Advances in Experimental Medicine and Biology

Volume 883

More information about this series at http://www.springer.com/series/5584

Nevan J. Krogan • Mohan Babu
Editors

Prokaryotic Systems Biology

Springer

Editors
Nevan J. Krogan
Cellular and Molecular Pharmacology
University of California, San Francisco
San Francisco, California, USA

Mohan Babu
Department of Biochemistry
University of Regina
Regina, Saskatchewan, Canada

ISSN 0065-2598 ISSN 2214-8019 (electronic)
Advances in Experimental Medicine and Biology
ISBN 978-3-319-79508-9 ISBN 978-3-319-23603-2 (eBook)
DOI 10.1007/978-3-319-23603-2

Springer Cham Heidelberg New York Dordrecht London

Softcover re-print of the Hardcover 1st edition 2015

Printed on acid-free paper

Springer International Publishing AG Switzerland is part of Springer Science+Business Media (www.springer.com)

Contents

Contributors

Ruedi Aebersold Institute of Molecular Systems Biology, Zurich, Switzerland

Faculty of Science, University of Zurich, Zurich, Switzerland

Mohan Babu Department of Biochemistry, University of Regina, Regina, SK, Canada

Vaibhav Bhandari Department of Biochemistry, University of Toronto, Toronto, ON, Canada

Steven Bowden Graduate School of Biological Sciences, Nara Institute of Science and Technology, Takayama, Ikoma, Nara, Japan

Cedoljub Bundalovic-Torma Department of Molecular Structure and Function, Hospital for Sick Children, Toronto, ON, Canada

Charles Calmettes Department of Biochemistry, University of Toronto, Toronto, ON, Canada

Catherine S. Chan Department of Biological Sciences, University of Calgary, Calgary, AB, Canada

Tatyana N. Chernikova School of Biological Sciences, Bangor University, Gwynedd, UK

Colin A. Cooper Department of Molecular and Cellular Biology, University of Guelph, Guelph, ON, Canada

Viktor Deineko Department of Biochemistry, University of Regina, Regina, SK, Canada

Karine Dufresne Department of Microbiology, Infectiology and Immunology, Université de Montréal, Montreal, QC, Canada

Andrew Emili Department of Molecular Genetics, University of Toronto, Toronto, ON, Canada

Donnelly Centre for Cellular and Biomolecular Research, University of Toronto, Toronto, ON, Canada

Peter N. Golyshin School of Biological Sciences, Bangor University, Gwynedd, UK

Alla Gagarinova Department of Molecular Genetics, University of Toronto, Toronto, ON, Canada

Donnelly Centre for Cellular and Biomolecular Research, University of Toronto, Toronto, ON, Canada

Olga V. Golyshina School of Biological Sciences, Bangor University, Gwynedd, UK

Walid A. Houry Department of Biochemistry, University of Toronto, Toronto, ON, Canada

Andrew Judd Department of Biochemistry, University of Toronto, Toronto, ON, Canada

Ashwani Kumar Department of Biochemistry, University of Regina, Regina, SK, Canada

M. Joanne Lemieux Department of Biochemistry, University of Alberta, Edmonton, AB, Canada

Igor Libourel The Biotechnology Institute, University of Minnesota, St. Paul, MN, USA

Iain L. Mainprize Department of Molecular and Cellular Biology, University of Guelph, Guelph, ON, Canada

Trevor F. Moraes Department of Biochemistry, University of Toronto, Toronto, ON, Canada

G. Moreno-Hagelsieb Department of Biology, Wilfrid Laurier University, Waterloo, ON, Canada

Hirotada Mori Graduate School of Biological Sciences, Nara Institute of Science and Technology, Takayama, Ikoma, Nara, Japan

Ai Muto Graduate School of Biological Sciences, Nara Institute of Science and Technology, Takayama, Ikoma, Nara, Japan

Nicholas N. Nickerson Department of Molecular and Cellular Biology, University of Guelph, Guelph, ON, Canada

Nicholas N. Nickerson Department of Infectious Diseases, Genentech Inc., South San Francisco, CA, USA

Yuta Otsuka Graduate School of Biological Sciences, Nara Institute of Science and Technology, Takayama, Ikoma, Nara, Japan

Rashmi Panigrahi Department of Biochemistry, University of Alberta, Edmonton, AB, Canada

Catherine Paradis-Bleau Department of Microbiology, Infectiology and Immunology, Université de Montréal, Montreal, QC, Canada

John Parkinson Department of Molecular Structure and Function, Hospital for Sick Children, Toronto, ON, Canada

Leopold Parts EMBL-EBI (South Building), Wellcome Trust Genome Campus, Saffron Walden, United Kingdom

Ana Popovic Department of Chemical Engineering and Applied Chemistry, University of Toronto, Toronto, ON, Canada

Seesandra Venkatappa Rajagopala J. Craig Venter Institute, Rockville, MD, USA

G. Santoya Instituo de Investigaciones Químico Biológicas, Universidad Michoacana de San Nicolás de Hidalgo, Morelia, Michoacán, Mexico

Olga T. Schubert Institute of Molecular Systems Biology, Zurich, Switzerland

Systems Biology Graduate School, Zurich, Switzerland

Rikiya Takeuchi Graduate School of Biological Sciences, Nara Institute of Science and Technology, Takayama, Ikoma, Nara, Japan

Anatoly Tchigvintsev Department of Chemical Engineering and Applied Chemistry, University of Toronto, Toronto, ON, Canada

Hai Tran School of Biological Sciences, Bangor University, Gwynedd, UK

Raymond J. Turner Department of Biological Sciences, University of Calgary, Calgary, AB, Canada

James Vlasblom Department of Biochemistry, University of Regina, Regina, SK, Canada

Joseph T. Wade New York State Department of Health, Wadsworth Center, Albany, NY, USA

Department of Biomedical Sciences, University at Albany, Albany, NY, USA

Omar Wagih EMBL-EBI (South Building), Wellcome Trust Genome Campus, Saffron Walden, United Kingdom

Barry L. Wanner Department of Biological Sciences, Purdue University, West Lafayette, IN, USA

Michail M. Yakimov Institute for Coastal Marine Environment, Messina, Italy

Alexander F. Yakunin Department of Chemical Engineering and Applied Chemistry, University of Toronto, Toronto, ON, Canada

Katsushi Yokoyama Graduate School of Biological Sciences, Nara Institute of Science and Technology, Takayama, Ikoma, Nara, Japan

Chapter 1
Metagenomics as a Tool for Enzyme Discovery: Hydrolytic Enzymes from Marine-Related Metagenomes

Ana Popovic, Anatoly Tchigvintsev, Hai Tran, Tatyana N. Chernikova, Olga V. Golyshina, Michail M. Yakimov, Peter N. Golyshin, and Alexander F. Yakunin

Abstract This chapter discusses metagenomics and its application for enzyme discovery, with a focus on hydrolytic enzymes from marine metagenomic libraries. With less than one percent of culturable microorganisms in the environment, metagenomics, or the collective study of community genetics, has opened up a rich pool of uncharacterized metabolic pathways, enzymes, and adaptations. This great untapped pool of genes provides the particularly exciting potential to mine for new biochemical activities or novel enzymes with activities tailored to peculiar sets of environmental conditions. Metagenomes also represent a huge reservoir of novel enzymes for applications in biocatalysis, biofuels, and bioremediation. Here we present the results of enzyme discovery for four enzyme activities, of particular industrial or environmental interest, including esterase/lipase, glycosyl hydrolase, protease and dehalogenase.

Keywords Metagenome • Gene library • Gene discovery • Enzyme screening • Hydrolase

1.1 Introduction to Metagenomics and Its Applications

Prokaryotes constitute the largest fraction of individual organisms on Earth, accounting for up to 10^8 separate genotypes, with conservative estimates of up

A. Popovic • A. Tchigvintsev • A.F. Yakunin (✉)
Department of Chemical Engineering and Applied Chemistry, University of Toronto, Toronto, ON M5G 1L6, Canada
e-mail: a.iakounine@utoronto.ca

H. Tran • T.N. Chernikova • O.V. Golyshina • P.N. Golyshin
School of Biological Sciences, Bangor University, Gwynedd, LL57 2UW, UK

M.M. Yakimov
Institute for Coastal Marine Environment, CNR, 98122 Messina, Italy

N.J. Krogan, M. Babu (eds.), *Prokaryotic Systems Biology*, Advances in Experimental Medicine and Biology 883, DOI 10.1007/978-3-319-23603-2_1

to 52,000 microbial species residing in just one gram of soil, and several hundred to several thousand species in just one millilitre of sea water (Simon and Daniel 2011; Roesch et al. 2007; Kemp and Aller 2004; Ravenschlag et al. 1999; Schloss and Handelsman 2005). Less than one percent of these microorganisms, however, are culturable in the laboratory, and amenable to traditional experimental studies (Giovannoni et al. 1990). Through the advent of metagenomics, we are just now starting to gain insight into the rich microbial worlds thriving within distinct habitats. One of the first heralded successes of metagenomics was the discovery of bacteriorhodopsin in marine bacterioplankton (Beja et al. 2000). This chapter gives an overview of the importance and applications of function-based metagenomic studies, and describes enzyme screening of metagenome libraries and findings to demonstrate what metagenomes have to offer.

1.1.1 Metagenomics and Its Approaches

Metagenomics is the study of community genetics through the extraction and direct analysis of environmental DNA, most often via creating large or small insert DNA libraries transformed into *E. coli* as a surrogate host. It allows us to circumvent the problems associated with culturing environmental bacteria and to study the biodiversity and biogeochemical roles of the communities through sequence analysis and function-based enzyme screens (Fig. 1.1). The increasingly more accessible and economical Next-Generation Sequencing platforms and continuous advances in computational biology allow us to analyse ever larger sets of sequence data, but prediction and annotation of new genes still relies on sequence similarity to already characterized genes and pathways in the public databases (GenBank, UniProt, KEGG, etc.). As a result, 40–50 % of genes in genomes are routinely labelled as "hypothetical" or proteins of unknown function (Koonin and Galperin 2003; Ferrer et al. 2007; Pelletier et al. 2008). In this scenario, function-based metagenomics is invaluable. The magnitude of microbial and protein diversity of marine metagenomes was demonstrated by two landmark papers by Venter et al., which together revealed over 500 new species, over 6 million protein encoding genes, and almost 2000 new protein families with unknown function (Venter et al. 2004; Yooseph et al. 2007). A more recent high-throughput metagenomics project identified over 27,000 putative carbohydrate-active genes in the cow rumen metagenome and demonstrated the presence of glycosyl hydrolase activity in 51 of 90 tested proteins.

The experimental approaches of functional metagenomics include developing new cultivation methods, meta-transcriptomics, meta-proteomics, meta-metabolomics, and enzyme screening (Rondon et al. 2000; Ferrer et al. 2007; Simon and Daniel 2011; Uchiyama and Miyazaki 2009). The enzyme screening approach involves gene expression and directly assaying metagenomic gene libraries for the ability to modify or hydrolyze a specific chemical substrate. Most often, this means expressing metagenomic enzymes from native or inducible promoters in *E. coli* and detecting enzymatic activity using chromogenic or insoluble substrates in

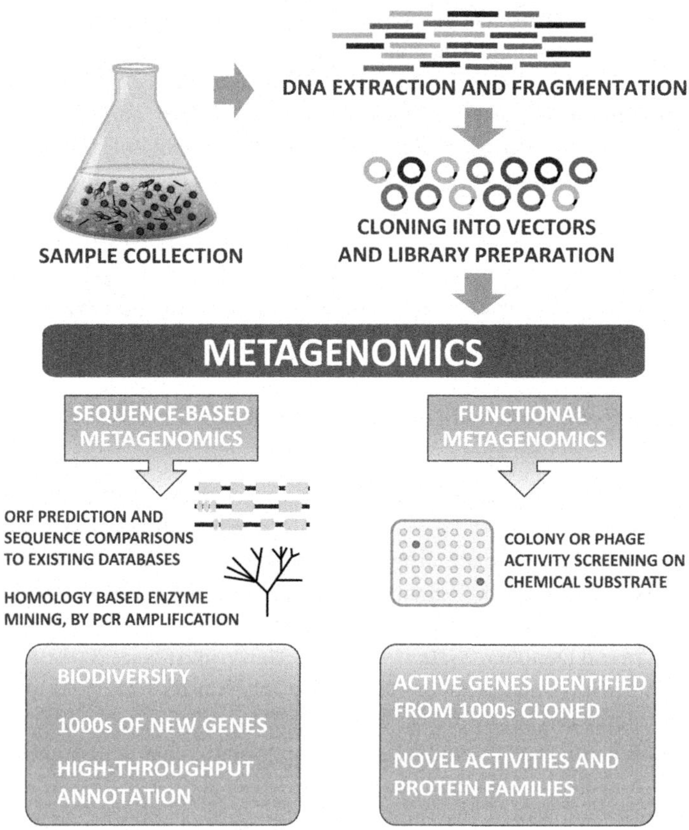

Fig. 1.1 Overview of sequence-based and functional metagenomics

agar (Rondon et al. 2000). An alternate approach is to clone environmental DNA fragments into a lambda phage-based expression vector and to screen for particular enzymatic activities directly on phage plaques (Ferrer et al. 2005). Enzymatic screening of metagenome libraries allows mining for new enzyme activities, and offers the possibility to discover novel families of enzymes with no sequence similarity to previously characterized enzymes found in BRENDA or Uniprot. It also offers an immense repository of new enzymes with an incredible variety of characteristics evolved to accommodate the unique environments that the microbes reside in.

Screening of metagenome gene libraries has greatly expanded the number of novel enzymes, including over 130 new nitrilases and many cellulases, carboxyl esterases, and laccases (Robertson et al. 2004; Lorenz and Eck 2005; Beloqui et al. 2006). Recently, metagenomes of several extreme environments have also been explored and revealed a rich biochemical diversity of enzymes adapted to function under extreme conditions, such as low or high temperatures, low or high pH, and high salt concentrations (Ferrer et al. 2007). Biochemical and structural characterization of these enzymes revealed different molecular mechanisms of adaptation to extreme environmental conditions (Feller and Gerday 2003; Olufsen et al. 2005; Siddiqui and Cavicchioli 2006). The potential for enzyme discovery in metagenomes has not gone unnoticed in the industrial sector, and several companies such as Diversa, Genencor International (now part of DuPont), Henkel, Degussa (now Evonik Industries), among others, have already made efforts in this area (reviewed by Lorenz and Eck 2005). The efforts will only increase as comparative studies have shown that replacing conventional industrial processes with enzymatic processes in a multitude of industries have indeed lead to savings and cleaner production (reviewed by Jegannathan and Nielsen 2013, of Novozymes).

1.1.2 Metagenomics and Enzyme Discovery

Examples of industries which employ enzymatic processes include pulp and paper production, household detergent production, the textile industry, food and beverage industries, animal feed production and biodiesel production, among others (Jimenez et al. 1999; Nguyen et al. 2008; Hemachander and Puvanakrishnan 2000; Saeki et al. 2007; Aly et al. 2004; Osma et al. 2010; Okamura-Matsui et al. 2003; Gado et al. 2009; Monsan and Donohue 2010; Hernández-Martín and Otero 2008). Novozymes, which holds a 48 % share of the global market for industrial enzymes, has reported enzyme business sales up by 5 % in 2013, at 1574 million euros, with the strongest sales to household care (including most importantly detergents) and bioenergy industries. They also calculated that customers saved approximately 52 million tons of CO_2 (editor Bedingfield 2013). In the trend toward alternative, cleaner, cheaper and more efficient processes, the potential for enzyme application is great.

Cold-active enzymes offer particular advantages in the household cleaner and food industries, including primarily energy savings compared to traditional high-temperature processes, both for consumers, in the case of detergents marketed for cold-water cleaning, and manufacturing processes (Cavicchioli et al. 2002). These enzymes maintain high levels of activity and specificity, and many have the convenient property of being thermolabile, allowing for easy inactivation prior to subsequent processing steps. As a result of their inherent flexibility, in order to remain active and mobile at low temperatures, they have been proposed as candidates for organic synthesis in non-aqueous or mixed aqueous-organic solvents, where the absence of water stabilizes and inhibits activity of many mesophilic enzymes. Similarly halophilic enzyme instability in aqueous low salt environments has made them attractive for non-aqueous synthesis (van den Burg 2003; Sellek and Chaudhuri 1999; Cavicchioli et al. 2002). Marine environments offer an ideal opportunity to sample microbial communities which have evolved to thrive at both cold temperatures and hypersaline conditions.

The Earth's biosphere is predominantly aqueous (70 % water) and cold (around 5 °C). Marine microorganisms that are able to grow at low temperatures have evolved different adaptation mechanisms to survive under these conditions.

Temperature is one of the most important factors for enzyme activity as the reaction rate can be reduced 30–80 times when the temperature drops from 37 to 0 °C (Lonhienne et al. 2000). To explore the biochemical diversity of marine metagenomes, we have screened twelve metagenomic libraries from a diverse set of marine-related environments for one or more of the following hydrolytic functions: esterase/lipase, glycosyl hydrolase and protease, which have applications in one or more of the industries listed above, and lastly dehalogenase, which together with the above mentioned enzyme activities, plays an important role in bioremediation and detoxification of halogenated organic pollutants (Brisson et al. 2012; Jegannathan and Nielsen 2013).

1.2 Metagenome Gene Library Preparation

1.2.1 Environments and Library Preparation

Immense sequence diversity has so far been documented in environmental metagenomes, suggesting significant metabolic and biochemical variety as well (Dinsdale et al. 2008; Yooseph et al. 2007). To exploit this diversity, we selected a set of marine-related metagenomes to search for cold-active and salt-tolerant esterases, lipases, proteases, glycosyl hydrolases and dehalogenases.

We sampled 12 marine-related environments, from communities thriving under extreme anoxic deep sea conditions (Urania, Kryolo, Medee, Rimicaris Gill and Gut) or extremely low pHs (Vulcano), to various regions associated with heavy industrialization and oil-contamination of the Mediterranean or Barents Sea (Messina, Milazzo, Priolo, Haven, Kolguev, Murmansk). As examples of the environmental diversity, the Rimicaris Gut and Gill libraries were prepared from dense bacterial communities inhabiting *Rimicaris exoculata* shrimp, which lives on chimney walls of hydrothermal vents in the Mid-Atlantic ridge, over 3 km beneath the surface (Williams and Rona 1986). The Haven library, though, is derived from samples of tar collected on the coast of Genoa, Italy, where the Amoco Milford Haven tanker exploded in 1991 releasing thousands of tonnes of crude oil. The temperatures of all of selected environments range from 3 to 15 °C and water salinity ranges from 3.1 to 3.8 %. Selected communities were treated for 1 month with phenanthrene and pyrene (Milazzo, Messina) or crude oil (Kolguev, Murmansk), prior to extraction of large molecular weight DNA.

Two types of metagenomic libraries were prepared from the marine samples—large DNA insert (40,000 bases) fosmid libraries and small DNA insert (4000–7000 bases) lambda phage libraries, using commercially available kits (Epicentre's CopyControl Fosmid and Stratagene's, now Agilent's, Lambda-ZAP). The names and descriptions of the 12 metagenomic libraries are provided in Table 1.1.

Table 1.1 Metagenomic libraries prepared and screened for enzyme activity

Metagenomic library	FosmidTotal clones	Lambda-ZAPTotal clones	Notes
Contaminated source			
Kolguev Island	N/A	100,000	Crude oil-degrading psychrophilic community; Barents Sea
Port of Murmansk	N/A	100,000	Crude oil-degrading psychrophilic community; Barents Sea
Milazzo enrichment	50,000	2400	Marine-based enrichment cultures with phenanthrene and pyrene; Mediterranean Sea
Messina enrichment	21,000	1000	Marine-based enrichment cultures with phenanthrene and pyrene; Mediterranean Sea
Priolo sediment	3000	40,000	Anoxic community, heavy industrialization and crude/refined oil contamination; harbor of Priolo Gargallo, Italy
Haven sediment	9200	25,000	Petroleum contamination; harbor of Arenzano, Italy
Specialized environment			
Rimicaris exoculata gut	11,100	150,000	Deep sea shrimp metagenome; Mid-Atlantic Ridge
Rimicaris exoculata gill	20,000	350,000	Deep sea shrimp metagenome; Mid-Atlantic Ridge
Urania basin interface	N/A	100,000	Deep hypersaline anoxic lake; Mediterranean Sea
Kryos brine interface	9300	620,000	Deep hypersaline anoxic lake; Mediterranean Sea
Medee basin interface	18,432	N/A	Deep hypersaline anoxic lake (salinity 170–190 g/l); Mediterranean Sea, Cycloclasticus naphthalene enrichment
Vulcano acidic pool	3456	N/A	Enrichment made from acidic pool sand/gravel; Mediterranean Sea

1.2.2 *Advantages and Disadvantages of Large and Small DNA Insert Libraries*

Each of the two types of metagenome gene libraries (large DNA insert fosmid libraries and small DNA insert phage libraries) offers particular advantages and disadvantages. Although the lambda phage libraries may be converted to phagemid clones and screened as colonies, one of the biggest advantages to the phage is lysis of *E. coli* cells at the end of the infection cycle and release of translated metagenomic proteins to the extracellular matrix, and consequently the substrate. Lambda phage are stable for long periods, both at −80 °C and at 4 °C, and library contamination with other laboratory strains is a smaller concern due to the specificity of the host–virus interaction. Perhaps most importantly, toxic effects of metagenomic enzymes generally do not pose problems as they would in cell-based libraries, since gene expression in cells is only driven during the short interval of viral infection. In addition, the lambda libraries, although containing small inserts of on average 4–8 kilobases (4–8 genes), have IPTG-driven expression which allows for higher concentrations of metagenomic proteins. This is important for phage screening, since there is only a short window of infection (approximately 50 min for a wildtype lambda virus) during which the metagenomic genes can be expressed prior to cell lysis. In screening phage, it is possible to screen large numbers of clones (600–2000 per 10 cm plate, depending on the substrate) very quickly and easily. Because the enzyme is released to the extracellular matrix upon cell lysis, the activity is often seen earlier than with fosmid libraries.

Alternatively, the large insert libraries have the advantage of just that—the presence of a longer metagenomic DNA insert. In some cases, more than one gene is required for activity or correct expression and folding of the enzyme of interest, and fosmid libraries offer the advantage of screening larger gene clusters (up to 50 genes), in many cases entire operons. In the particular cases of two esterases we have identified during functional screening, we have also found predicted lipase chaperones immediately downstream of the active genes. Sequencing the genetic neighbourhood of an active fosmid clone can also offer hints about the native metabolic role of the enzyme in question, and more accurately predict the taxonomy of the source organism. Finally, vectors used for cloning large-insert DNA fragments are typically low copy which, unless induced by engineered copy-control, minimize effects of toxic genes (Taupp et al. 2011) and allow basal expression levels from native promoters, avoiding inclusion body formation often associated with overexpression in *E. coli*.

1.3 Agar-Based Enzyme Screens for Hydrolytic Enzymes

Agar plate-based screening of metagenomic libraries provides a direct and simple approach to mine for industrially useful enzymes that function under diverse conditions—low temperature, extreme pH, nonaqueous, anoxic, hypersaline, among

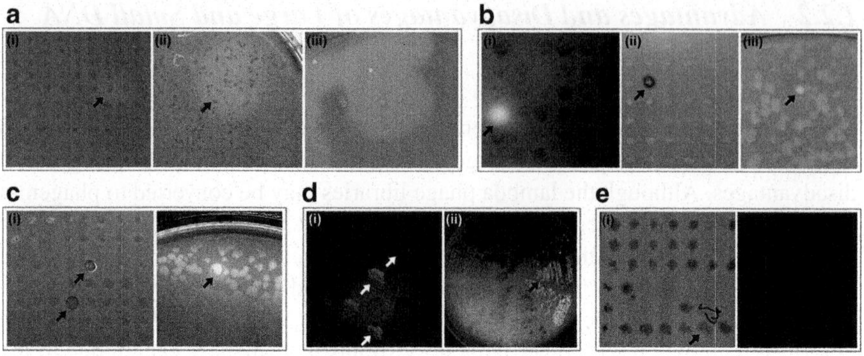

Fig. 1.2 Agar-based enzymatic screening of metagenomic libraries. Panels show positive hits on fosmid or phage library screens as follows: (**a**) Glycosyl hydrolase positive fosmid (i) and phage (ii and iii) clones screened on carboxymethyl cellulose. (**b**) Skim milk screens for protease and glycosyl hydrolase activities, on fosmid libraries with and without a pH indicator (i and ii), and on phage libraries (iii). (**c**) Esterase or lipase positive fosmid (i) and phage (ii) hits screened on tributyrin. (**d**) Lipase positive fosmid (i) and excised phagemid (ii) clones screened on olive oil. (**e**) Fosmid clones positive for dehalogenase activity

others. This method has successfully produced a large number of novel esterases, lipases, proteases, glycosyl hydrolases, laccases, and other enzymes (Lorenz and Eck 2005; Ferrer et al. 2009; Steele et al. 2009). Below we describe the five screens we have used to screen up to 100,000 clones per experiment using Lambda-ZAP or over 6000 clones for all activities using fosmid libraries. Examples of enzyme screens and positive clones are shown in Fig. 1.2.

1.3.1 Esterases and Lipases

We use two substrates to screen for esterase and lipase activity. A commonly used substrate for detection of esterase or lipase activity is the simple ester tributyrin. To specifically detect lipase activity, we use olive oil, which is comprised primarily of the long carbon chain (C >16) triglyceride esters oleic acid, linoleic acid and palmitic acid. The screens are adapted from previously published protocols (Kok et al. 1993; Kouker and Jaeger 1987).

In fosmid screens, the substrates, 1 % tributyrin and 3 % olive oil, are emulsified with 0.5 % gum arabic in standard Luria-Bertani (LB) broth and agar plate media containing appropriate antibiotics. The clones, containing metagenomic DNA, are grown in LB in microtiter plates for several hours, plated on substrate plates using 96-pin or 384-pin replicators and incubated overnight at 37 °C. Emulsified tributyrin gives a turbid appearance to the plates, and hydrolysis by esterases or lipases is seen as a clearing or halo around the colony, or plaque in phage screening (Fig. 1.2c). In the case of olive oil plates, 0.001 % w/v Rhodamine B dye is also added to the plates, and activity is detected as orange fluorescence under UV light (Fig. 1.2d), presumably caused by formation of complexes of Rhodamine B molecules and the hydrolyzed fatty acids (Kouker and Jaeger 1987). During phage screening, phage clones are preincubated with host cells, added to several millilitres of soft LB-agar containing emulsified tributyrin and gum arabic, and plated on LB-agar plates, containing 1 mM IPTG, overnight at 37 °C.

All plates are kept for an additional 2–4 days at room temperature to 37 °C, and monitored for positive clones.

1.3.2 Glycosyl Hydrolases

One rather simple color-based method to screen for glycosyl hydrolase activity uses carboxymethyl cellulose as a substrate and Congo red dye as an indicator (adapted from Teather and Wood 1982). In this assay, 0.3 % carboxymethyl cellulose is added to conventional plate media (for fosmid screening) or to soft LB-agar (for phage screening), and cells or phage are plated as described in the screen above. After 2–4 days, cellulose is stained with a 0.1 % Congo red solution, and unstained haloes are observed around positive colonies (Fig. 1.2a).

1.3.3 Proteases

The skim milk-based agar screen has been proposed for detecting proteases in soil metagenome libraries (Rondon et al. 2000). In these screens, 1 % skim milk is added to LB-agar plates for colony screening, and 3–4 % skim milk is added to soft LB-agar for phage library screening, as described for screens above (adapted from Rondon et al. 2000). Activity is detected as a clearing in the turbidity (Fig. 1.2b), as the protease degrades milk proteins, mostly caseins. Later, Jones et al. showed that skim milk screens can also detect the metagenomic clones expressing glycosyl hydrolases or releasing acid (Jones et al. 2007). Therefore, to some screens we have added pH indicator dyes (0.5 mM phenol red and/or bromothymol blue) to increase sensitivity and detect the acidic shift during hydrolysis of casein by proteases or lactose by glycosyl hydrolases (Jones et al. 2007). In order to distinguish proteases from glycosyl hydrolases, we rescreened positive clones from the skim milk assay on plates containing X-gal. In pH based assays, including also the assay described below for dehalogenases, however, we have found a higher tendency for false positives, and positive clone genes must be further tested to ensure the enzymatic activity is present.

1.3.4 Dehalogenases

Detection of dehalogenase activity has been described using a variety of haloalkane and haloacid substrates (Holloway et al. 1998). The basic concept involves freeing the halogen (e.g. chloride or bromine) from the substrate and detecting the ensuing pH change with an indicator dye, which results from haloacid formation (HCl, HBr, etc.). The described agar based screen is set up in a similar way to those of the other activities above with the following changes. Colonies or phage are plated on LB-agar plates containing 0.5 mM phenol red and/or bromothymol blue pH indicator dyes, but no substrates. After overnight incubation, or in the case of fosmid libraries, possibly after 2–3 days of incubation, a mixture of melted 0.4 % agarose, 20 mM EPPS buffer (pH 8.0) and a 2.5 mM combination of haloalkanes and haloacids, such as bromoacetic acid, 3-bromo-2-methylpropanoic acid, iodoacetic acid, ethyl-3-(bromomethyl)propanoate, 3-dibromopropanol, 1-iodopropane, is layered on top of the colonies or phage. The plates are incubated at 30 °C, and checked every few minutes for colour change (Fig. 1.2e). The pH change on solid media is short lived as a result of diffusion, therefore requires close monitoring.

In the above described screens, we have found phage particularly useful for screens involving turbid substrates (tributyrin or skim milk), as well as color-based end-point assays (such as the glycosyl hydrolase screen described above). Positive plaques are difficult to spot in screens involving transient colour changes, such as the dehalogenase screen. For these screens, liquid-based microplate assays with fosmid or excised phagemid libraries are perhaps best.

1.4 Screening and Sequencing Results for Marine Metagenome Libraries

1.4.1 General Functional Screening Statistics

Over 1.3 million clones from the 12 described marine-related metagenomic libraries were screened for esterase/lipase, glycosyl hydrolase, protease and dehalogenase activities, yielding 545 positive hits. Over half of these were putative esterases or lipases identified with tributyrin screens, the quickest and most successful assay, followed by glycosyl hydrolases, putative proteases or glycosyl hydrolases from skim milk screens, and finally dehalogenases (Table 1.2). The tributyrin plate screen for esterases and lipases is the most common metagenomic screen, which depending on the metagenomic library has been reported to have a hit rate in the range of 1 positive per 5 to 4000 Mb of DNA screened (Lorenz and Eck 2005; Steele et al. 2009; Uchiyama and Miyazaki 2009). In our work, this screen produced 1 hit per 9 Mb of screened DNA (Table 1.3), which is within the reported range. The glycosyl hydrolase and dehalogenase screens produced comparable frequencies of positive hits (1 hit per 28.4 and 23.9 Mb DNA, respectively). The highest frequency

Table 1.2 Enzyme screening statistics for marine metagenomic libraries

	Esterase/lipase		Glycosyl hydrolase		Protease		Dehalogenase	
Metagenomic library	Screened	Positive	Screened	Positive	Screened	Positive	Screened	Positive
Contaminated source								
Kolguev (phage)	154,000	30	111,500	5	8200	6	–	–
Murmansk (phage)	108,000	41	103,000	9	–	–	–	–
Milazzo (phage)	20,000	8	15,000	0	24,000	0	–	–
Messina (phage)	24,000	18	–	–	25,000	3	–	–
Priolo (phage)	118,500	5	54,000	0	–	–	–	–
Haven (phage)	36,800	11	94,500	4	16,000	2	–	–
Specialized environment								
Rimicaris Gill (fosmid)	–	–	8400	27	8400	38	8400	17
Rimicaris Gut (phage)	137,500	7	–	–	–	–	–	–
Rimicaris Gill (phage)	21,100	8	–	–	–	–	–	–
Urania (phage)	90,800	36	130,000	5	–	–	–	–
Kryos (fosmid)	–	–	3456	49	3456	16	3456	17
Medee (fosmid)	4992	147	4992	0	4992	1	4992	0
Vulcano (fosmid)	1920	32	3456	1	3456	2	3456	0
Total	717,612	343	528,304	100	93,504	68	20,304	34

Table 1.3 Frequency of positive hits (hit per Mb of DNA screened)[a]

Metagenomic library	Esterase/lipase	Glycosyl hydrolase	Protease	Dehalogenase
Contaminated environment				
Kolguev (phage)	1/20.5	1/89.2	1/5.5	–
Murmansk (phage)	1/10.5	1/45.7	–	–
Milazzo (phage)	1/10	0	0	–
Messina (phage)	1/5.3	–	1/33.3	–
Priolo (phage)	1/94.8	0	–	–
Haven (phage)	1/13.3	1/94.5	1/32	–
Specialized environment				
Rimicaris Gill (fosmid)	–	1/12.4	1/8.8	1/19.8
Rimicaris Gut (phage)	1/78.6	–	–	–
Rimicaris Gill (phage)	1/10.5	–	–	–
Urania (phage)	1/10.1	1/104	–	–
Kryos (fosmid)	–	1/2.8	1/8.6	1/8.1
Medee (fosmid)	1/1.4	0	1/200	0
Vulcano (fosmid)	1/2.4	1/138	1/69	0
Total	1/9.1	1/28.4	1/16.2	1/23.9

[a]For calculation of hit frequency, an average of 4000 bp/phage clone and 40,000 bp/fosmid clone were used

of positive esterase hits (1 hit per 1.4 Mb DNA) was obtained from the Medee fosmid library, whereas the Kryos fosmid library produced the highest frequencies of positive glycosyl hydrolase and dehalogenase clones (1 hit per 2.8 and 8.1 Mb DNA, respectively) and the Rimicaris Gill and Kryos fosmid libraries revealed the highest frequencies of positive hits on skim milk screens (1 hit per 8.8 and 8.6 Mb DNA, respectively) (Table 1.3). In some libraries, the high frequency of positive hits can be explained by the presence of multiple copies of a limited number of positive genes (redundancy) as was revealed by DNA sequencing (see below), or a high propensity for false positives in enzyme activity screens using transient pH changes.

1.4.2 Clone Sequencing

Due to the size differences between phagemids (on average 4–7 kilobases) and fosmids (on average 40 kilobases), we used different strategies for sequencing the positive clones. Simple gene walking was used to sequence phagemids. This approach, however, would have been tedious and impractical for fosmids. Fosmid DNA samples were pooled into mixtures of 40–100 clones and submitted for Next Generation sequencing by Illumina. The massive amount of sequence data obtained, the substantial decrease in cost of this sequencing platform, and the increasing number of companies offering the service have made this approach practical. In theory, it is possible to obtain 1000× sequence coverage for samples of 500 pooled fosmids, assuming 200 million reads per lane (Genome Quebec, personal communication), however one must keep in mind inevitable downstream problems of contig assembly, including overlapping clones, underrepresented sequences, short-sequence DNA repeats, or stretches of DNA inherently difficult to sequence. For these reasons, we were relatively conservative in our DNA pool sizes. There are a variety of assemblers for Illumina data (Velvet, ABySS, DNASTAR). We used Geneious, a powerful sequence analysis program, which uses a Velvet based algorithm for sequence assembly. In order to trace the assembled contigs to particular clones, we also sequenced the ends of each fosmid. In lieu of end sequencing, it is also possible to barcode the fosmids, in house or at added expense by the sequencing companies.

We obtained sequences for most isolated phage and phagemid positives and 50 % of our fosmids. Open reading frames were predicted and annotated using a combination of Geneious' gene prediction, Glimmer, BLASTx and MG-RAST. Gene annotation is laborious and requires significant manual curation to accurately resolve and annotate predicted overlapping open reading frames. Genes with predicted enzyme activities of interest were cloned and rescreened in standard *E. coli* expression vectors. Where the putative enzyme could not be identified through sequence-based searches, multiple, if not all, genes were cloned and retested.

Sequencing revealed a high proportion of marine hydrocarbon degrading bacteria in the tested metagenomic libraries. Murmansk and Kolguev libraries both had a large proportion of the bacterium *Alcanivorax borkumensis*, an oil-degrading bacterium. The deep sea Urania library had predominantly *Marinobacter aquaeolei* and *Marinobacter hydrocarbonoclasticus*, the most abundant sea dwelling bacteria capable of degrading oil or other hydrocarbons, or species with high nucleotide similarity to these. Finally, a majority of positive clones isolated from the Medee deep sea basin were predicted to contain DNA from one or more of the marine polycyclic aromatic hydrocarbon degrading *Cycloclasticus* species. Positive hits isolated from libraries with geographical proximity showed some overlap, as expected, with seven enzymes recovered from both Murmansk and Kolguev screening, two enzymes from Milazzo and Messina libraries, and the same *Cycloclasticus* enzyme isolated from Messina, Vulcano as well as the Medee libraries.

In our experience, one quarter of all initially cloned metagenomic genes could be expressed in *E. coli* and purified sufficiently for biochemical analysis. Quite often, it became necessary to clone protein fragments and remove predicted N-terminal signal peptides or transmembrane domains, identified using prediction programs such as TMHMM or SignalP. Based on sequence, nearly one half (45 %) of the experimentally confirmed metagenomic esterases were predicted to have N-terminal signal or transmembrane sequences. A combination of removing these sequences and using chemical chaperones, such as sorbitol or glycerol, for recombinant protein expression (Prasad et al. 2011), greatly aided in obtaining soluble proteins.

1.4.3 Sequence Analysis of Esterase Positive Genes

Thirty-nine so far cloned and confirmed esterases from marine-related metagenomes belong to a diverse set of protein families (Fig. 1.3). As expected, a majority (70 %) is predicted as α/β-hydrolase family proteins, the largest proportion belonging

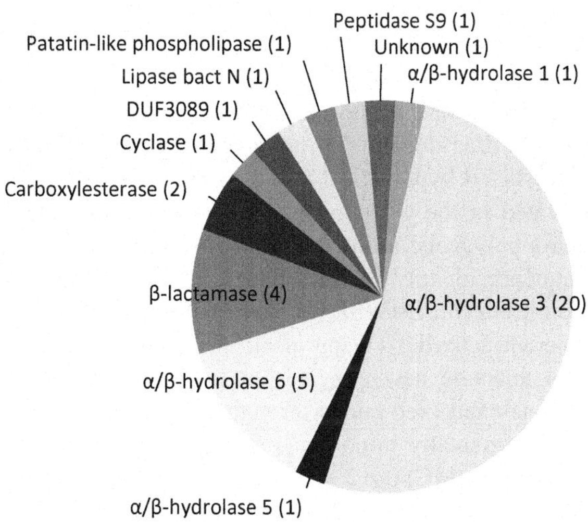

Fig. 1.3 Protein family prediction for 39 confirmed esterases

to the α/β-hydrolase 3 family. Four enzymes are predicted β-lactamases, two carboxylesterases, and the remaining are a predicted lipase, patatin, and interestingly a predicted cyclase, prolyl oligopeptidase, DUF3089 family protein and one completely unknown.

Most of the biochemically characterized α/β-hydrolases contain a Ser-His-Asp catalytic triad in the active site. Sequence analysis of the identified α/β-hydrolases and carboxylesterases reveals conserved GxSxG motifs, suggesting the positions of the catalytic Ser residues, as well as several conserved candidates for the His and Asp residues of the catalytic triad. The β-lactamases each have the typical N-terminal SxxK motif, as well as a conserved Tyr and an additional GxSxG or GxSxx near the C-terminus described for Family VIII esterases (reviewed by Hausmann and Jaeger 2010). The predicted lipase, patatin and DUF3089 proteins also show conserved GxSxG motifs and several conserved His and Asp residues suggesting that these proteins also belong to serine hydrolases. The enzyme of unknown protein family (internally identified as MGS0084) shares low sequence similarity (20–23 % of sequence identity) with only four predicted proteins in Genbank and Uniprot. These proteins have several conserved Ser residues suggesting that these sequences might represent another family of Ser-dependent hydrolases. Finally, both esterase and peptidase activities have been previously detected in select serine proteases, specifically prolyl oligopeptidases predicted to contain an α/β hydrolase fold (Wang et al. 2006). It was found that a single conserved Arg residue can discriminate between the two enzymatic activities. The active residues of the latter enzymes will need to be confirmed through mutagenesis.

Thus, while a majority of the lipolytic enzymes so far mined belong to one of the well-known esterase families, a significant 4 (10 %) are either unknown or predicted to have alternate activities. For esterases from characterized families, we have the opportunity to compare and study particular amino acid substitutions or structural changes that confer psychro or halophilicity.

1.4.3.1 Case Example of Venturing into Novel Sequence Space

MGS0012, one of several enzymes isolated in our functional assays that were predicted as domains of unknown function, shares 99 % sequence identity with a predicted hypothetical protein from *Kordiimonas gwangyangensis*, an organism isolated in the Gwangyang Bay of Korea capable of degrading high-molecular-mass polycyclic aromatic hydrocarbons (Kwon et al. 2005). It also shares 40 % sequence identity with another recently and independently isolated soil metagenomic enzyme EstWSD (Uniprot K9RYB0), which exhibits salt tolerant and solvent tolerant activities (Wang et al. 2013). It is particularly useful that as we isolate and annotate new enzymes in functional metagenomics, we can extend these findings and predictions through protein sequence space to other interesting and environmentally important organisms. A protein sequence alignment (Fig. 1.4) shows that MGS0012 and EstWSD, predicted members of the DUF3089 family, share sequence similarity with uncharacterized proteins from *Dehalococcoides*

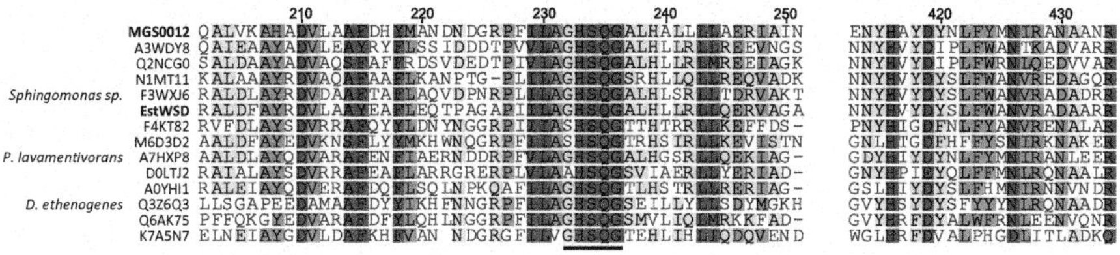

Fig. 1.4 Sequence alignment of the conserved Ser motif in identified metagenomic carboxyl esterase MGS0012 and DUF3089 family proteins

ethenogenes (Q3Z6Q3), an organism known to reductively dechlorinate the groundwater pollutants tetrachloroethene and trichloroethene (Maymo-Gatell et al. 1997), *Parvibaculum lavamentivorans* (A7HXP8), a microbe which degrades the commercial laundry surfactant linear alkylbenzenesulfonate (Schleheck et al. 2000), and proteins from various species of *Sphingomonas* (F3WXJ6), some of which are known to degrade aromatic compounds (Fredrickson et al. 1991, 1995; Baraniecki et al. 2002). In fact, MGS0012 has >30 % sequence identity to 197 proteins in Uniprot. Thus slowly, but surely, we are venturing into novel sequence space, and making progress into those 40 % hypothetical or uncharacterized proteins currently in the database.

1.5 Biochemical Properties of Carboxyl Esterases from Marine Metagenomes

Eight genes encoding metagenomic carboxyl esterases were cloned for overexpression in *E. coli* and purified using affinity chromatography. Detailed biochemical characterization of these enzymes revealed a high prevalence of cold-adapted activity and salt tolerance, in accordance with the microbial environment. The enzyme activities of purified proteins were assayed using a set of *p*-nitrophenyl ester substrates with different acyl chain lengths (C2–C16). They were also tested for temperature optimum, pH preference, as well as salt and organic solvent tolerance, in addition to the substrate specificity.

The metagenomic carboxyl esterases were active over a broad pH range at 30 °C (pH 7–9). These enzymes display a wide range of chain length preferences, ranging from acyl chain length C2 (*p*NP-acetate for MGS0109) to C8 (*p*NP-octanoate for MGS0010). The preference for short acyl chain substrates is typical of carboxyl esterases (Hausmann and Jaeger 2010). However, two enzymes, MGS0012 from the Messina library and ABO1197 from the oil-degrading *Alcanivorax borkumensis* of the Murmansk library, exhibit high activity (up to 30 μmoles/min per mg protein) against the C16 *p*NP-palmitate, a model substrate used for lipase activity. However, the preferred substrates for these two enzymes are α-naphthyl acetate (61 μmoles/min per mg protein) and *p*NP-valerate (nearly 300 μmoles/min per mg

Table 1.4 Biochemical characteristics of purified metagenomic carboxyl esterases

Clone	Metagenomic library	Predicted Pfam	Optimal temp	Activity at 4 °C	Optimal salt conc	Activity	
						0 M NaCl	2 M NaCl
MGS0006	Messina	α/β-hydrolase 3	30 °C	55 %	0 M	100 %	3 %
MGS0010	Messina	β-lactamase	30 °C	32 %	>3.5 M	100 %	144 %
MGS0012	Messina	DUF3089	40 °C	9 %	2 M	100 %	187 %
MGS0018	Rimicaris Gill	α/β-hydrolase 3	30 °C	37 %	0 M	100 %	9 %
MGS0105	Kolguev	β-lactamase	15 °C	41 %	0 M	100 %	0 %
MGS0109	Kolguev	α/β-hydrolase 6	30 °C	50 %	0 M	100 %	0 %
ABO1197	Murmansk	α/β-hydrolase 6	30 °C	66 %	0 M	100 %	40 %
ABO1251	Murmansk	Carboxylesterase	35 °C	46 %	0 M	100 %	23 %

protein), respectively. A weak lipase activity was also detected for MGS0012 in an agar-based olive oil assay, which contains 50–80 % oleic acid (C18).

Temperature optima for seven out of eight enzymes were found to range from 15 to 35 °C, with 32–66 % activity at 4 °C (Table 1.4) suggesting that these proteins are cold-adapted (psychrophilic) enzymes. MGS00012 alone showed an optimal temperature of 40 °C, and only 9 % activity at 4 °C, reminiscent of a mesophilic enzyme. As intracellular salt concentrations vary species by species, and can differ from external salt concentrations (Oren 2002; Christian and Waltho 1961, 1962), we would expect enzymes recovered from different species to reflect these differences. For the eight characterized metagenomic enzymes (Table 1.4), we found a range of salt effects on activity including stimulation and various levels of inhibition. As shown in Table 1.4, three enzymes were completely inhibited by 2 M salt (MGS0006, MGS0105 and MGS0109), while the activity of two other enzymes was stimulated by addition of NaCl, and in fact have optimal activity at 2 M or over 3.5 M concentrations (MGS0012 and MGS0010, respectively). The remaining enzymes show intermediate salt tolerance and lie on various points along this spectrum. Thus, many carboxyl esterases from marine metagenomes are cold-adapted enzymes showing different levels of salt resistance.

1.6 Conclusion

In summary, enzymatic screening of metagenomic marine libraries identifies genes from diverse organisms and protein families with enzymatic properties that reflect the environmental conditions of the microbial community. Most of the esterases we have biochemically characterized are halotolerant or halophilic, cold adapted enzymes, as one would expect for proteins from a marine environment. We have identified several esterases which belong to uncharacterized families or proteins annotated to have alternate functions, which could not have been identified through sequence analysis alone.

Since only 40 % of enzymes from environmental DNA have been suggested to express in *E. coli*, (Gabor et al. 2004), this leaves a large proportion of the environmental gene pool unsampled during a standard enzyme activity screen, missing some potentially very exciting enzymes. In order to increase physiologic and metabolic diversity and therefore close this expression gap, multi-host shuttle vectors have already been designed for expression in *Bacillus subtilis*, *Pseudomonas putida*, *Streptomyces lividans*, and *Rhizobium leguminosarum* (Martinez et al. 2004; Staskawicz et al. 1987; Troeschel et al. 2012; Li et al. 2005; Wexler et al. 2005). Enzymatic screening of metagenomic libraries expressed in these hosts revealed different gene expression profiles indicating that additional metagenomic enzymes can be identified using this approach. For example, these hosts have successfully identified a novel alcohol/aldehyde dehydrogenase in *R. leguminosarum*, and multiple hemolytic clones that were active in *S. lividans* but not *E. coli*.

Presently, DNA sequencing and sequence-based studies of metagenomes far outpace functional metagenomic studies. As the computational load has now fallen on improving high-throughput annotation of thousands of metagenomes, functional studies offer us a targeted approach to find the needles in the haystack. Activity based studies remain crucial in identifying novel enzymes and enzyme families, and allow us to isolate and clone those genes which we know can be expressed in industrially-relevant hosts from gigabases of environmental DNA. Selecting appropriate environmental communities enriched in the activities we wish to mine for, and being creative in designing screens to identify enzymes with desired characteristics are key in the field.

Acknowledgments This work was supported in part by the Government of Canada through Genome Canada and the Ontario Genomics Institute (2009-OGI-ABC-1405 and the research agreement OGI-055), Ontario Research Fund (ORF-GL2-01-004), the EU FP7 project MAMBA (FP7-KBBE-2008-226977) and MicroB3 (OCEAN-2011-287589). We thank all members of the Structural Proteomics in Toronto (SPiT) Centre for help in conducting the experiments and Dr. Manuel Ferrer (Institute of Catalysis, Madrid) for helpful discussions.

References

Aly AS, Moustafa AB, Hebeish A (2004) Bio-technological treatment of cellulosic textiles. J Clean Prod 12(7):697–705

Baraniecki CA, Aislabie J, Foght JM (2002) Characterization of Sphingomonas sp. Ant 17, an aromatic hydrocarbon-degrading bacterium isolated from Antarctic soil. Microb Ecol 43(1):44–54

Bedingfield J (ed) (2013) The Novozymes Report. Novozymes. http://report2013.novozymes.com Accessed 24 Feb 2014

Beja O et al (2000) Bacterial rhodopsin: evidence for a new type of phototrophy in the sea. Science 289(5486):1902–1906

Beloqui A et al (2006) Novel polyphenol oxidase mined from a metagenome expression library of bovine rumen: biochemical properties, structural analysis, and phylogenetic relationships. J Biol Chem 281(32):22933–22942

Brisson VL et al (2012) Metagenomic analysis of a stable trichloroethene-degrading microbial community. ISME J 6(9):1702–1714
Cavicchioli R et al (2002) Low-temperature extremophiles and their applications. Curr Opin Biotechnol 13(3):253–261
Christian JH, Waltho JA (1961) The sodium and potassium content of non-halophilic bacteria in relation to salt tolerance. J Gen Microbiol 25:97–102
Christian JH, Waltho JA (1962) Solute concentrations within cells of halophilic and non-halophilic bacteria. Biochim Biophys Acta 65:506–508
Dinsdale EA et al (2008) Functional metagenomic profiling of nine biomes. Nature 452(7187):629–632
Feller G, Gerday C (2003) Psychrophilic enzymes: hot topics in cold adaptation. Nat Rev Microbiol 1(3):200–208
Ferrer M, Martínez-Abarca F, Golyshin PN (2005) Mining genomes and "metagenomes" for novel catalysts. Curr Opin Biotechnol 16(6):588–593
Ferrer M et al (2007) Mining enzymes from extreme environments. Curr Opin Microbiol 10(3):207–214
Ferrer M et al (2009) Metagenomics for mining new genetic resources of microbial communities. J Mol Microbiol Biotechnol 16(1–2):109–123
Fredrickson JK et al (1991) Isolation and characterization of a subsurface bacterium capable of growth on toluene, naphthalene, and other aromatic compounds. Appl Environ Microbiol 57(3):796–803
Fredrickson JK et al (1995) Aromatic-degrading Sphingomonas isolates from the deep subsurface. Appl Environ Microbiol 61(5):1917–1922
Gabor EM, Alkema WBL, Janssen DB (2004) Quantifying the accessibility of the metagenome by random expression cloning techniques. Environ Microbiol 6(9):879–886
Gado HM et al (2009) Influence of exogenous enzymes on nutrient digestibility, extent of ruminal fermentation as well as milk production and composition in dairy cows. Anim Feed Sci Technol 154(1–2):36–46
Giovannoni SJ, Britschgi TB, Moyer CL, Field KG (1990) Genetic diversity in Sargasso Sea bacterioplankton. Nature 345:60–63
Hausmann S, Jaeger EK (2010) Lipolytic enzymes from bacteria. In: Timmis KN (ed) Handbook of hydrocarbon and lipid microbiology. Springer, Berlin/Heidelberg, pp 1100–1126
Hemachander C, Puvanakrishnan R (2000) Lipase from Ralstonia pickettii as an additive in laundry detergent formulations. Process Biochem 35(8):809–814
Hernández-Martín E, Otero C (2008) Different enzyme requirements for the synthesis of biodiesel: Novozym 435 and Lipozyme TL IM. Bioresour Technol 99(2):277–286
Holloway P, Trevors JT, Lee H (1998) A colorimetric assay for detecting haloalkane dehalogenase activity. J Microbiol Methods 32(1):31–36
Jegannathan KR, Nielsen PH (2013) Environmental assessment of enzyme use in industrial production – a literature review. J Clean Prod 42:228–240
Jimenez L et al (1999) Biobleaching of cellulose pulp from wheat straw with enzymes and hydrogen peroxide. Process Biochem 35:149–157
Jones BV, Sun F, Marchesi JR (2007) Using skimmed milk agar to functionally screen a gut metagenomic library for proteases may lead to false positives. Lett Appl Microbiol 45(4):418–420
Kemp PF, Aller JY (2004) Bacterial diversity in aquatic and other environments: what 16S rDNA libraries can tell us. FEMS Microbiol Ecol 47(2):161–177
Kok RG et al (1993) Growth-phase-dependent expression of the lipolytic system of Acinetobacter calcoaceticus BD413: cloning of a gene encoding one of the esterases. J Gen Microbiol 139:2329–2342
Koonin EV, Galperin MY (2003) Sequence - evolution - function: computational approaches in comparative genomics. Kluwer Academic, Boston
Kouker G, Jaeger KE (1987) Specific and sensitive plate assay for bacterial lipases. Appl Environ Microbiol 53(1):211–213

Kwon KK et al (2005) Kordiimonas gwangyangensis gen. nov., sp. nov., a marine bacterium isolated from marine sediments that forms a distinct phyletic lineage (Kordiimonadales ord. nov.) in the “Alphaproteobacteria”. Int J Syst Evol Microbiol 55(Pt 5):2033–2037

Li Y et al (2005) Screening a wide host-range, waste-water metagenomic library in tryptophan auxotrophs of Rhizobium leguminosarum and of Escherichia coli reveals different classes of cloned trp genes. Environ Microbiol 7(12):1927–1936

Lonhienne T, Gerday C, Feller G (2000) Psychrophilic enzymes: revisiting the thermodynamic parameters of activation may explain local flexibility. Biochim Biophys Acta 1543(1):1–10

Lorenz P, Eck J (2005) Metagenomics and industrial applications. Nat Rev Microbiol 3(June):510–516

Martinez A et al (2004) Genetically modified bacterial strains and novel bacterial artificial chromosome shuttle vectors for constructing environmental libraries and detecting heterologous natural products in multiple expression hosts. Appl Environ Microbiol 70(4):2452–2463

Maymo-Gatell X et al (1997) Isolation of a bacterium that reductively dechlorinates tetrachloroethene to ethene. Science 276(5318):1568–1571

Monsan P, Donohue MJO (2010) Industrial biotechnology in the food and feed sector. In: Soetaert W, Vandamme EJ (eds) Industrial biotechnology: sustainable growth and economic success. Wiley-VCH Verlag GmbH & Co. KGaA, Weinheim, p 350

Nguyen D et al (2008) Bleaching of kraft pulp by a commercial lipase: accessory enzymes degrade hexenuronic acids. Enzyme Microb Technol 43(2):130–136

Okamura-Matsui T et al (2003) Discovery of alcohol dehydrogenase from mushrooms and application to alcoholic beverages. J Mol Catal B: Enzym 23(2–6):133–144

Olufsen M et al (2005) Increased flexibility as a strategy for cold adaptation: a comparative molecular dynamics study of cold- and warm-active uracil DNA glycosylase. J Biol Chem 280(18):18042–18048

Oren A (2002) Halophilic microorganisms and their environments. Kluwer Academic, Dordrecht

Osma JF, Toca-Herrera JL, Rodríguez-Couto S (2010) Uses of Laccases in the Food Industry. Enzyme Research 2010:918761. doi: 10.4061/2010/918761

Pelletier E et al (2008) “Candidatus Cloacamonas Acidaminovorans”: genome sequence reconstruction provides a first glimpse of a new bacterial division. J Bacteriol 190(7):2572–2579

Prasad S, Khadatare PB, Roy I (2011) Effect of chemical chaperones in improving the solubility of recombinant proteins in Escherichia coli. Appl Environ Microbiol 77(13):4603–4609

Ravenschlag K et al (1999) High bacterial diversity in permanently cold marine sediments. Appl Environ Microbiol 65(9):3982–3989

Robertson DE et al (2004) Exploring nitrilase sequence space for enantioselective catalysis. Appl Environ Microbiol 70(4):2429–2436

Roesch LFW et al (2007) Pyrosequencing enumerates and contrasts soil microbial diversity. ISME J 1(4):283–290

Rondon MR et al (2000) Cloning the soil metagenome: a strategy for accessing the genetic and functional diversity of uncultured microorganisms. Appl Environ Microbiol 66(6):2541–2547

Saeki K et al (2007) Detergent alkaline proteases: enzymatic properties, genes, and crystal structures. J Biosci Bioeng 103(6):501–508

Schleheck D et al (2000) An α-proteobacterium converts linear alkylbenzenesulfonate surfactants into sulfophenylcarboxylates and linear alkyldiphenyletherdisulfonate surfactants into sulfodiphenylethercarboxylates. Appl Environ Microbiol 66(5):1911–1916

Schloss PD, Handelsman J (2005) Introducing DOTUR, a computer program for defining operational taxonomic units and estimating species richness. Appl Environ Microbiol 71(3):1501–1506

Sellek GA, Chaudhuri JB (1999) Biocatalysis in organic media using enzymes from extremophiles. Enzyme Microb Technol 25(6):471–482

Siddiqui KS, Cavicchioli R (2006) Cold-adapted enzymes. Annu Rev Biochem 75:403–433

Simon C, Daniel R (2011) Metagenomic analyses: past and future trends. Appl Environ Microbiol 77(4):1153–1161

Staskawicz B et al (1987) Molecular characterization of cloned avirulence genes from race 0 and race 1 of Pseudomonas syringae pv. glycinea. J Bacteriol 169(12):5789–5794
Steele HL et al (2009) Advances in recovery of novel biocatalysts from metagenomes. J Mol Microbiol Biotechnol 16(1–2):25–37
Taupp M, Mewis K, Hallam SJ (2011) The art and design of functional metagenomic screens. Curr Opin Biotechnol 22(3):465–472
Teather RM, Wood PJ (1982) Use of Congo red-polysaccharide interactions in enumeration and characterization of cellulolytic bacteria from the bovine rumen. Appl Environ Microbiol 43(4):777–780
Troeschel SC et al (2012) Novel broad host range shuttle vectors for expression in Escherichia coli, Bacillus subtilis and Pseudomonas putida. J Biotechnol 161(2):71–79
Uchiyama T, Miyazaki K (2009) Functional metagenomics for enzyme discovery: challenges to efficient screening. Curr Opin Biotechnol 20(6):616–622
Van den Burg B (2003) Extremophiles as a source for novel enzymes. Curr Opin Microbiol 6(3):213–218
Venter JC et al (2004) Environmental genome shotgun sequencing of the Sargasso Sea. Science 304(5667):66–74
Wang Q et al (2006) Discrimination of esterase and peptidase activities of acylaminoacyl peptidase from hyperthermophilic Aeropyrum pernix K1 by a single mutation. J Biol Chem 281(27):18618–18625
Wang S et al (2013) Isolation and characterization of a novel organic solvent-tolerant and halotolerant esterase from a soil metagenomic library. J Mol Catal B: Enzym 95:1–8
Wexler M et al (2005) A wide host-range metagenomic library from a waste water treatment plant yields a novel alcohol/aldehyde dehydrogenase. Environ Microbiol 7(12):1917–1926
Williams AB, Rona PA (1986) Two new caridean shrimps (Bresiliidae) from a hydrothermal field on the Mid-Atlantic Ridge. J Crustac Biol 6(3):446–462
Yooseph S et al (2007) The Sorcerer II Global Ocean Sampling expedition: expanding the universe of protein families. PLoS Biol 5(3):e16

Chapter 2
Investigating Bacterial Protein Synthesis Using Systems Biology Approaches

Alla Gagarinova and Andrew Emili

Abstract Protein synthesis is essential for bacterial growth and survival. Its study in *Escherichia coli* helped uncover features conserved among bacteria as well as universally. The pattern of discovery and the identification of some of the longest-known components of the protein synthesis machinery, including the ribosome itself, tRNAs, and translation factors proceeded through many stages of successively more refined biochemical purifications, finally culminating in the isolation to homogeneity, identification, and mapping of the smallest unit required for performing the given function. These early studies produced a wealth of information. However, many unknowns remained. Systems biology approaches provide an opportunity to investigate protein synthesis from a global perspective, overcoming the limitations of earlier ad hoc methods to gain unprecedented insights. This chapter reviews innovative systems biology approaches, with an emphasis on those designed specifically for investigating the protein synthesis machinery in *E. coli*.

Keywords Bacteria • *E. coli* • Protein synthesis • Translation • Ribosome biogenesis • Systems biology • Ribosome profiling • Synthetic genetic array • Genetic interaction • Proteomics

Bacterial protein synthesis is an adaptive, multi-step process performed by the ribonucleoprotein machinery, consisting of the ribosome and a multitude of factors and accessory components (e.g. tRNA) that translate nucleic acid sequences encoded by different messenger RNAs (mRNAs) into the corresponding sequences of cognate amino acids in response to changing cellular demands. Due to evolutionary conservation of many aspects of protein synthesis, the enteric microbe *Escherichia coli* played a key role in understanding both prokaryotic and eukaryotic translation. For instance, mRNA was first identified as the template for protein synthesis in *E. coli* (Astrachan and Volkin 1958; Brenner et al. 1961; Gale and

A. Gagarinova (✉) • A. Emili
Department of Molecular Genetics, University of Toronto, Toronto, ON, Canada, M5S 1A8

Donnelly Centre for Cellular and Biomolecular Research, University of Toronto, Toronto, ON, Canada, M5S 3E1
e-mail: alla.gagarinova@mail.utoronto.ca

N.J. Krogan, M. Babu (eds.), *Prokaryotic Systems Biology*, Advances in Experimental Medicine and Biology 883, DOI 10.1007/978-3-319-23603-2_2

Folkes 1955) and *E. coli* was the first organism for which the genetic code was deciphered (Adams and Capecchi 1966; Brenner et al. 1965; Bretscher et al. 1965; Clark and Marcker 1966; Ganoza and Nakamoto 1966; Khorana et al. 1966; Lamborg and Zamecnik 1960; Last et al. 1967; Lengyel et al. 1961; Weigert and Garen 1965). This chapter will focus on *E. coli* protein synthesis because this is the best-studied bacterium and because of the important role this model organism has played in elucidating bacterial protein synthesis components, concepts, and mechanisms.

Early studies of bacterial protein synthesis relied on low throughput biochemical approaches and forward genetic screens to identify components of the protein synthesis machinery and investigate pertinent mechanisms (e.g. see Capecchi 1967a, b; Fakunding and Hershey 1973; Helser et al. 1971, 1972; Noll and Noll 1972; Sabol and Ochoa 1971; Sabol et al. 1970; Scolnick et al. 1968; Sparling 1970). Although many pivotal discoveries and breakthroughs were made using these conventional approaches, many unknowns about bacterial protein synthesis components and mechanisms remain (see Kaczanowska and Ryden-Aulin 2007 and below). At the same time, in the best annotated and studied bacterium, *E. coli*, about a third of its genes still lack experimental annotations (Hu et al. 2009). It is not surprising, therefore, that with the advent of systems biology approaches they are increasingly frequently being applied to the study bacterial protein synthesis, effectively complementing earlier approaches and circumventing their limitations.

Because protein synthesis is a complex, dynamic, and adaptive process, with many unknowns, diverse systems biology approaches are used to tackle this system from various perspectives. These approaches can be roughly categorized based on the kinds of information they collect and thus how they approach the study of bacterial protein synthesis. Methods in the first category approach protein synthesis by looking at translated messages (see Sect. 2.1); the second category includes genetic and proteomic methods that are aimed at identifying new components of the protein synthesis machinery and exploring component functions (see Sect. 2.2). These categories are not exclusive and can be combined to gain even greater insights about bacterial protein synthesis, as will be discussed below.

2.1 Systematic Investigation of Translated Messages Provides Insights into the Functioning of Bacterial Protein Synthesis Machinery

Taniguchi et al. (2010) used a fluorescent protein fusion library to measure mRNA and protein abundances for over 1000 proteins in *E. coli* on single cell level. The study revealed that mRNA and protein levels within a cell did not correlate well for any given gene (Taniguchi et al. 2010). Since most cellular processes are accomplished by proteins, this finding emphasizes the insufficiency of transcript abundance studies and the concurrent need to understand translation and translational regulation of gene expression for understanding cellular dynamics and processes.

Ingolia et al. (2009, 2012) developed ribosome profiling for investigating translation by looking at mRNA translated *in vivo*. The approach was first developed in yeast and then adapted to *E. coli*. Ribosome profiling, briefly, involves capturing translating ribosomes along with their target mRNAs; then nuclease footprinting is performed, during which all mRNAs not protected by the ribosomes are degraded; the short mRNA fragments protected by the ribosomes are then isolated and used in creating next generation deep sequencing libraries, which are sequenced (Ingolia et al. 2009, 2012). The obtained sequencing reads identify regions occupied by the ribosomes with nucleotide resolution and provide quantitative occupancy measurements, while preserving coding strand information (Ingolia et al. 2009, 2012). Oh et al. applied ribosome profiling to *E. coli* (Oh et al. 2011). The resulting map was reported to permit monitoring gene expression and defining protein coding sequence boundaries (Oh et al. 2011). Furthermore, it revealed nonuniform ribosome occupancy between and within genes, with average differences in gene expression levels spanning staggering five orders of magnitude (Oh et al. 2011). The authors also used affinity purification in combination with ribosome profiling (Fig. 2.1) to describe quantitatively and for the first time the substrates of co-translational chaperone trigger factor (Oh et al. 2011). For example, it was found that trigger factor recruitment is delayed until approximately 100 amino acids have been added to the growing polypeptide (Oh et al. 2011). Furthermore, trigger factor exhibited preference for interacting with nascent outer-membrane proteins (Oh et al. 2011).

Li et al. (2012) applied ribosome profiling to *E. coli* and evolutionarily distinct *Bacillus subtilis*. The authors combined ribosomal profiling with the use of differing growth conditions, an engineered ribosome and a reporter gene system to address a long-standing question about why translation rates differ between genes (Li et al. 2012). The genetic code is degenerate, meaning that multiple codons encode the same amino acid. Some codons and tRNAs are present in the cell at lower frequency. Proteins containing rare codons are less frequent, leading to speculations that ribosomes pause at these rare codons because of lower tRNA frequencies, which in turn leads to lower translational efficiency (see Gingold and Pilpel 2011). Many models where developed to quantify and describe these effects (see Gingold and Pilpel 2011 and references therein). However, surprisingly, Li et al. (2012) found that under nutrient-rich conditions ribosomes do not slow down at rare codons. Instead, approximately 70 % of the strong pauses occur at Shine-Dalgarno-like sequences within the coding regions of mRNA (Li et al. 2012). Bacterial translation typically initiates with the interaction between the Shine-Dalgarno sequence of mRNA upstream of the protein-coding region and the anti-Shine-Dalgarno sequence of 16S ribosomal RNA (rRNA), which places the codon for the first amino acid in the position suitable for translation initiation (Li et al. 2012; Shine and Dalgarno 1975). Li et al. (2012) demonstrated increased ribosomal occupancy at Shine-Dalgarno-like sequences within protein-coding regions of mRNA and that it was not due to the instances of internal initiation of translation. The authors proposed that the rare, Shine-Dalgarno-like codons were evolutionarily disfavored and the tRNA abundances adjusted through evolution to these codon adaptations

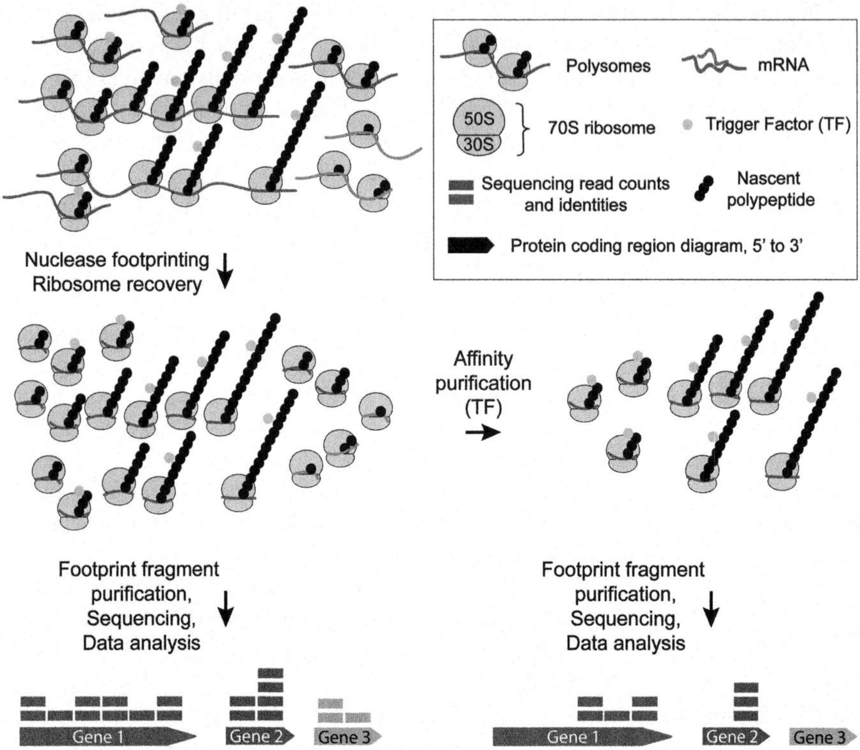

Fig. 2.1 Schematic of ribosome profiling procedure for quantifying translated mRNA by deep sequencing. In this example of selective ribosome profiling, mRNAs from all ribosomes or from ribosomes with nascent polypeptides bound to trigger factor (TF) chaperone are analyzed. Subsequently, the two sets of sequences are compared to characterize TF function. See main text for references and details

(Li et al. 2012). While frequent appearances of Shine-Dalgarno-like sequences would hinder growth, strategically placed translational pause sites can help ensure proper nascent peptide folding, targeting, translation, and transcription. Indeed, Fluman et al. (2014) analyzed the same data and found that Shine-Dalgarno-like elements trigger strategic pauses during the translation of membrane proteins. Fluman et al. (2014) further demonstrated that these pauses correlate with better folding of overexpressed membrane proteins, further reinforcing the shift in the perception about codon adaptation and evolution.

Ribosome profiling was similarly essential for dispelling the misconception about the mechanism through which macrolide antibiotics inhibit protein synthesis. Macrolides are large antibiotics binding the large ribosomal subunit inside the peptide exit tunnel (Wilson 2009). It has long been thought that macrolide antibiotics inhibit translation by blocking the peptide exit tunnel so that several initial amino acids are translated but once the growing polypeptide reaches the macrolide, which blocks its exit, no further elongation can take place, with peptide and macrolide occupancy of the exit tunnel being mutually exclusive (Wilson 2009). If this were the mode of macrolide action, ribosome profiling experiments following maximal inhibition of translation by a macrolide antibiotic would reveal arrest of translation near the nascent peptide's N-terminus. Kannan et al. (2014) found three common patterns of macrolide action in their ribosome profiling experiments for each of the two macrolides they tested. The effect of macrolide action matched the expected pattern for one group of genes (Kannan et al. 2014). However, the patterns observed for two other groups could not be explained by the existing model (Kannan et al. 2014). Translation of the second group of genes remained unaffected by macrolides; site-specific arrest of translation at positions past the N-terminal region was observed for the third group of genes (Kannan et al. 2014). Search for sequence specificity in arrest sites of this third group revealed specific and distinct features (Kannan et al. 2014). These features were roughly placed near the peptidyl transferase center rather than at the peptide exit tunnel (Kannan et al. 2014). Biochemical mapping and in vitro experiments confirmed that the sites of macrolide-induced translation arrest were defined primarily by the nature of amino acids present in the peptidyl transferase center (Kannan et al. 2014). Since some such macrolide-dependent stall sites were apparently missed, and since in vitro and *in vivo* inhibition patterns did not always mirror each other (Kannan et al. 2014), additional, possibly yet undiscovered, factors may come into play during *in vivo* protein synthesis.

Ribosome profiling was also used in a number of other studies. For example, it was used to investigate translation following bacteriophage lambda infection (Liu et al. 2013), to examine the consequences of the loss of poorly characterized elongation factor 4 (Balakrishnan et al. 2014), and to investigate the function of elongation factor P (EF-P) (Elgamal et al. 2014; Hersch et al. 2014; Woolstenhulme et al. 2015). Biochemically, EF-P was shown to be required for efficient elongation of translation across polyproline stretches (Doerfel et al. 2013; Ude et al. 2013). However, EF-P loss results in a minor growth defect despite many essential genes containing such motifs (Zou et al. 2012). Ribosome profiling helped reconcile and explain these seemingly opposing observations, expanded the understanding of the role of EF-P, and improved the understanding of translation (Elgamal et al. 2014; Hersch et al. 2014; Woolstenhulme et al. 2015).

Although only a limited number of publications used ribosome profiling so far, this approach clearly demonstrated its vast potential for future discoveries. For example, the approach offers a unique opportunity to query the functions of some conserved, yet dispensable rRNA or ribosomal protein (r-protein) modifications, whose functions thus far could not be successfully discerned (Kaczanowska and Ryden-Aulin 2007; Sergiev et al. 2012). When combined with appropriate genetic and environmental perturbations, ribosome profiling may represent a particularly powerful tool not only for investigating and understanding translation, but also for examining other biological phenomena, such as stress adaptation.

Continued and expanded application of ribosome profiling requires addressing the limitations of the approach as it relates to the system at hand. For example, elongation in *E. coli* is extremely fast, making capturing bacterial polysomes challenging without the rapid filtration and flash freezing before cell lysis (Li et al. 2012; Oh et al. 2011), which should therefore continue to be used in *E. coli* ribosome profiling studies. The limitation still present in all herein discussed ribosome profiling studies (Balakrishnan et al. 2014; Elgamal et al. 2014; Hersch et al. 2014; Li et al. 2012; Liu et al. 2013; Oh et al. 2011; Woolstenhulme et al. 2015) is that highly-translated sequences were the only ones that could be reliably mapped and analyzed. In the future, less abundant translated sequences may be captured and investigated by ribosome profiling if highly abundant sequences are depleted (Labaj et al. 2011) or if specific low-abundance sequences are selected (Levin et al. 2009), for instance, before the nuclease footprinting step.

2.2 Identifying New Components of the Bacterial Protein Synthesis Machinery and Investigating the Functions of Known Components

E. coli is the best annotated bacterium (Hu et al. 2009). Nonetheless, about one third of the protein-coding genes, including those that are essential or broadly conserved, lack experimental annotations (Hu et al. 2009). Given the many unknowns about protein synthesis, at least some of uncharacterized genes may affect this process. In an effort to facilitate gene function discovery for these 'orphan' unannotated *E. coli* genes, Hu et al. (2009) performed a thorough survey of the affinity-purification/mass spectrometry data encompassing nearly six thousand physical associations. In addition, the authors generated a comprehensive, high-quality map of genome context associations, encompassing nearly 75 thousand links and derived from gene co-conservation, gene fusion, and other data (Hu et al. 2009). Integration and analysis of the data helped assign orphans to specific biological processes, including protein synthesis (Hu et al. 2009). The effect of deleting two such orphans on protein synthesis was experimentally tested and verified (Hu et al. 2009), highlighting the utility of high-throughput datasets for investigating bacterial protein synthesis.

Sergiev et al. (2012) compiled data from various high-throughput datasets, including gene co-conservation and co-expression, phenotypic similarity between single gene deletion mutants across a variety of growth conditions, and protein physical association data with the aim of identifying uncharacterized genes whose products may affect or participate in rRNA modification. Each of the datasets contributed new and unique associations to the rRNA modification machinery of *E. coli* (Sergiev et al. 2012), which likely attests to the complementarity of different kinds of systems biology data in facilitating the understanding of bacterial protein synthesis. Vlasblom et al. (2015) compiled 48 biological networks from various literature sources to facilitate gene function discovery and exploration in *E. coli* in an accessible and user-friendly GeneMANIA (Mostafavi et al. 2008) format.

The compilation was experimentally verified to be useful for predicting genes affecting protein synthesis (Vlasblom et al. 2015). This resource will likely continue to be extremely useful. However, additional systematic systems biology studies need to be undertaken to create a more comprehensive framework for continued discoveries in the field of bacterial protein synthesis. Among published datasets, interactions and relationships most relevant to the study of protein synthesis are underrepresented, as will be described in the following two sections.

2.2.1 Genetic Interactions for Investigating Bacterial Protein Synthesis

Only 303 of the 4453 protein coding genes of *E. coli* are essential for growth under standard laboratory conditions (Baba et al. 2006). This low percentage (7 %), relative, for example, to yeast, a eukaryote with approximately 19 % gene essentiality (Giaever et al. 2002), suggests a high degree of functional buffering against loss-of-function mutations. Genetic buffering complicates the study of molecular functions: when a functionally redundant or a genetically buffered gene is deleted, only a muted phenotype indicative of a biological role may be observed in the absence of a secondary 'modifier' mutation (e.g. a growth defect may be apparent only in a sensitized genetic background). In such circumstances, mutations in two or more functionally overlapping genes may be necessary to reveal participation in a biological process that is subject to extensive functional redundancy.

The fact that several protein synthesis machinery components were discovered as genetic determinants in sensitized backgrounds suggests that the *E. coli* protein synthesis machinery is functionally buffered. For example, *ad hoc* forward genetic screens were used in the discovery of post-transcriptional modifications of 16S rRNA and the corresponding enzyme-encoding genes under conditions of inhibited translation in the presence of the antibiotic kasugamycin (Helser et al. 1971, 1972; Sparling 1970); the discovery of ribosomal proteins S4, S5, and S12 as being important for translational fidelity in the presence of spectinomycin (Carter et al. 2001; Ogle et al. 2002; Spahn and Prescott 1996); and the discovery of ribosome biogenesis factors that become essential at low and/or high temperatures in certain genetic backgrounds. For instance, *rimN*—encoding a ribosome maturation factor—was discovered as a suppressor of a temperature sensitive mutation in release factor 1, which is involved in the termination of translation (Kaczanowska and Ryden-Aulin 2005). RbfA and RimJ—two other ribosome biogenesis factors—were discovered as suppressors of temperature sensitive mutations in 16S rRNA and ribosomal protein S5, respectively (Dammel and Noller 1995; Roy-Chaudhuri et al. 2008). These examples demonstrate that genetic interactions can identify new protein synthesis genes. Genetic interactions can also help explore and define functions for translation genes. For example, Campbell and Brown (2008) combined *yjeQ* deletion with single mutations in each of pre-selected protein synthesis genes.

Based on the observed phenotypes, the authors concluded that *yjeQ* was likely involved in the biogenesis of the small ribosomal subunit of *E. coli* (Campbell and Brown 2008). This prediction was later verified by orthogonal approaches (Jomaa et al. 2011; Leong et al. 2013).

Synthetic genetic interaction screens, investigating pair-wise interactions between genes, systematically combine mutations and identify interacting genes by comparing the observed double mutant fitness measurements to expected measurements, which are determined based on the fitness of the corresponding single mutants (Mani et al. 2008). For example, according to the multiplicative model of genetic interactions (Mani et al. 2008), the fitness of the double mutant is expected to equal the product of the corresponding single mutant measurements (Fig. 2.2a). When the observed fitness is significantly less than or greater than expected, the two genes are said to be linked by aggravating or alleviating genetic interactions, respectively (Fig. 2.2a). Genetic interaction maps can help discern gene functions and pathway relationships (see example in Fig. 2.2b–e).

One systematic genetic interaction screening approach, termed eSGA (for *E. coli* Synthetic Genetic Arrays (Butland et al. 2008)), produced most of the currently available genetic interaction data for *E. coli*. The approach exploits bacterial mating, or conjugation, as well as successive rounds of robotic pinning, colony growth, and selection to construct arrays of double mutants from defined arrays of marked *E. coli* single gene deletion mutants (Butland et al. 2008). The fitness of single and double mutants is measured by measuring colony sizes and the interacting genes are identified (Butland et al. 2008). This approach has been successful in defining pathways, predicting gene functions, and describing the rewiring of genetic interactions in response to changing environments (Babu et al. 2011a, b, 2014; Butland et al. 2008). The latest and largest eSGA-based study investigated genome-wide genetic interactions of 163 *E. coli* genes and identified a new gene affecting protein synthesis despite the fact that protein synthesis genes were underrepresented among the selected 163 queries, with only three ribosome biogenesis factors, one elongation factor, and one tRNA synthetase having been screened (Babu et al. 2014). This highlights the potential of systematic genetic interaction studies for making new discoveries about bacterial translation. A variety of systems biology approaches for investigating genetic interactions in *E. coli* and other bacteria exist (see Gagarinova and Emili 2012 for review). Their systematic application in a variety of relevant conditions should continue improving the understanding of bacterial protein synthesis.

2.2.2 *Proteomic Approaches for Investigating Bacterial Protein Synthesis*

Most cellular processes are accomplished by proteins, which rarely act alone. The translation process and the ribosome itself are amazing examples of dynamic

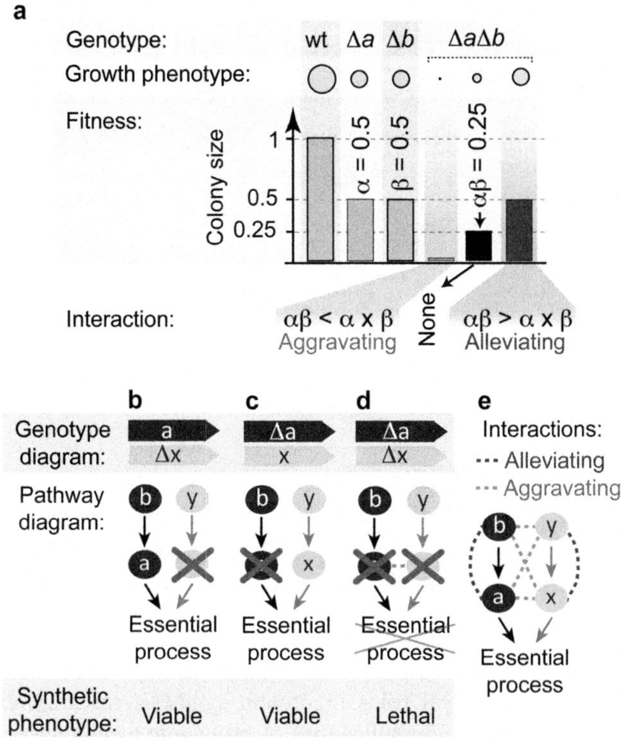

Fig. 2.2 The use of genetic interactions for investigating functional relationships. (**a**) Multiplicative model for scoring genetic interactions. The extent to which mutations in two genes jointly impact growth rate may be evaluated by measuring single and double mutant fitness. One way to estimate fitness is to measure the colony sizes of mutants arrayed on plates. Each of the colonies is quantified, with the wild type strain being assigned an arbitrary fitness of one unit. In the example shown, the single mutants, Δa and Δb, have the fitnesses α and β, respectively, equaling 0.5 relative to the wild type. The double mutants are denoted as $\Delta a\Delta b$, with fitness αβ. When two mutations are combined, according to the multiplicative model of genetic interactions, the double mutant is expected to have the αβ fitness of 0.25 ($\alpha\beta = \alpha \times \beta$) if the two genes do not interact (i.e. are functionally unrelated, or neutral). However, when a functional relationship connects the two genes, either a worse than expected ($\alpha\beta < 0.25$) or better than expected ($\alpha\beta > 0.25$) double mutant fitness is likely to be observed. The two genes are then said to have aggravating or alleviating genetic interactions, respectively. (**b**–**e**) Genetic interactions can help predict gene functions and delineate pathway relationships. Pathways 'ab' (consisting of genes 'a' and 'b') and 'xy' (consisting of genes 'x' and 'y') buffer each other and contribute to the same essential process. (**b**, **c**) Inactivating either pathway by a single mutation may slow growth but will not abolish viability. (**d**) However, when both pathways are abolished, the cell can no longer survive. The two inactivated genes will be said to show an extreme example of an aggravating genetic interaction—i.e. synthetic lethality. (**e**) When all combinations of pair-wise mutations for the two pathways are interrogated, the resulting pattern of genetic interactions will delineate pathways. Aggravating interactions will be observed between buffering pathways. Since perturbing two or more components in the same pathway may not result in further fitness reduction than seen upon single gene loss, alleviating interactions between genes within the same pathway will be observed. If the function of one of the genes (e.g. gene 'x') in this pathway map were unknown, its function and participation in pathway 'xy' could be predicted based on the aforementioned pattern of genetic interactions. See main text for references and details

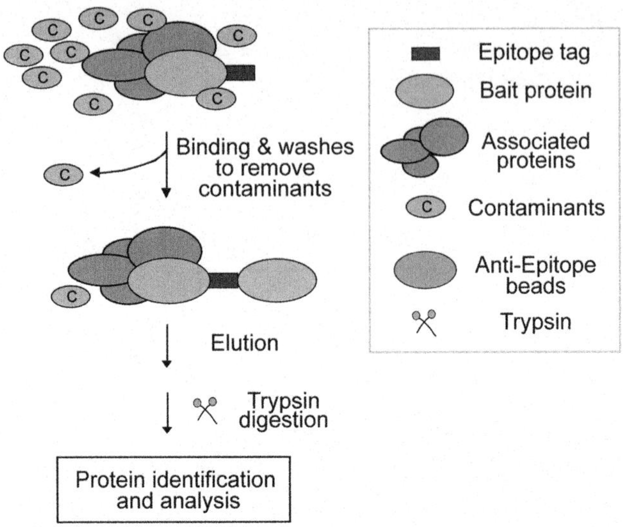

Fig. 2.3 A diagram showing the use of affinity purification with subsequent mass spectrometry for mapping protein–protein interactions. See main text for references and details

molecular cooperation. Correspondingly, biochemical protein-centric approaches have played key and pivotal roles in protein synthesis research. For example, successive improvements and applications of proteomic techniques were essential for identifying the 21 and 34 r-proteins of the small 30S and the large 50S ribosomal subunits, respectively (Kaltschmidt and Wittmann 1970). They were also essential for identifying and characterizing the various translation factors that transiently or stably interact with various components of the translation machinery to synthesize proteins (Capecchi 1967a, b; Fakunding and Hershey 1973; Noll and Noll 1972; Sabol and Ochoa 1971; Sabol et al. 1970; Scolnick et al. 1968). These early studies benefited from the high abundance of the protein synthesis machinery components they investigated. Newer methods that are continually being developed permit deeper querying of the protein synthesis machinery.

Affinity purification with subsequent mass spectrometry (AP/MS) can be used to map protein–protein interactions (PPI) corresponding to high- and low-abundance endogenous protein complexes (Fig. 2.3). From among all prokaryotic organisms, this approach has been most extensively applied to *E. coli* (Babu et al. 2009; Butland et al. 2005; Hu et al. 2009). For typical *E. coli* AP/MS studies, briefly, strains with individually epitope-tagged proteins are constructed; each epitope-tagged bait protein, along with its stable PPI partners, is affinity purified; all purified proteins are then identified by mass spectrometry and interactions between each bait protein and its interacting partners are mapped (Babu et al. 2009; Butland et al. 2005; Hu et al. 2009; Zeghouf et al. 2004). Such typical AP/MS studies are limited to stable complexes. Since the purification procedures are not perfect in eliminating all high-abundance co-purifying proteins, false positive identifications may be made (Nesvizhskii 2012). The strength of systematic genome-wide AP/MS studies compared to single purifications is that common contaminants can be identified and filtered out (Nesvizhskii 2012).

While the analysis of high throughput *E. coli* AP/MS data allowed identifying new, previously uncharacterized proteins affecting protein synthesis (Hu et al. 2009), these data likely do not capture the entire repertoire of physical associations underlying translation. One reason for this is that many known PPIs, required for efficient translation, are transient and therefore may be overlooked. Cross-linking interacting proteins before in AP/MS should help recover transient interactions (Kim et al. 2012). Another limitation is due to the confounding abundance of many core translation components, which typically represent the most abundant proteins during exponential growth (Taniguchi et al. 2010). Exponentially growing *E. coli* contain about 90,000 ribosomes per cell, which constitute up to approximately half of the microbial dry mass (Kjeldgaard and Gausing 1974; Tissieres and Watson 1958). Such highly abundant proteins may frequently co-purify with unrelated baits and distinguishing their true interactions from irrelevant co-purifications can be difficult. Therefore, new ribosome biogenesis or translation factors may be obscured in standard AP/MS studies.

Biochemical pre-fractionation of ribosomal components can help overcome some of the AP/MS limitations, providing additional insights. Briefly, fractions are separated by sucrose gradient ultracentrifugation and fractionation (Fig. 2.4a); then quantitative mass spectrometry approaches identify and determine protein abundances in each of the fractions (Fig. 2.4b). Fractionation can separate the peaks consisting primarily of the small 30S and the large 50S subunits of the ribosome, the 70S ribosome, as well as the polysomes (Fig. 2.4a). Since ribosomal proteins are sequentially incorporated into each of the ribosomal subunits during ribosome biogenesis, analysis of sequential fractions, along with relative or absolute quantitation, can allow for molecular reconstruction of the *in vivo* assembly process (Chen et al. 2012; Chen and Williamson 2013; Jiang et al. 2007; Sykes et al. 2010). For example, *in vivo* maps of normal ribosome biogenesis, generated by using fractionation and mass spectrometry, are largely consistent with the in vitro assembly maps, which were produced and refined over decades (Chen and Williamson 2013; Guo et al. 2013; Kaczanowska and Ryden-Aulin 2007; Xu and Culver 2010). At the same time, analysis of fractions from cultures grown in low temperature revealed that several late-binding proteins were underrepresented in subunit peaks but were present in 70S ribosomes, likely indicating the delayed addition of these proteins at low temperature (Jiang et al. 2007). Furthermore, combining fractionation and quantitative proteomic analysis with pulse-labeling helped investigate not only the incorporation of ribosomal proteins but also their exchange during normal bacterial growth (Bunner et al. 2008; Chen et al. 2012; Chen and Williamson 2013) as well as during growth in the presence of neomycin, which causes the accumulation of 30S precursor particles (Sykes et al. 2010).

Extending the analysis beyond ribosomal proteins helped find new ribosome-associated proteins in addition to recovering known ribosome biogenesis and translation factors (Chen and Williamson 2013; Jiang et al. 2007). For example,

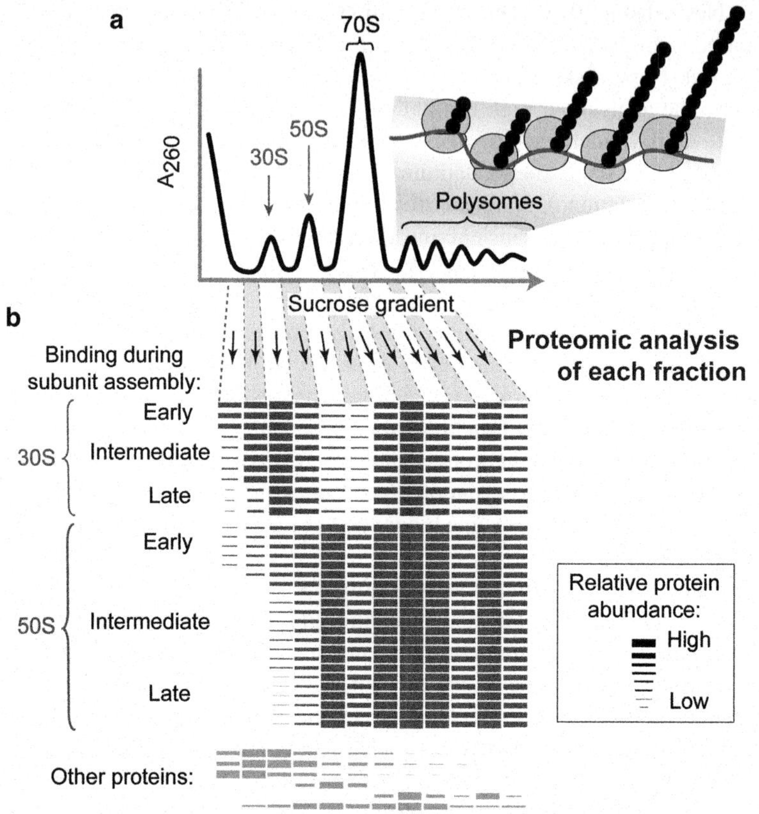

Fig. 2.4 A diagram of fractionation and quantitative proteomics approach for investigating protein synthesis. (**a**) Ribosomal particles are separated in sucrose gradient by ultracentrifugation and fractionation. A hypothetical absorbance trace is shown; 30S and 50S subunit, 70S ribosome, and polysomal peaks are highlighted. (**b**) Quantitative or semi-quantitative analyses of the protein contents of each of the selected fractions can help identify and quantify ribosome-associated factors and ribosomal proteins. The figure highlights the groups of ribosomal proteins that are added at early, intermediate, and late stages of ribosome biogenesis. Not all ribosomal proteins or ribosome-associated factors are shown. See main text for references and details

the protein YbeB was identified as a novel ribosome-associating factor by Jiang et al. (2007). Subsequent work by Hauser et al. (2012) demonstrated that it is a ribosome silencing factor, necessary for efficient adaptation to stationary growth and to the shift from rich to poor growth medium. Despite the demonstrated usefulness of the aforementioned studies for investigating protein synthesis in normal and altered growth conditions, none identified the full complement of known translation or ribosome biogenesis factors (Chen and Williamson 2013; Jiang et al. 2007), indicating incomplete coverage. This may be due to the transient nature of many of the interactions, essential for protein synthesis (see Kaczanowska and Ryden-Aulin 2007 for review). An additional complication is that large, unrelated complexes co-sedimented with ribosomal components (Jiang et al. 2007) and the corresponding proteins would need to be excluded in studies aiming to identify new translation and ribosome biogenesis factors. The increasing sensitivity and detection range of new mass spectrometry instruments and selective application of cross-linking with improved fractionation or affinity purifications can help recover all known factors to characterize their functions in addition to discovering new ones, while eliminating unrelated contaminants. Standard and modified proteomic approaches, applied to the study of protein synthesis in wild type and mutant strains, grown in standard and altered conditions will provide new insights about translation.

2.3 Outlook and Conclusions

Bacterial protein synthesis has been the subject of intense investigations since the middle of the last century. Plethora of information about this system has since been accumulated. However, much still remains unknown. Ribosome biogenesis and translational fidelity are two of the many areas, where the new systems biology approaches will likely be essential for finding answers to the multitude of pending questions.

In vivo, ribosome biogenesis starts concomitantly with the synthesis the three *E. coli* rRNA molecules—5S, 16S, and 23S, which are transcribed together as large precursor molecules from each of seven operons (Kaczanowska and Ryden-Aulin 2007). Each precursor is then cleaved, modified, and processed in a series of enzymatic steps coupled to ribosomal subunit assembly facilitated by ribosome biogenesis factors, wherein over 50 r-proteins are sequentially incorporated to form the functional ribosome (Kaczanowska and Ryden-Aulin 2007; Srivastava and Schlessinger 1990).

Despite this apparent complexity, *E. coli* ribosome maturation is highly efficient *in vivo*, being accomplished in less than 2 min during fast growth (Lindahl 1975). Although far less efficient, in vitro assembly is also possible (Xu and Culver 2010), indicating that all information needed for ribosome assembly is contained within its components. However, unlike *in vivo* assembly, in vitro assembly depends on pre-processed ribosomal components and requires long incubation steps and non-physiological conditions (Guo et al. 2013). While in vitro experiments have generated ribosomal subunit assembly maps and insights into the cooperative nature of r-protein binding (Hamacher et al. 2006; Mizushima and Nomura 1970; Nierhaus and Dohme 1974), it is clear that *in vivo* assembly involves not only known but also yet unknown factors (Shajani et al. 2011). Similarly, while many aspects of rRNA processing have been meticulously and carefully elucidated, the enzyme(s) responsible for the processing of the $3'$ end of 16S rRNA and the $5'$ ends of 23S and 5S rRNA remain unknown (Kaczanowska and Ryden-Aulin 2007). Taken together, these observations point to the existence of yet unidentified cellular components playing a role in ribosome biogenesis.

Another aspect of protein synthesis—translational fidelity—has gathered some attention but is still poorly understood. Translational fidelity can vary significantly even within the range of normal physiological conditions—for example, depending on the nutritional status of the cell (O'Farrell 1978). Various sources of 'error' may contribute to the overall accuracy of translational fidelity. These include tRNA mischarging (Cramer et al. 1991) and errors made by the ribosome, such as (1) non-canonical interpretation of the genetic code (Farabaugh 1996; Jorgensen et al. 1993); (2) loss of the reading frame through frameshift (Baranov et al. 2002) or slippage (Gallant and Lindsley 1998; Herr et al. 2000; Huang et al. 1988; Weiss et al. 1990), whereby ribosome skips a normally translated region of mRNA without translating it; and (3) processivity errors, whereby the nascent polypeptide is released prematurely (Janssen and Hayes 2012; Jorgensen and Kurland 1990; Keiler and Lee 2010; Keiler et al. 1996; Menninger 1976).

Instances of programmed translational errors and observations of sequence context affecting the frequencies of 'misreading' of the mRNA by the ribosome have been proposed to contribute to proteome diversity, particularly for low-abundance polypeptides (Alam et al. 1999). Thus, non-canonical interpretations by the ribosome are sometimes referred to as 'recoding' (Baranov et al. 2002). Recoding types were investigated to various depths, with specific environmental or sequence determinants having been discovered for each type of recoding. For example, ribosomes can slip over "hungry codons", which encode amino acids whose availability is limited, and continue translating at a cognate downstream codon (Gallant and Lindsley 1998). In an extreme example of programmed slippage, during the synthesis of a topoisomerase encoded by the bacteriophage T4, 50 nucleotides are omitted in order to synthesize the correct full-length protein product (Herr et al. 2000; Huang et al. 1988; Weiss et al. 1990). The sequence of amino acids joined prior to the slippage event (Weiss et al. 1990), nucleic acid sequence context (Farabaugh 1996; Gallant and Lindsley 1998; Herr et al. 2000; Huang et al. 1988; Stahl et al. 2002; Weiss et al. 1990), and physiological context (Gallant and Lindsley 1998) can affect the likelihood of slippage. For all of these errors, the mechanistic dependencies and potentially new regulators remain unclear.

New factors affecting ribosome biogenesis and translation may be identified and their roles investigated by systematic genetic interaction screens (see Sect. 2.2.1) or by proteomic approaches (see Sect. 2.2.2). Ribosome biogenesis or translation factors, associating with pre-ribosomal or ribosomal particles, are best targeted by proteomic approaches, separating the various ribosomal particles by fractionation. The capture of these factors may be further improved, for example, by purifying the particles of interest by AP/MS after fractionation and by using cross-linking to retain transient interactions (see Sect. 2.2.2). Some modifications of ribosomal RNA and proteins as well as some ribosomal features have currently no known roles in protein synthesis (Golovina et al. 2012; Kaczanowska and Ryden-Aulin 2007). To investigate the roles of these modifications in ribosome biogenesis, null mutants can be constructed and used in the fractionation-based proteomics experiments (see Sect. 2.2.2), while the effects of the same mutations on translation can be examined using ribosome profiling experiments (see Sect. 2.1). The individual and combined roles of translation and ribosome biogenesis factors can be examined in similar detail by examining mutants with single and multiple mutations in genetic, proteomic, and ribosome profiling experiments.

The potential effects of physiologically-relevant and stress conditions on the aforementioned and other aspects of protein synthesis are even less well understood. Since the rate of peptide bond formation, temperature, and growth rate exhibit mutual dependencies (Asato 2005), better understanding of protein synthesis requires investigating it in a variety of relevant conditions. Given that many aspects of the protein synthesis machinery are retained through evolution (Carter et al. 2001; Liebman et al. 1995; Ogle et al. 2002; Spahn and Prescott 1996; Watanabe 2010), the discoveries made in *E. coli* should help understand translation in other species, including those that cannot be easily manipulated or cultured in the laboratory settings. Better understanding of translation will help harness it for biotechnological applications (Han et al. 2004; Hochkoeppler 2013; Lee and Lee 2005; Waegeman and Soetaert 2011) and may provide additional avenues of developing antimicrobials (Comartin and Brown 2006), which is particularly relevant in view of the increasing drug resistance in the clinic.

References

Adams JM, Capecchi MR (1966) N-formylmethionyl-sRNA as the initiator of protein synthesis. Proc Natl Acad Sci U S A 55:147–155

Alam SL, Atkins JF, Gesteland RF (1999) Programmed ribosomal frameshifting: much ado about knotting! Proc Natl Acad Sci U S A 96:14177–14179

Asato Y (2005) Control of ribosome synthesis during the cell division cycles of E. coli and Synechococcus. Curr Issues Mol Biol 7:109–117

Astrachan L, Volkin E (1958) Properties of ribonucleic acid turnover in T2-infected Escherichia coli. Biochim Biophys Acta 29:536–544

Baba T, Ara T, Hasegawa M, Takai Y, Okumura Y, Baba M, Datsenko KA, Tomita M, Wanner BL, Mori H (2006) Construction of Escherichia coli K-12 in-frame, single-gene knockout mutants: the Keio collection. Mol Syst Biol 2, 2006.0008

Babu M, Butland G, Pogoutse O, Li J, Greenblatt JF, Emili A (2009) Sequential peptide affinity purification system for the systematic isolation and identification of protein complexes from Escherichia coli. Methods Mol Biol 564:373–400

Babu M, Aoki H, Chowdhury WQ, Gagarinova A, Graham C, Phanse S, Laliberte B, Sunba N, Jessulat M, Golshani A, Emili A, Greenblatt JF, Ganoza MC (2011a) Ribosome-dependent ATPase interacts with conserved membrane protein in Escherichia coli to modulate protein synthesis and oxidative phosphorylation. PLoS One 6:e18510

Babu M, Diaz-Mejia JJ, Vlasblom J, Gagarinova A, Phanse S, Graham C, Yousif F, Ding H, Xiong X, Nazarians-Armavil A, Alamgir M, Ali M, Pogoutse O, Pe'er A, Arnold R, Michaut M, Parkinson J, Golshani A, Whitfield C, Wodak SJ, Moreno-Hagelsieb G, Greenblatt JF, Emili A (2011b) Genetic interaction maps in Escherichia coli reveal functional crosstalk among cell envelope biogenesis pathways. PLoS Genet 7:e1002377

Babu M, Arnold R, Bundalovic-Torma C, Gagarinova A, Wong KS, Kumar A, Stewart G, Samanfar B, Aoki H, Wagih O, Vlasblom J, Phanse S, Lad K, Yeou Hsiung Yu A, Graham C, Jin K, Brown E, Golshani A, Kim P, Moreno-Hagelsieb G, Greenblatt J, Houry WA, Parkinson J, Emili A (2014) Quantitative genome-wide genetic interaction screens reveal global epistatic relationships of protein complexes in Escherichia coli. PLoS Genet 10:e1004120

Balakrishnan R, Oman K, Shoji S, Bundschuh R, Fredrick K (2014) The conserved GTPase LepA contributes mainly to translation initiation in Escherichia coli. Nucleic Acids Res 42:13370–13383
Baranov PV, Gesteland RF, Atkins JF (2002) Recoding: translational bifurcations in gene expression. Gene 286:187–201
Brenner S, Jacob F, Meselson M (1961) An unstable intermediate carrying information from genes to ribosomes for protein synthesis. Nature 190:576–581
Brenner S, Stretton AO, Kaplan S (1965) Genetic code: the 'nonsense' triplets for chain termination and their suppression. Nature 206:994–998
Bretscher MS, Goodman HM, Menninger JR, Smith JD (1965) Polypeptide chain termination using synthetic polynucleotides. J Mol Biol 14:634–639
Bunner AE, Trauger SA, Siuzdak G, Williamson JR (2008) Quantitative ESI-TOF analysis of macromolecular assembly kinetics. Anal Chem 80:9379–9386
Butland G, Peregrin-Alvarez JM, Li J, Yang W, Yang X, Canadien V, Starostine A, Richards D, Beattie B, Krogan N, Davey M, Parkinson J, Greenblatt J, Emili A (2005) Interaction network containing conserved and essential protein complexes in Escherichia coli. Nature 433:531–537
Butland G, Babu M, Diaz-Mejia JJ, Bohdana F, Phanse S, Gold B, Yang W, Li J, Gagarinova AG, Pogoutse O, Mori H, Wanner BL, Lo H, Wasniewski J, Christopolous C, Ali M, Venn P, Safavi-Naini A, Sourour N, Caron S, Choi JY, Laigle L, Nazarians-Armavil A, Deshpande A, Joe S, Datsenko KA, Yamamoto N, Andrews BJ, Boone C, Ding H, Sheikh B, Moreno-Hagelseib G, Greenblatt JF, Emili A (2008) eSGA: E. coli synthetic genetic array analysis. Nat Methods 5:789–795
Campbell TL, Brown ED (2008) Genetic interaction screens with ordered overexpression and deletion clone sets implicate the Escherichia coli GTPase YjeQ in late ribosome biogenesis. J Bacteriol 190:2537–2545
Capecchi MR (1967a) Polypeptide chain termination in vitro: isolation of a release factor. Proc Natl Acad Sci U S A 58:1144–1151
Capecchi MR (1967b) A rapid assay for polypeptide chain termination. Biochem Biophys Res Commun 28:773–778
Carter AP, Clemons WM Jr, Brodersen DE, Morgan-Warren RJ, Hartsch T, Wimberly BT, Ramakrishnan V (2001) Crystal structure of an initiation factor bound to the 30S ribosomal subunit. Science 291:498–501
Chen SS, Williamson JR (2013) Characterization of the ribosome biogenesis landscape in E. coli using quantitative mass spectrometry. J Mol Biol 425:767–779
Chen SS, Sperling E, Silverman JM, Davis JH, Williamson JR (2012) Measuring the dynamics of E-coli ribosome biogenesis using pulse-labeling and quantitative mass spectrometry. Mol Biosyst 8:3325–3334
Clark BF, Marcker KA (1966) N-formyl-methionyl-sigma-ribonucleic acid and chain initiation in protein biosynthesis. Polypeptide synthesis directed by a bacteriophage ribonucleic acid in a cell-free system. Nature 211:378–380
Comartin DJ, Brown ED (2006) Non-ribosomal factors in ribosome subunit assembly are emerging targets for new antibacterial drugs. Curr Opin Pharmacol 6:453–458
Cramer F, Englisch U, Freist W, Sternbach H (1991) Aminoacylation of tRNAs as critical step of protein biosynthesis. Biochimie 73:1027–1035
Dammel CS, Noller HF (1995) Suppression of a cold-sensitive mutation in 16S rRNA by overexpression of a novel ribosome-binding factor, RbfA. Genes Dev 9:626–637
Doerfel LK, Wohlgemuth I, Kothe C, Peske F, Urlaub H, Rodnina MV (2013) EF-P is essential for rapid synthesis of proteins containing consecutive proline residues. Science 339:85–88
Elgamal S, Katz A, Hersch SJ, Newsom D, White P, Navarre WW, Ibba M (2014) EF-P dependent pauses integrate proximal and distal signals during translation. PLoS Genet 10:e1004553
Fakunding JL, Hershey JW (1973) The interaction of radioactive initiation factor IF-2 with ribosomes during initiation of protein synthesis. J Biol Chem 248:4206–4212
Farabaugh PJ (1996) Programmed translational frameshifting. Annu Rev Genet 30:507–528

Fluman N, Navon S, Bibi E, Pilpel Y (2014) mRNA-programmed translation pauses in the targeting of E. coli membrane proteins. eLife 3, e03440

Gagarinova A, Emili A (2012) Genome-scale genetic manipulation methods for exploring bacterial molecular biology. Mol Biosyst 8:1626–1638

Gale EF, Folkes JP (1955) The assimilation of amino acids by bacteria. 21. The effect of nucleic acids on the development of certain enzymic activities in disrupted staphylococcal cells. Biochem J 59:675–684

Gallant JA, Lindsley D (1998) Ribosomes can slide over and beyond "hungry" codons, resuming protein chain elongation many nucleotides downstream. Proc Natl Acad Sci U S A 95:13771–13776

Ganoza MC, Nakamoto T (1966) Studies on the mechanism of polypeptide chain termination in cell-free extracts of E. coli. Proc Natl Acad Sci U S A 55:162–169

Giaever G, Chu AM, Ni L, Connelly C, Riles L, Veronneau S, Dow S, Lucau-Danila A, Anderson K, Andre B, Arkin AP, Astromoff A, El-Bakkoury M, Bangham R, Benito R, Brachat S, Campanaro S, Curtiss M, Davis K, Deutschbauer A, Entian KD, Flaherty P, Foury F, Garfinkel DJ, Gerstein M, Gotte D, Guldener U, Hegemann JH, Hempel S, Herman Z, Jaramillo DF, Kelly DE, Kelly SL, Kotter P, LaBonte D, Lamb DC, Lan N, Liang H, Liao H, Liu L, Luo C, Lussier M, Mao R, Menard P, Ooi SL, Revuelta JL, Roberts CJ, Rose M, Ross-Macdonald P, Scherens B, Schimmack G, Shafer B, Shoemaker DD, Sookhai-Mahadeo S, Storms RK, Strathern JN, Valle G, Voet M, Volckaert G, Wang CY, Ward TR, Wilhelmy J, Winzeler EA, Yang Y, Yen G, Youngman E, Yu K, Bussey H, Boeke JD, Snyder M, Philippsen P, Davis RW, Johnston M (2002) Functional profiling of the Saccharomyces cerevisiae genome. Nature 418:387–391

Gingold H, Pilpel Y (2011) Determinants of translation efficiency and accuracy. Mol Syst Biol 7:481

Golovina AY, Dzama MM, Osterman IA, Sergiev PV, Serebryakova MV, Bogdanov AA, Dontsova OA (2012) The last rRNA methyltransferase of E. coli revealed: the yhiR gene encodes adenine-N6 methyltransferase specific for modification of A2030 of 23S ribosomal RNA. RNA 18:1725–1734

Guo Q, Goto S, Chen Y, Feng B, Xu Y, Muto A, Himeno H, Deng H, Lei J, Gao N (2013) Dissecting the *in vivo* assembly of the 30S ribosomal subunit reveals the role of RimM and general features of the assembly process. Nucleic Acids Res 41:2609–2620

Hamacher K, Trylska J, McCammon JA (2006) Dependency map of proteins in the small ribosomal subunit. PLoS Comput Biol 2:e10

Han MJ, Park SJ, Park TJ, Lee SY (2004) Roles and applications of small heat shock proteins in the production of recombinant proteins in Escherichia coli. Biotechnol Bioeng 88:426–436

Hauser R, Pech M, Kijek J, Yamamoto H, Titz B, Naeve F, Tovchigrechko A, Yamamoto K, Szaflarski W, Takeuchi N, Stellberger T, Diefenbacher ME, Nierhaus KH, Uetz P (2012) RsfA (YbeB) proteins are conserved ribosomal silencing factors. PLoS Genet 8:e1002815

Helser TL, Davies JE, Dahlberg JE (1971) Change in methylation of 16S ribosomal RNA associated with mutation to kasugamycin resistance in Escherichia coli. Nat New Biol 233:12–14

Helser TL, Davies JE, Dahlberg JE (1972) Mechanism of kasugamycin resistance in Escherichia coli. Nat New Biol 235:6–9

Herr AJ, Gesteland RF, Atkins JF (2000) One protein from two open reading frames: mechanism of a 50 nt translational bypass. EMBO J 19:2671–2680

Hersch SJ, Elgamal S, Katz A, Ibba M, Navarre WW (2014) Translation initiation rate determines the impact of ribosome stalling on bacterial protein synthesis. J Biol Chem 289:28160–28171

Hochkoeppler A (2013) Expanding the landscape of recombinant protein production in Escherichia coli. Biotechnol Lett 35:1971–1981

Hu P, Janga SC, Babu M, Diaz-Mejia JJ, Butland G, Yang W, Pogoutse O, Guo X, Phanse S, Wong P, Chandran S, Christopoulos C, Nazarians-Armavil A, Nasseri NK, Musso G, Ali M, Nazemof N, Eroukova V, Golshani A, Paccanaro A, Greenblatt JF, Moreno-Hagelsieb G, Emili A (2009) Global functional atlas of Escherichia coli encompassing previously uncharacterized proteins. PLoS Biol 7:e96

Huang WM, Ao SZ, Casjens S, Orlandi R, Zeikus R, Weiss R, Winge D, Fang M (1988) A persistent untranslated sequence within bacteriophage T4 DNA topoisomerase gene 60. Science 239:1005–1012

Ingolia NT, Ghaemmaghami S, Newman JR, Weissman JS (2009) Genome-wide analysis *in vivo* of translation with nucleotide resolution using ribosome profiling. Science 324:218–223

Ingolia NT, Brar GA, Rouskin S, McGeachy AM, Weissman JS (2012) The ribosome profiling strategy for monitoring translation *in vivo* by deep sequencing of ribosome-protected mRNA fragments. Nat Protoc 7:1534–1550

Janssen BD, Hayes CS (2012) The tmRNA ribosome-rescue system. Adv Protein Chem Struct Biol 86:151–191

Jiang M, Sullivan SM, Walker AK, Strahler JR, Andrews PC, Maddock JR (2007) Identification of novel Escherichia coli ribosome-associated proteins using isobaric tags and multidimensional protein identification techniques. J Bacteriol 189:3434–3444

Jomaa A, Stewart G, Martin-Benito J, Zielke R, Campbell TL, Maddock JR, Brown ED, Ortega J (2011) Understanding ribosome assembly: the structure of *in vivo* assembled immature 30S subunits revealed by cryo-electron microscopy. RNA 17:697–709

Jorgensen F, Kurland CG (1990) Processivity errors of gene expression in Escherichia coli. J Mol Biol 215:511–521

Jorgensen F, Adamski FM, Tate WP, Kurland CG (1993) Release factor-dependent false stops are infrequent in Escherichia coli. J Mol Biol 230:41–50

Kaczanowska M, Ryden-Aulin M (2005) The YrdC protein—a putative ribosome maturation factor. Biochim Biophys Acta 1727:87–96

Kaczanowska M, Ryden-Aulin M (2007) Ribosome biogenesis and the translation process in Escherichia coli. Microbiol Mol Biol Rev 71:477–494

Kaltschmidt E, Wittmann HG (1970) Ribosomal proteins. XII. Number of proteins in small and large ribosomal subunits of Escherichia coli as determined by two-dimensional gel electrophoresis. Proc Natl Acad Sci U S A 67:1276–1282

Kannan K, Kanabar P, Schryer D, Florin T, Oh E, Bahroos N, Tenson T, Weissman JS, Mankin AS (2014) The general mode of translation inhibition by macrolide antibiotics. Proc Natl Acad Sci U S A 111:15958–15963

Keiler K, Lee D (2010) trans-Translation. In: Atkins JF, Gesteland RF (eds) Recoding: expansion of decoding rules enriches gene expression. Springer, New York, pp 383–405

Keiler KC, Waller PR, Sauer RT (1996) Role of a peptide tagging system in degradation of proteins synthesized from damaged messenger RNA. Science 271:990–993

Khorana HG, Buchi H, Ghosh H, Gupta N, Jacob TM, Kossel H, Morgan R, Narang SA, Ohtsuka E, Wells RD (1966) Polynucleotide synthesis and the genetic code. Cold Spring Harb Symp Quant Biol 31:39–49

Kim KM, Yi EC, Kim Y (2012) Mapping protein receptor–ligand interactions via *in vivo* chemical crosslinking, affinity purification, and differential mass spectrometry. Methods 56:161–165

Kjeldgaard NO, Gausing K (1974) Regulation of Biosynthesis of Ribosomes. In Ribosomes, P. Lengyel, M. Nomura, and A. Tissières, eds. (Cold Spring Harbor, New York: Cold Spring Harbor Laboratory), pp. 369–392

Labaj PP, Leparc GG, Linggi BE, Markillie LM, Wiley HS, Kreil DP (2011) Characterization and improvement of RNA-Seq precision in quantitative transcript expression profiling. Bioinformatics 27:i383–i391

Lamborg MR, Zamecnik PC (1960) Amino acid incorporation into protein by extracts of E. coli. Biochim Biophys Acta 42:206–211

Last JA, Stanley WM Jr, Salas M, Hille MB, Wahba AJ, Ochoa S (1967) Translation of the genetic message, IV. UAA as a chain termination codon. Proc Natl Acad Sci U S A 57:1062–1067

Lee PS, Lee KH (2005) Engineering HlyA hypersecretion in Escherichia coli based on proteomic and microarray analyses. Biotechnol Bioeng 89:195–205

Lengyel P, Speyer JF, Ochoa S (1961) Synthetic polynucleotides and the amino acid code. Proc Natl Acad Sci U S A 47:1936–1942

Leong V, Kent M, Jomaa A, Ortega J (2013) Escherichia coli rimM and yjeQ null strains accumulate immature 30S subunits of similar structure and protein complement. RNA 19:789–802
Levin JZ, Berger MF, Adiconis X, Rogov P, Melnikov A, Fennell T, Nusbaum C, Garraway LA, Gnirke A (2009) Targeted next-generation sequencing of a cancer transcriptome enhances detection of sequence variants and novel fusion transcripts. Genome Biol 10:R115
Li GW, Oh E, Weissman JS (2012) The anti-Shine-Dalgarno sequence drives translational pausing and codon choice in bacteria. Nature 484:538–541
Liebman SW, Chernoff YO, Liu R (1995) The accuracy center of a eukaryotic ribosome. Biochem Cell Biol 73:1141–1149
Lindahl L (1975) Intermediates and time kinetics of the *in vivo* assembly of Escherichia coli ribosomes. J Mol Biol 92:15–37
Liu X, Jiang H, Gu Z, Roberts JW (2013) High-resolution view of bacteriophage lambda gene expression by ribosome profiling. Proc Natl Acad Sci U S A 110:11928–11933
Mani R, St Onge RP, Hartman JL 4th, Giaever G, Roth FP (2008) Defining genetic interaction. Proc Natl Acad Sci U S A 105:3461–3466
Menninger JR (1976) Peptidyl transfer RNA dissociates during protein synthesis from ribosomes of Escherichia coli. J Biol Chem 251:3392–3398
Mizushima S, Nomura M (1970) Assembly mapping of 30S ribosomal proteins from E. coli. Nature 226:1214
Mostafavi S, Ray D, Warde-Farley D, Grouios C, Morris Q (2008) GeneMANIA: a real-time multiple association network integration algorithm for predicting gene function. Genome Biol 9(Suppl 1):S4
Nesvizhskii AI (2012) Computational and informatics strategies for identification of specific protein interaction partners in affinity purification mass spectrometry experiments. Proteomics 12:1639–1655
Nierhaus KH, Dohme F (1974) Total reconstitution of functionally active 50S ribosomal subunits from Escherichia coli. Proc Natl Acad Sci U S A 71:4713–4717
Noll M, Noll H (1972) Mechanism and control of initiation in the translation of R17 RNA. Nat New Biol 238:225–228
O'Farrell PH (1978) The suppression of defective translation by ppGpp and its role in the stringent response. Cell 14:545–557
Ogle JM, Murphy FV, Tarry MJ, Ramakrishnan V (2002) Selection of tRNA by the ribosome requires a transition from an open to a closed form. Cell 111:721–732
Oh E, Becker AH, Sandikci A, Huber D, Chaba R, Gloge F, Nichols RJ, Typas A, Gross CA, Kramer G, Weissman JS, Bukau B (2011) Selective ribosome profiling reveals the cotranslational chaperone action of trigger factor *in vivo*. Cell 147:1295–1308
Roy-Chaudhuri B, Kirthi N, Kelley T, Culver GM (2008) Suppression of a cold-sensitive mutation in ribosomal protein S5 reveals a role for RimJ in ribosome biogenesis. Mol Microbiol 68:1547–1559
Sabol S, Ochoa S (1971) Ribosomal binding of labelled initiation factor F3. Nat New Biol 234:233–236
Sabol S, Sillero MA, Iwasaki K, Ochoa S (1970) Purification and properties of initiation factor F3. Nature 228:1269–1273
Scolnick E, Tompkins R, Caskey T, Nirenberg M (1968) Release factors differing in specificity for terminator codons. Proc Natl Acad Sci U S A 61:768–774
Sergiev PV, Golovina AY, Sergeeva OV, Osterman IA, Nesterchuk MV, Bogdanov AA, Dontsova OA (2012) How much can we learn about the function of bacterial rRNA modification by mining large-scale experimental datasets? Nucleic Acids Res 40:5694–5705
Shajani Z, Sykes MT, Williamson JR (2011) Assembly of bacterial ribosomes. Annu Rev Biochem 80:501–526
Shine J, Dalgarno L (1975) Determinant of cistron specificity in bacterial ribosomes. Nature 254:34–38
Spahn CM, Prescott CD (1996) Throwing a spanner in the works: antibiotics and the translation apparatus. J Mol Med (Berl) 74:423–439

Sparling PF (1970) Kasugamycin resistance: 30S ribosomal mutation with an unusual location on the Escherichia coli chromosome. Science 167:56–58

Srivastava AK, Schlessinger D (1990) Mechanism and regulation of bacterial ribosomal RNA processing. Annu Rev Microbiol 44:105–129

Stahl G, McCarty GP, Farabaugh PJ (2002) Ribosome structure: revisiting the connection between translational accuracy and unconventional decoding. Trends Biochem Sci 27:178–183

Sykes MT, Shajani Z, Sperling E, Beck AH, Williamson JR (2010) Quantitative proteomic analysis of ribosome assembly and turnover *in vivo*. J Mol Biol 403:331–345

Taniguchi Y, Choi PJ, Li GW, Chen H, Babu M, Hearn J, Emili A, Xie XS (2010) Quantifying E. coli proteome and transcriptome with single-molecule sensitivity in single cells. Science 329:533–538

Tissieres A, Watson JD (1958) Ribonucleoprotein particles from Escherichia coli. Nature 182:778–780

Ude S, Lassak J, Starosta AL, Kraxenberger T, Wilson DN, Jung K (2013) Translation elongation factor EF-P alleviates ribosome stalling at polyproline stretches. Science 339:82–85

Vlasblom J, Zuberi K, Rodriguez H, Arnold R, Gagarinova A, Deineko V, Kumar A, Leung E, Rizzolo K, Samanfar B, Chang L, Phanse S, Golshani A, Greenblatt JF, Houry WA, Emili A, Morris Q, Bader G, Babu M (2015) Novel function discovery with GeneMANIA: a new integrated resource for gene function prediction in Escherichia coli. Bioinformatics 31:306–310

Waegeman H, Soetaert W (2011) Increasing recombinant protein production in Escherichia coli through metabolic and genetic engineering. J Ind Microbiol Biotechnol 38:1891–1910

Watanabe K (2010) Unique features of animal mitochondrial translation systems. The non-universal genetic code, unusual features of the translational apparatus and their relevance to human mitochondrial diseases. Proc Jpn Acad Ser B 86:11–39

Weigert MG, Garen A (1965) Base composition of nonsense codons in E. coli. Evidence from amino-acid substitutions at a tryptophan site in alkaline phosphatase. Nature 206:992–994

Weiss RB, Huang WM, Dunn DM (1990) A nascent peptide is required for ribosomal bypass of the coding gap in bacteriophage T4 gene 60. Cell 62:117–126

Wilson DN (2009) The A–Z of bacterial translation inhibitors. Crit Rev Biochem Mol Biol 44:393–433

Woolstenhulme CJ, Guydosh NR, Green R, Buskirk AR (2015) High-precision analysis of translational pausing by ribosome profiling in bacteria lacking EFP. Cell Rep 11:13–21

Xu Z, Culver GM (2010) Differential assembly of 16S rRNA domains during 30S subunit formation. RNA 16:1990–2001

Zeghouf M, Li J, Butland G, Borkowska A, Canadien V, Richards D, Beattie B, Emili A, Greenblatt JF (2004) Sequential Peptide Affinity (SPA) system for the identification of mammalian and bacterial protein complexes. J Proteome Res 3:463–468

Zou SB, Hersch SJ, Roy H, Wiggers JB, Leung AS, Buranyi S, Xie JL, Dare K, Ibba M, Navarre WW (2012) Loss of elongation factor P disrupts bacterial outer membrane integrity. J Bacteriol 194:413–425

Chapter 3
Biology and Assembly of the Bacterial Envelope

Karine Dufresne and Catherine Paradis-Bleau

Abstract All free-living bacterial cells are delimited and protected by an envelope of high complexity. This physiological barrier is essential for bacterial survival and assures multiple functions. The molecular assembly of the different envelope components into a functional structure represents a tremendous biological challenge and is of high interest for fundamental sciences. The study of bacterial envelope assembly has also been fostered by the need for novel classes of antibacterial agents to fight the problematic of bacterial resistance to antibiotics. This chapter focuses on the two most intensively studied classes of bacterial envelopes that belong to the phyla *Firmicutes* and *Proteobacteria*. The envelope of *Firmicutes* typically has one membrane and is defined as being monoderm whereas the envelope of *Proteobacteria* contains two distinct membranes and is referred to as being diderm. In this chapter, we will first discuss the multiple roles of the bacterial envelope and clarify the nomenclature used to describe the different types of envelopes. We will then define the architecture and composition of the envelopes of *Firmicutes* and *Proteobacteria* while outlining their similarities and differences. We will further cover the extensive progress made in the field of bacterial envelope assembly over the last decades, using *Bacillus subtilis* and *Escherichia coli* as model systems for the study of the monoderm and diderm bacterial envelopes, respectively. We will detail our current understanding of how molecular machines assure the secretion, insertion and folding of the envelope proteins as well as the assembly of the glycosidic components of the envelope. Finally, we will highlight the topics that are still under investigation, and that will surely lead to important discoveries in the near future.

Keywords Bacterial envelope assembly • *Firmicutes* • Monoderm • *Bacillus subtilis* • *Proteobacteria* • Diderm • *Escherichia coli* • Molecular machines • Envelope proteins • Glycosidic components of the envelope

---

K. Dufresne • C. Paradis-Bleau (✉)
Department of Microbiology, Infectiology and Immunology, Université de Montréal, Pavillon Roger-Gaudry, room S-640, Montreal, QC, Canada, H3T 1J4
e-mail: karine.dufresne@umontreal.ca; Catherine.Paradis-Bleau@umontreal.ca

N.J. Krogan, M. Babu (eds.), *Prokaryotic Systems Biology*, Advances in Experimental Medicine and Biology 883, DOI 10.1007/978-3-319-23603-2_3

3.1 Introduction

3.1.1 The Multiple Functions of the Bacterial Envelope

The bacterial envelope is a complex structure that surrounds and delimits all free-living bacterial cells. Its integrity is essential for bacterial survival, division, morphogenesis, adaptation and pathogenesis. The envelope provides the bacterial shape and protects bacterial cells against variations in osmotic pressure and external insults (Silhavy et al. 2010; Holtje 1998). It enables influx of nutrients and cofactors required for bacterial cell metabolism, and efflux of toxins (Silhavy et al. 2010; Piddock 2006). It drives energy production through the generation of an electrochemical proton gradient via the electron transport chain. This allows the formation of the proton motive force in the cytoplasmic membrane (Nelson 1994; Taylor 1983) and the production of cytoplasmic ATP by oxidative phosphorylation (Harold 1972; Maloney et al. 1974). The envelope acts as an interface between the bacterial cells and the environment, and is the first site of contact with the hosts. It allows signal sensing and transduction, thus mediating cell signaling for stress response and adaptation to continuously changing extracellular conditions (Jordan et al. 2008; Raivio 2005; Utsumi 2008). It also confers adherence, motility and effector secretion for bacteria-host interactions (Jordan et al. 2008; Wooldridge 2009; Harshey 2003; Thanassi et al. 2012; Haiko and Westerlund-Wikstrom 2013). Bacterial envelope components are in turn recognized by host innate immune receptors as signs of invasion (Li et al. 2013; Ramos et al. 2004; Raetz and Whitfield 2002; Dziarski and Gupta 2005; Royet and Dziarski 2007).

3.1.2 Nomenclature and Types of Bacterial Envelopes

This chapter focuses on the architecture, components and assembly of the two most intensively studied classes of envelopes that belong to the phyla *Firmicutes* and *Proteobacteria*. The members of those phyla are commonly referred to as being Gram-positive and Gram-negative, respectively. This classification is based on the capacity of the bacteria to retain a primary dye in their envelope after a decolorization step according to the staining procedure developed by Christian Gram and published in 1884 (Bartholomew and Mittwer 1952). The envelope of *Firmicutes* typically contains a thick layer of cell wall that retains the primary dye and is thus referred to as being Gram-positive. The envelope of *Proteobacteria* includes a thin layer of cell wall that is protected by an additional membrane named the outer membrane (Fig. 3.1). The decolorizant used in the Gram staining protocol disrupts the outer membrane and removes the primary dye from the envelope of *Proteobacteria* that is therefore classified as being Gram-negative (Beveridge 2001). Even though the Gram stain procedure has been extremely useful for bacterial identification and diagnostic for over a century, it is more relevant to define bacterial envelope architecture with the number of membranes than with the staining properties of bacterial cells (Sutcliffe 2010). In this chapter, the envelopes of *Firmicutes* and *Proteobacteria* that respectively have one and two membranes will be referred to as being monoderm and diderm (Fig. 3.1). For an overview of the architecture and components of the envelope in the different bacterial phyla, we refer the readers to the excellent review by Iain C. Sutcliffe (Sutcliffe 2010) and to the Bergey's Manual of Systematic Bacteriology (Boone et al. 2001).

3.2 Architecture and Components of the Envelope in *Firmicutes* and *Proteobacteria*

In the following sub-sections, we will define the architecture of the envelopes in *Firmicutes* and *Proteobacteria*, detail the nature of their core components and explain their importance in bacterial physiology. The assembly of the described core components into functional envelopes will be addressed in Sect. 3.3. The organization and assembly of motility devices such as the flagella (Chevance and Hughes 2008) and of virulence factors such as secretion systems (Wooldridge 2009) in the envelope as well as the formation of a protective capsule (Whitfield 2006; Fagan and Fairweather 2014; Beveridge and Graham 1991) is beyond the scope of this chapter.

3.2.1 The Envelope of Firmicutes (Monoderm)

Most members of the phylum *Firmicutes* such as *Bacillus subtilis* have an envelope composed of a cytoplasmic or inner membrane surrounded by a thick layer of cell wall facing the external environment (Fig. 3.1a). The inner membrane delimits the cytoplasmic content of bacterial cells and is formed of a symmetrical bilayer of phospholipids. In the model system *B. subtilis*, this bilayer is typically made of 30 % of the zwitterionic phospholipid phosphatidylethanolamine and 70 % of the anionic phospholipid phosphatidylglycerol (van der Does et al. 2000). It also contains the polyisoprenoid lipid carrier named undecaprenyl-phosphate (undecaprenyl-P) required for envelope assembly, saccharidic linkage units for teichoic acids as well as integral α-helical proteins and lipoproteins (Silhavy et al. 2010). These proteins assure many functions such as the formation of a proton gradient via the electron transport chain, the production of cytoplasmic ATP by oxidative phosphorylation, the transport of nutrients inside the cytoplasm, the detection of extracellular signals and the export of envelope components across the inner membrane (Facey and Kuhn 2010). In *Firmicutes*, the cell wall is mainly composed of peptidoglycan and teichoic acids. The peptidoglycan layer is a network of long glycan chains made of alternating units of *N*-acetylglucosamine (GlcNAc) and *N*-

acetylmuramic acid (MurNAc) cross-linked by short peptide bridges (Holtje 1998). This macromolecular network is organized as an exoskeleton around the inner membrane and constitutes the major structural component of the bacterial envelope. It confers the bacterial shape and provides rigidity, flexibility and strength necessary for all free-living bacterial cells to grow and divide while withstanding their high internal osmotic pressure (Typas et al. 2012). Peptidoglycan is the site of attachment of surface proteins that interact with the extracellular environment (Dramsi et al. 2008) (Fig. 3.1a). Members of the phylum *Firmicutes* contain two types of teichoic acids in their envelope: wall teichoic acids (WTAs) and lipoteichoic acids (LTAs) that are respectively anchored in the peptidoglycan layer (Brown et al. 2013) and in the outer leaflet of the inner membrane (Reichmann and Grundling 2011) (Fig. 3.1a). These glycopolymers typically consist of a disaccharide linkage unit and an anionic chain of polyglycerolphosphate or polyribitolphosphate decorated with saccharides and positively charged D-alanyl esters. WTAs are a major constituent of the envelope of *Firmicutes* and account for about 50 % of the cell wall material (Hancock 1997; Sewell and Brown 2014). They are critical determinants of the bacterial surface charge and hydrophobicity (Brown et al. 2013). WTAs are involved in the regulation of cell division and peptidoglycan biosynthesis, and are particularly important for morphogenesis in rod-shaped bacteria like *B. subtilis* (Sewell and Brown 2014; Swoboda et al. 2010; Schirner et al. 2009). LTAs are important for membrane physiology and are needed to maintain divalent cation homoeostasis. They are involved in morphogenesis and are crucial for cell division (Schirner et al. 2009; Percy and Grundling 2014).

3.2.2 The Envelope of Proteobacteria (Diderm)

Members of *Proteobacteria* such as *Escherichia coli* have an envelope made of an inner and an outer membrane with a thin layer of cell wall sandwiched in between them (Fig. 3.1b). The inner membrane is formed of a symmetrical bilayer of phospholipids. In the model system *E. coli*, this bilayer is typically made of 75 % of the zwitterionic phospholipid phosphatidylethanolamine, 20 % of the anionic phospholipid phosphatidylglycerol and 5 % of the anionic phospholipid cardiolipin (Raetz 1978). The inner membrane of the diderm model system *E. coli* thus has a smaller net negative charge than the inner membrane of the monoderm model system *B. subtilis* that consists mainly of phosphatidylglycerol (van der Does et al. 2000). As for the *Firmicutes*, members of the phylum *Proteobacteria* have the polyisoprenoid lipid carrier undecaprenyl-P, integral α-helical proteins and lipoproteins in their inner membrane. However, the envelope of *Proteobacteria* does not contain techoic acids. (Silhavy et al. 2010). In *Proteobacteria*, the inner and outer membranes delimit the periplasm; an oxidizing environment densely packed with proteins and containing the cell wall (Silhavy et al. 2010; Mullineaux et al. 2006) (Fig. 3.1b). The cell wall of *Proteobacteria* is composed mainly of a thin layer of peptidoglycan. *Firmicutes* and *Proteobacteria* have peptidoglycan of similar composition, although the peptidoglycan layer of *Proteobacteria* is thinner and less cross-linked than for *Firmicutes* (Vollmer et al. 2008a). The outer membrane of *Proteobacteria* is an asymmetrical lipid bilayer; its inner leaflet is made of phospholipids while its outer leaflet is composed of anionic glycolipids, mainly lipopolysaccharides (LPS) (Muhlradt and Golecki 1975; Kamio and Nikaido 1976) (Fig. 3.1b). The inner leaflet of the outer membrane contains the same phospholipids as the inner membrane but the phospholipid ratio differs. In *E. coli*, the inner membrane and the inner leaflet of the outer membrane are respectively made of 75 % and 90 % of the zwitterionic phospholipid phosphatidylethanolamine. Furthermore, phospholipids of the outer membrane contain relatively more saturated fatty acids than phospholipids of the inner membrane. The inner leaflet of the outer membrane consequently has a smaller net negative charge and is more rigid than the leaflets of the inner membrane (Lugtenberg and Peters 1976). LPS in the outer leaflet of the outer membrane typically consist of a lipid A (also referred to as the endotoxin), a negatively charged core oligosaccharide and a distal chain of repeating oligosaccharides called O-antigen (Raetz and Whitfield 2002). The negative charges of LPS are balanced and stabilized by divalent cations such as magnesium and calcium in the outer leaflet of the outer membrane (Nikaido 2003; Holst 2007). As for WTAs in *Firmicutes*, LPS are critical determinants of the bacterial surface charge and hydrophobicity in *Proteobacteria*. Finally, the outer membrane harbors a unique set of lipoproteins as well as integral β-barrel proteins such as the porins for transport across the outer membrane (Silhavy et al. 2010; Doerrler 2006). Like the other envelope layers, the outer membrane is essential for bacterial survival. It reduces the overall cell envelope permeability and acts as a selective barrier that protects bacterial cells from harmful extracellular compounds (Delcour 2009; Nikaido 1989; Bos et al. 2007).

3.3 Biosynthesis and Assembly of the Bacterial Envelope

The building blocks for the construction of bacterial envelopes, notably the phospholipids and polyisoprenoid lipid carrier, the proteins, the peptidoglycan precursor lipid II, the LPS precursors lipid A-core and O-antigen units, WTAs and the precursors for LTAs are synthesized in the cytoplasm and the inner leaflet of the inner membrane. These components then have to be transported across the inner membrane for their assembly and/or incorporation at their correct locations in the envelope. This process named envelope assembly is required for the formation, maintenance and reorganization of a physiologically active structure (Silhavy et al. 2010). Envelope assembly is performed in a potentially hostile environment that lacks ATP and other high-energy molecular carriers (Silhavy et al. 2010; Oliver 1996). The energy required to assemble the envelope is either provided by exergonic reactions with substrates that have been energized before their transport across the inner membrane or by protein machineries in the inner membrane that exploit and translocate the energy from the proton motive force or ATP hydrolysis in the cytoplasm (Polissi and Sperandeo 2014).

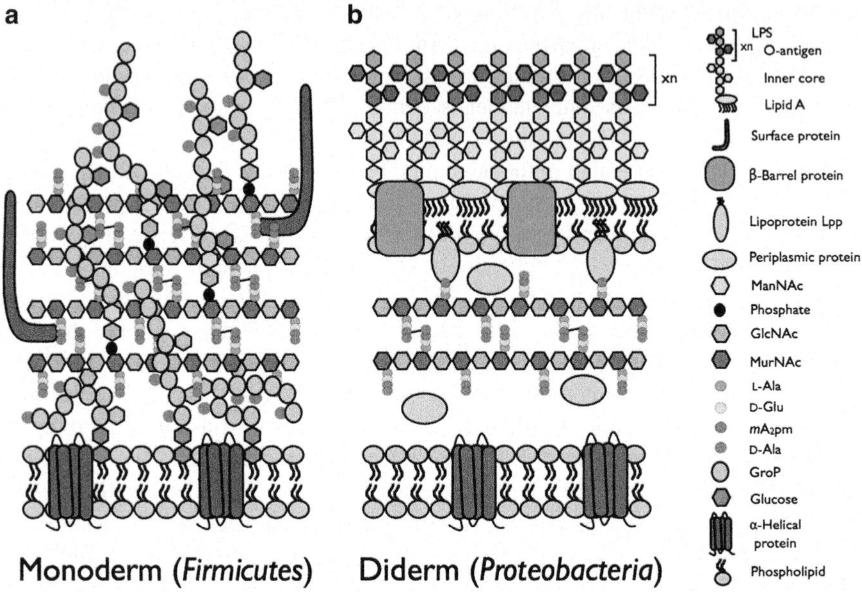

Fig. 3.1 Schematic representation of the architecture and components of the envelope in *Firmicutes* and *Proteobacteria*. (**a**) The monoderm envelope of the members of the phylum *Firmicutes* is typically made of an inner membrane and a thick layer of cell wall. The inner membrane is composed of a bilayer of phospholipids, it contains integral α-helical proteins and LTAs anchored in the outer leaflet. The membrane also includes the lipid carrier undecaprenyl-P and a few lipoproteins in the outer leaflet that are not represented in the figure. The cell wall is composed of peptidoglycan, surface proteins and WTAs. Peptidoglycan is made of long glycan chains containing alternating units of GlcNAc and MurNAc cross-linked by short peptide bridges. It is the site of attachment of surface proteins and WTAs. In the model system *B. subtilis* strain 168, LTAs and WTAs are composed of a disaccharide linkage unit attached to a chain of polyglycerolphosphate decorated with monosaccharides and D-alanine residues. (**b**) The diderm envelope of the members of the phylum *Proteobacteria* is made of an inner membrane, a thin layer of cell wall and an outer membrane. The inner membrane is the same as described for the *Firmicutes* but does not contain LTAs. The inner and outer membranes delimit an environment called the periplasm that contains the cell wall and proteins. The cell wall is mainly composed of peptidoglycan, which is the site of attachment of the lipoprotein Lpp that links the peptidoglycan layer with the outer membrane. The inner leaflet of the outer membrane is composed of phospholipids whereas the outer leaflet of the outer membrane is made of LPS composed of a lipid A anchor, an inner core and an O-antigen. The outer membrane contains lipoproteins and integral β-barrel proteins

While the biochemical pathways for the biosynthesis of the cytoplasmic and lipid-linked bacterial envelope precursors are well characterized (Silhavy et al. 2010; Raetz and Whitfield 2002; Brown et al. 2013; Reichmann and Grundling 2011; Barreteau et al. 2008; Bouhss et al. 2008; Manat et al. 2014), their assembly represents an extremely complex biological process that is still not well understood (Silhavy et al. 2010). However, extensive progress has been made in the field of bacterial envelope assembly over the last decades using model systems of the phyla *Firmicutes* and *Proteobacteria*. We will resume our current understanding of the transport, assembly and incorporation systems that have been identified.

3.3.1 Transport, Insertion and Folding of Envelope Proteins

3.3.1.1 Insertion of Inner Membrane Proteins and Secretion of Envelope Proteins Across the Inner Membrane by the Sec Translocase

All bacterial proteins are made in the cytoplasm by the translation machinery that converts the genetic information from messenger RNAs into chains of amino acids via ribosomes and transfer RNAs (Clark 2010). The vast majority of envelope proteins require the Sec translocase for their insertion in the inner membrane or their translocation across the inner membrane. These proteins have hydrophobic N-terminal sequences that act as a signal for their recruitment to the Sec translocase; a highly conserved protein-conducting channel in the inner membrane composed of the heterotrimeric protein complex SecYEG (Driessen and Nouwen 2008; du Plessis et al. 2011) (Fig. 3.2).

Integral inner membrane proteins contain transmembrane domains made of hydrophobic α-helices that have to be inserted perpendicular to the membrane (du Plessis et al. 2011) (Fig. 3.2a). The signal sequences of inner membrane proteins are located in their N-terminal transmembrane domains. For the insertion of proteins in the inner membrane, a ribonucleoprotein named signal recognition particle (SRP) recognizes the highly hydrophobic N-terminal sequences of ribosome-bound nascent polypeptide chains and directs them to the translocase (du Plessis et al. 2011; Dalbey et al. 2011) (Fig. 3.2a). SRP interacts with the SRP receptor FtsY, a cytoplasmic protein that associates with phospholipids in the inner leaflet of the membrane and interacts with the translocase (Angelini et al. 2005, 2006; Millman et al. 2001). The binding of SRP to FtsY triggers the GTPase activity of both protein partners. The energy from GTP hydrolysis drives the transfer of the ribosome-nascent chain complexes to the translocase and the dissociation of the SRP/FtsY complex in the cytoplasm (Dalbey et al. 2011; Egea et al. 2004; Focia et al. 2004). The lateral insertion of proteins in the phospholipid bilayer from the translocase gate is driven by polypeptide chain elongation from the ribosomes. The inner membrane protein complex SecDF\YajC and the insertase YidC transiently interact with the translocase to assist the insertion and folding of inner membrane proteins (du Plessis et al. 2011; Dalbey et al. 2011) (Fig. 3.2a). The insertion of inner membrane proteins containing large periplasmic domains involves an additional factor. The SecA protein acts as an ATP-dependent motor for the transport of large hydrophilic domains across the inner membrane (du Plessis et al. 2011; Andersson and von Heijne 1993; Koch and Muller 2000). A subset of inner membrane proteins is inserted in a Sec-independent fashion. In this case, the complexes formed by the ribosome-bound nascent polypeptide chain, SRP and FtsY directly interact with YidC independently of the Sec translocase. The ribosome-bound nascent polypeptide chains are transferred from the SRP/FtsY complex to the YidC integrase that directs their insertion in the inner membrane (du Plessis et al. 2011; Welte et al. 2012; Samuelson et al. 2000) (Fig. 3.2a). However, the insertion of inner membrane proteins containing large periplasmic domains cannot be performed by

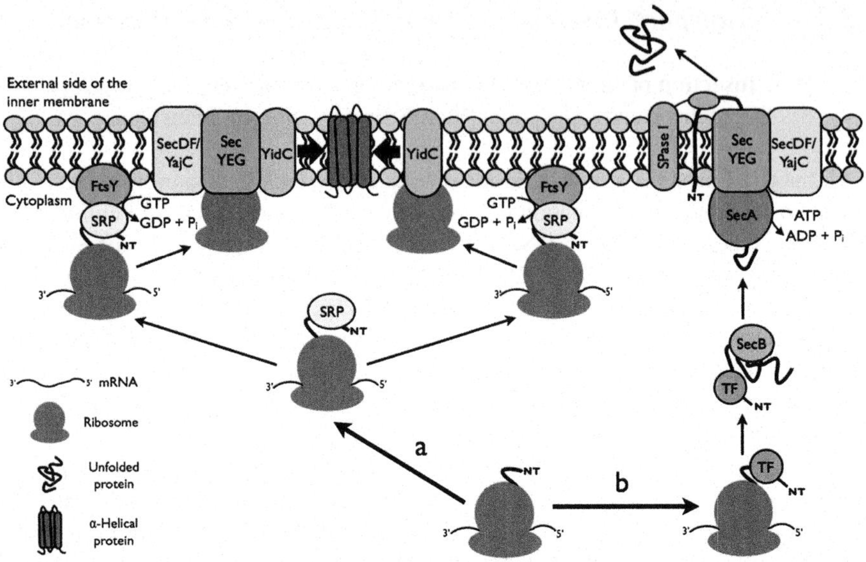

Fig. 3.2 Insertion of inner membrane proteins and secretion of envelope proteins across the inner membrane by the Sec translocase. Most envelope proteins are inserted in the inner membrane or translocated across the inner membrane by the Sec translocase: a protein-conducting channel composed of the SecYEG protein complex. (**a**) Co-translational insertion of inner membrane proteins by the Sec translocase. SRP first recognizes the highly hydrophobic N-terminal sequences of ribosome-bound nascent polypeptide chains. The interaction of SRP with its receptor FtsY triggers the GTPase activity of both protein partners and GTP hydrolysis drives the transfer of the ribosome-nascent chain complexes to the Sec translocase. Protein insertion in the membrane is energized by polypeptide chain elongation from the ribosomes. The SecDF\YajC protein complex and YidC interact with the translocase and contribute to the insertion and folding of inner membrane proteins. A subset of inner membrane proteins is inserted by the integrase YidC in a Sec-independent fashion. The ribosome-bound nascent polypeptide chains are then transferred from the SRP/FtsY complex to YidC for membrane protein insertion. (**b**) Post-translational secretion of envelope proteins by the Sec translocase. TF binds to the N-terminal signal sequences of secretory proteins when they exit the ribosomes. For protein secretion, the N-terminal signal sequences of unfolded preproteins are further recognized by SecA. In *Proteobacteria*, preproteins are generally maintained in an unfolded state by the chaperone SecB that recruits them to SecA. The ATP-dependent motor protein SecA drives the transport of preproteins across the translocase and the SecDF\YajC complex contributes to protein secretion. The N-terminal signal sequences of preproteins enter of the translocase first and are inserted in the inner membrane from the external side of the inner membrane. The signal peptidase SPase I cleaves the N-terminal sequences of the preproteins and allows the release of mature unfolded proteins on the external side of the inner membrane

the YidC-only pathway, indicating that the SecA motor protein required for the translocation of large hydrophilic domains across the inner membrane can only interact with the Sec translocase (Welte et al. 2012).

The Sec translocase is also responsible for the secretion of envelope proteins across the inner membrane (Fig. 3.2b). While the insertion of inner membrane proteins is a co-translational process, the translocation of envelope proteins occurs post-translationally (du Plessis et al. 2011; Ulbrandt et al. 1997). Secretory proteins are synthesized as preproteins in the cytoplasm and possess N-terminal signal sequences with a short positively charged region followed by a central hydrophobic region and a C-terminal polar region containing a type-I peptidase cleavage site for their subsequent maturation in the envelope compartment (Driessen and Nouwen 2008; Palmer and Berks 2012). When the signal sequences of secretory proteins exit the ribosome, they are recognized by a chaperone named trigger factor (TF) that prevents the binding by SRP (Beck et al. 2000; Eisner et al. 2003; Ferbitz et al. 2004). For protein secretion, the N-terminal signal sequences of unfolded preproteins are further recognized by SecA (Driessen and Nouwen 2008). In most *Proteobacteria* such as *E. coli*, preproteins are maintained in an unfolded state by the secretion-dedicated chaperone SecB that recruits them to SecA (Driessen and Nouwen 2008; van der Sluis and Driessen 2006). The motor protein SecA drives the transport of preproteins across the translocase in the inner membrane by its ATPase activity and the N-terminal signal sequences of secretory proteins enter the translocase first (Lycklama and Driessen 2012). On the external side of the inner membrane, the central hydrophobic region of the secreted N-terminal signal sequences are predicted to be inserted in the inner membrane. This would position the short positively charged N-terminal region back in the cytoplasm, thus retaining the preproteins in the membrane and exposing the type-I peptidase cleavage sites of the N-terminal signal sequences to the signal peptidase I (SPase I, also named Lep). This enzyme is integrated in the membrane with its catalytic domain facing the envelope compartment (Lycklama and Driessen 2012; Paetzel et al. 2002; Briggs et al. 1986) (Fig. 3.2b). To complete protein translocation across the inner membrane, SecA continues to drive the transport of secretory proteins through the translocase and the protein complex SecDF\YajC associates transiently with the translocase to stimulate the transit. This process directly involves SecDF and the membrane proton motive force (Lycklama and Driessen 2012). The combined action of the Sec secretory system and SPase I that cleaves the N-terminal sequences of preproteins allows the release of mature unfolded proteins on the external side of the inner membrane (Lycklama and Driessen 2012; Paetzel et al. 2002) (Fig. 3.2b).

3.3.1.2 Folding of the Secretory Sec Substrates in the Envelope Compartment

After their transport across the inner membrane and their processing by SPase I, mature unfolded secretory proteins must adopt their functional conformations. Protein folding involves the formation of the secondary structure elements, the α-helices and β-sheets, and their non-covalent interactions for the adoption of active tertiary structures (Merdanovic et al. 2011). In *Proteobacteria*, periplasmic molecular chaperones participate in the folding of mature secreted proteins and in the maintenance of protein quality under stress conditions by reducing protein aggregation and degradation (Silhavy et al. 2010; Merdanovic et al. 2011; Duguay and Silhavy 2004). The folding of some envelope proteins also requires folding catalysts; oxidoreductases for the formation of disulfide bonds between cysteine residues and peptidyl-prolyl isomerases (PPIases) for cis-trans isomerization of peptidyl-prolyl bonds (Merdanovic et al. 2011; Duguay and Silhavy 2004; Goemans et al. 2014; Depuydt et al. 2011). While the formation of disulfide bonds in the periplasm of *Proteobacteria* is an extremely important physiological process (Hatahet et al. 2014; Kadokura and Beckwith 2010), its role is very limited in *Firmicutes* (Dutton et al. 2008; van Wely et al. 2001). In *Firmicutes*, secretory proteins are released directly in the extracellular environment and the folding of mature proteins is assisted by folding catalysts integrated in the inner membrane such as the predicted PPIase PrsA present in many species such as *B. subtilis* (van Wely et al. 2001; Sarvas et al. 2004).

3.3.1.3 Translocation of Folded Proteins Across the Inner Membrane by the Tat System

Some envelope proteins are translocated across the inner membrane in their folded and/or cofactor-containing forms by the twin-arginine translocation (Tat) system (Palmer and Berks 2012; Natale et al. 2008). Secretory Tat substrates have hydrophobic N-terminal signal sequences with a tripartite structure and a type-I peptidase cleavage site like the signal sequences of secretory Sec substrates (see Sect. 3.3.1.1). However, the N-terminal positively charged region of the Tat signal sequence is longer than for the Sec signal sequence and the Tat signal sequence includes two invariable arginine residues at its C-terminal end (Palmer and Berks 2012; Goosens et al. 2014). For Tat substrates, the presence of the signal sequence is not sufficient for translocation, and the proteins must adopt the proper folded conformation and/or incorporate the correct cofactor in the cytoplasm to be transported across the inner membrane (Goosens et al. 2014). Some Tat substrates have to incorporate complex cofactors such as metal–sulphur clusters or nucleotide-based molecules that are present only in the cytoplasm while others need to fold around the correct cytoplasmic cofactor to avoid competition of periplasmic metal ions for their active sites. A subset of Tat substrates associates with cytoplasmic protein cofactors that do not possess signal sequences for their translocation across the inner membrane. In all cases, preprotein maturation in the cytoplasm is required for proper folding, stability and functionality of the Tat substrates in the envelope compartment (Palmer and Berks 2012).

The minimal Tat system for translocation of folded and/or cofactor-containing substrates is composed of the TatA and TatC proteins (Goosens et al. 2014). This system is used in *Firmicutes* such as *B. subtilis* that only has a few predicted Tat substrates (Palmer and Berks 2012; Dilks et al. 2003). In *Proteobacteria* such as *E. coli* that has 27 predicted Tat substrates, translocation also requires the TatA-like protein TatB (Palmer and Berks 2012; Goosens et al. 2014; Palmer et al. 2010) (Fig. 3.3). The Tat system is functionally divided in the pore-forming unit and the docking unit for substrate recognition. The small inner membrane protein

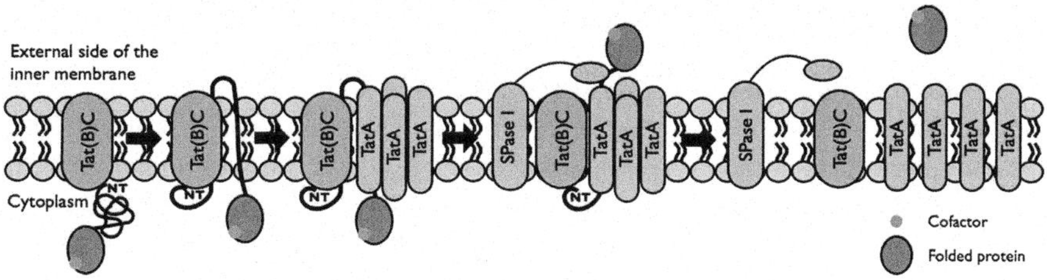

Fig. 3.3 Translocation of folded proteins across the inner membrane by the Tat system. The Tat system allows the transport of folded and/or cofactor-containing proteins through a large protein pore in the inner membrane. The N-terminal sequences of Tat substrates are recognized by the docking unit formed by TatC in *Firmicutes* such as *B. subtilis* and of the additional TatB protein in *Proteobacteria* such as *E. coli*. The formation of Tat(B)C-substrate complex triggers the insertion of N-signal sequences in the inner membrane and the recruitment of TatA that oligomerizes to form a large pore in the inner membrane for protein transport. The assembly of the Tat system is energized by the proton motive force but the dynamics of protein transport is not understood. After the translocation of Tat substrates, their N-terminal signal sequences are cleaved by the signal peptidase SPase I. Mature proteins are then released on the external side of the inner membrane and the Tat complex disassembles

TatA is responsible of forming the pore by oligomerization. The large integral inner membrane protein TatC in *B. subtilis* with the additional TatA-like protein TatB in *E. coli* constitutes the docking unit. For protein translocation, the docking unit recognizes the N-terminal signal sequences of folded and/or cofactor-containing Tat substrates. TatC binds to the twin-arginine consensus motif in the N-terminal positively charged region of the Tat signal sequences (Palmer and Berks 2012; Goosens et al. 2014) (Fig. 3.3). In *E. coli*, it has been shown that TatC inserts the following regions of the Tat signal sequences in the inner membrane. This step requires the membrane proton motive force, is regulated by TatB and results in the positioning of the signal sequences in a binding pocket formed by TatB and TatC (Frobel et al. 2012) (Fig. 3.3). The formation of the Tat(B)C-substrate complex triggers the recruitment of TatA monomers that oligomerize to form a large pore in the inner membrane, a process energized by the proton motive force (Palmer and Berks 2012; Goosens et al. 2014; Frobel et al. 2011). The mechanistic and energy source required for the passage of folded and/or cofactor-containing Tat substrates across the pore is not yet understood, but it is known that the twin-arginine consensus motif of the signal sequences remain attached to TatC during the transport. After substrate translocation across the inner membrane, the protease cleavage site of the signal sequences is exposed at the external side of the inner membrane. The signal sequences are then cleaved by SPase I for the release of mature proteins on the external side of the inner membrane, and the Tat complex disassembles (Palmer and Berks 2012; Goosens et al. 2014; Frobel et al. 2012) (Fig. 3.3).

3.3.1.4 Biogenesis of Lipoproteins and Transport by the Lol System

Some Sec and Tat substrates are designed to become lipoproteins; lipidated proteins embedded in a phospholipid leaflet (Fig. 3.4). These proteins are produced as preprolipoproteins in the cytoplasm and contain a Sec or Tat N-terminal targeting signal sequence (see Sects. 3.3.1.1 and 3.1.3, respectively). However, the C-terminal polar region of their signal sequences consists of a lipoprotein box including a type-II peptidase cleavage site instead of a type-I peptidase cleavage site and contains an invariable cysteine residue after the cleavage site (du Plessis et al. 2011; Palmer and Berks 2012). After the translocation of preprolipoproteins across the inner membrane by the Sec or Tat pathway (see Sects. 3.3.1.1 and 3.3.1.3, and Figs. 3.2 and 3.3), their signal sequences retain them in the inner membrane. The central hydrophobic region of the signal sequences stays embedded in the membrane while the positively charged N-terminal region remains in the cytoplasm and the C-terminal lipoprotein box faces the envelope compartment (Fig. 3.4). The cysteine residue at the C-terminal end of the lipoprotein box is lipidated by the Lgt enzyme in the inner membrane that transfers a diacylglycerol moiety from the phospholipid phosphatidylglycerol to the preprolipoproteins (Pailler et al. 2012; Okuda and Tokuda 2011; Sankaran and Wu 1994). Prolipoproteins are then embedded in the outer leaflet of the inner membrane by their lipidated cysteine residues. Their N-terminal signal peptides are subsequently cleaved by the signal peptidase SPase II (also named Lsp), and the membrane-anchored cysteine residues become the first residues of the lipoproteins (Fig. 3.4). This typically completes the maturation of lipoproteins in *Firmicutes* (Okuda and Tokuda 2011; Hutchings et al. 2009; Nakayama et al. 2012). In *Proteobacteria*, an additional fatty acid is attached to the N-terminal cysteine residues by the phospholipid/apolipoprotein transacylase Lnt in the inner membrane (Okuda and Tokuda 2011; Hutchings et al. 2009; Buddelmeijer and Young 2010; Robichon et al. 2005) (Fig. 3.4). The lipoprotein biosynthesis enzymes are essential in *Proteobacteria* such as *E. coli* whereas they are not essential in *Firmicutes* such as *B. subtilis* (Nakayama et al. 2012).

In *Proteobacteria*, most lipoproteins are localized in the inner leaflet of the outer membrane (Nakayama et al. 2012). Depending on the residues following the lipid-modified cysteine residues, mature lipoproteins stay anchored in the outer leaflet of the inner membrane or are transported to the outer membrane by the localization of lipoproteins (Lol) system. Typically, the presence of an aspartate residue at position 2 in mature lipoproteins functions as a Lol avoidance signal for lipoprotein retention in the inner membrane (Okuda and Tokuda 2011). In the absence of an avoidance signal, mature lipoproteins are released from the outer leaflet of the inner membrane by the LolCDE complex: an ATP-binding cassette (ABC) transporter in the inner membrane that requires ATP hydrolysis for its function (Yakushi et al. 2000). The LolCDE complex transfers mature lipoproteins to the soluble chaperone LolA that assures their transport across the hydrophilic periplasm to the receptor protein LolB located in the inner leaflet of the outer membrane. Finally, LolB incorporates lipoproteins in the inner leaflet of the outer membrane by an unknown mechanism (Okuda and Tokuda 2011; Narita and Tokuda 2010) (Fig. 3.4).

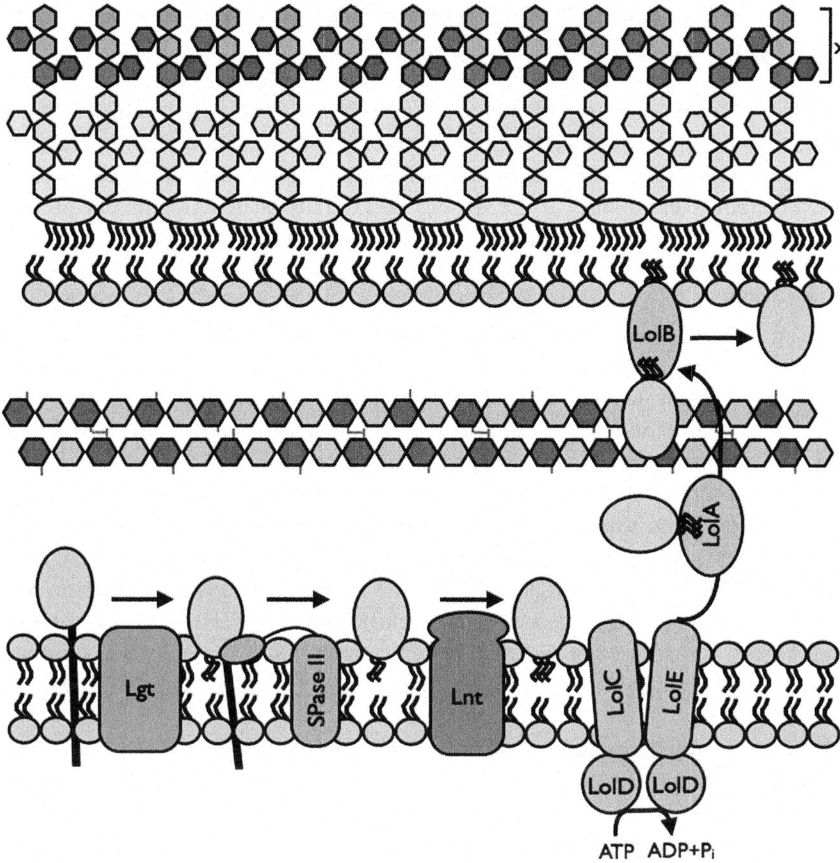

Fig. 3.4 Biogenesis of lipoproteins and transport by the Lol system. Lipoproteins are lipidated proteins embedded in a phospholipid leaflet. These proteins are produced as preprolipoproteins in the cytoplasm and are translocated across the inner membrane by the Sec or Tat pathway. The N-terminal sequences retain the preprolipoproteins in the inner membrane after secretion and position them for their lipidation with diacylglycerol by the Lgt enzyme. The N-terminal signal sequences of prolipoproteins are then cleaved by the signal peptidase SPase II and this completes the biogenesis of lipoproteins in *Firmicutes*. In *Proteobacteria*, a fatty acid is further added to the apolipoproteins by the transacylase Lnt. In *Proteobacteria*, most mature lipoproteins are transported to the inner leaflet of the outer membrane by the Lol system. For lipoprotein transport, the LolCDE complex forms an ABC transporter in the inner membrane and uses energy from ATP hydrolysis to transfer lipoproteins to LolA. This soluble chaperone transports lipoproteins across the periplasm and transfers them to the lipoprotein LolB for lipoprotein incorporation in the inner leaflet of the outer membrane

3.3.1.5 Insertion of Outer Membrane Proteins by the Bam Complex in *Proteobacteria*

In addition to lipoproteins, the outer membrane of the diderm envelope contains integral proteins. Most outer membrane proteins contain antiparallel β-sheets that form transmembrane domains for the adoption of a closed cylindrical β-barrel structure (Ricci and Silhavy 2012; Wimley 2003). Integral outer membrane proteins are translocated from the cytoplasm to the periplasm as unfolded polypeptides by the Sec translocase and are processed by the protease SPase I (Driessen and Nouwen 2008; du Plessis et al. 2011) (see Sect. 3.3.1.1 and Fig. 3.2b). Nascent outer membrane proteins associate with soluble molecular chaperones that prevent their aggregation, facilitate their transport across the periplasm in an unfolded but folding-competent state and target them to the β-barrel assembly machinery (Bam) in the outer membrane (Ricci and Silhavy 2012; Denoncin et al. 2012; Hagan et al. 2011). The molecular chaperones involved in the transport and assembly of outer membrane proteins include SurA, Skp and DegP in *E. coli* (Goemans et al. 2014). SurA transports most nascent outer membrane proteins across the periplasm while Skp and DegP form a partially redundant backup pathway (Denoncin et al. 2012; Rizzitello et al. 2001; Sklar et al. 2007) (Fig. 3.5). Nascent outer membrane proteins can also be the substrates of periplasmic folding catalysts such as oxidoreductases and PPIases as discussed in Sect. 3.1.2 on the maturation of secretory Sec substrates (Merdanovic et al. 2011; Duguay and Silhavy 2004; Goemans et al. 2014; Depuydt et al. 2011).

The most important and conserved component of the Bam complex for outer membrane protein insertion and folding is BamA: an integral β-barrel outer membrane protein with a large periplasmic portion containing five structurally homologous polypeptide translocation associated (POTRA) domains (Kim et al. 2007) (Fig. 3.5). In *E. coli*, the Bam system is composed of BamA and of the four outer membrane lipoproteins BamBCDE that form an oligomeric complex (Ricci and Silhavy 2012; Hagan et al. 2011) (Fig. 3.5). The POTRA domains of BamA are required for the interaction with the lipoprotein components of the complex, for proper BamA assembly in the outer membrane as well as for the binding and folding of the Bam substrates. Nascent outer membrane proteins in their unfolded but folding-competent states probably bind to the POTRA domains of BamA via C-terminal signal sequences. This interaction is though to initiate protein folding by a process called β-strand augmentation (Ricci and Silhavy 2012; Hagan et al. 2011; Kim et al. 2007). The mechanism of protein insertion and complete folding of β-barrel proteins in the outer membrane by the BAM complex is not yet understood. However, the reconstitution of the Bam system in liposomes by the Kahne lab demonstrated that SurA can deliver unfolded substrates to the Bam complex that facilitates outer membrane protein incorporation without an external energy source (Hagan et al. 2010).

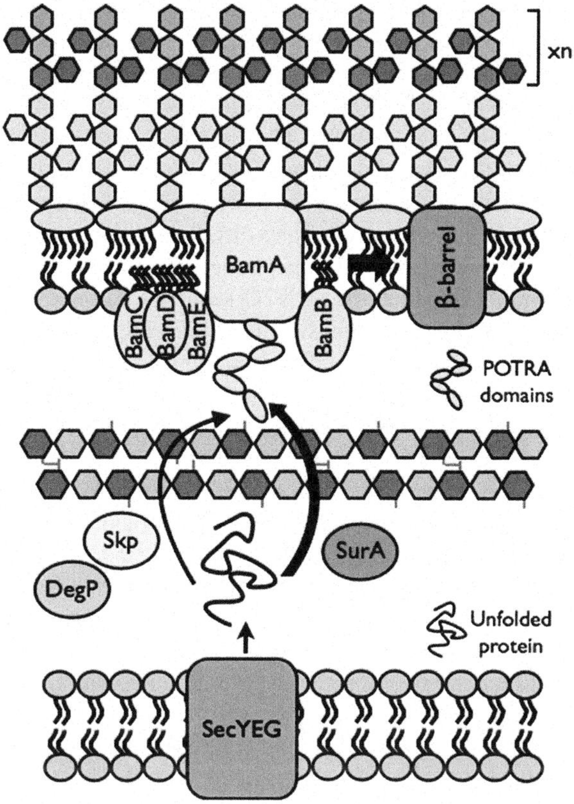

Fig. 3.5 Insertion of outer membrane proteins by the Bam complex in *Proteobacteria*. The outer membrane of *Proteobacteria* contains integral proteins that typically adopt a β-barrel conformation. These proteins are translocated from the cytoplasm to the periplasm as unfolded polypeptides by the Sec translocase. They associate with chaperones that assure their transport across the periplasm in an unfolded but folding-competent state and target them to the Bam complex in the outer membrane. In *E. coli*, the chaperone SurA transports most nascent outer membrane proteins across the periplasm while Skp and DegP form a backup pathway. The Bam complex of the model system *E. coli* is composed of the integral β-barrel outer membrane protein BamA that contains five POTRA domains and of the four lipoproteins BamBCDE. The mechanism of insertion and folding of β-barrel proteins by the Bam system is currently not understood but involves the interaction of nascent outer membrane proteins with the POTRA domains of BamA

3.3.1.6 Attachment of Proteins to Peptidoglycan

Some envelope proteins in *Firmicutes* and *Proteobacteria* are associated non-covalently with the peptidoglycan layer via cell wall binding domains (Visweswaran et al. 2011) while other envelope proteins are attached covalently to peptidoglycan by the enzymes described below.

In *Firmicutes*, many surface proteins are attached covalently to the thick peptidoglycan layer that faces the extracellular environment (Fig. 3.1a). These proteins are synthesized as preproteins in the cytoplasm and have N-terminal signal sequences for their translocation across the inner membrane by the Sec translocase and their processing by Spase I (see Sect. 3.1.1 and Fig. 3.2b). They also have C-terminal cell wall sorting signals for their attachment to peptidoglycan by sortase enzymes (Dramsi et al. 2008; Spirig et al. 2011). These signal sequences typically consist of the N-terminal LPXTG sorting motif where X can be any amino acid, followed by a central hydrophobic domain and a short positively charged C-terminal tail (Schneewind et al. 1992). However, many different classes of sortases have now been involved in protein attachment to peptidoglycan and some of them recognize distinct sorting motifs in which the LPXTG sequence varies (Spirig et al. 2011). Similarly to the N-terminal secretion signals of preprolipoproteins (see Sect. 3.1.4 and Fig. 3.4), the C-terminal cell wall sorting signals delay protein secretion by retaining the preproteins in the inner membrane. This allows proper preprotein positioning for enzymatic processing. The central hydrophobic region of the cell wall sorting signals stays embedded in the membrane while the positively charged C-terminal tail remains in the cytoplasm and the N-terminal sorting motif is exposed at the external side of the inner membrane for recognition by sortases. These enzymes are embedded in the inner membrane with their catalytic domains for cysteine transpeptidase activity facing the extracellular environment (Schneewind and Missiakas 2012). They typically function by first breaking the peptide bond between the threonine and glycine residues of the LPXTG sorting motif. They then form a peptide bond between the C-terminal threonine residue of the target proteins and the terminal amino group of the residue in the third position of the stem peptide of the peptidoglycan lipid II assembly unit (Dramsi et al. 2008; Spirig et al. 2011). In *B. subtilis* (and in *E. coli*), this residue in lipid II is *meso*-diaminopimelic acid ($m\mathrm{A_2pm}$) (Barreteau et al. 2008; Kouidmi et al. 2014). The lipid II units with attached proteins are then incorporated to the peptidoglycan layer in the process of peptidoglycan assembly described in Sect. 3.2.2 and Fig. 3.6. The structure of the peptidoglycan layer of the monoderm envelope of *Firmicutes* containing covalently bound proteins is pictured in Fig. 3.1a.

Sortases have been identified in a few species of *Proteobacteria*, but their function remains unknown (Spirig et al. 2011; Comfort and Clubb 2004; Pallen et al. 2003). However, the L,D-transpeptidases YbiS, YcfS and ErfK are redundant enzymes that attach the outer membrane protein Lpp (also named Braun lipoprotein, Braun and Rehn 1969) to the peptidoglycan layer in the diderm model system *E. coli* (Magnet et al. 2007a). These enzymes also have a cysteine transpeptidase activity. To link Lpp to peptidoglycan, they cleave the peptide bond between the third and fourth residue of the peptidoglycan stem peptide to form a peptide bond between the $m\mathrm{A_2pm}$ residue at the third position of the stem peptide and the side chain amine of the C-terminal lysine residue of Lpp. It is not known whether Lpp is attached to the peptidoglycan lipid II assembly unit or to mature peptidoglycan (Dramsi et al. 2008; Magnet et al. 2007a). The attachment of the outer membrane protein Lpp to the peptidoglycan layer physically links these two envelope layers as depicted in Fig. 3.1b.

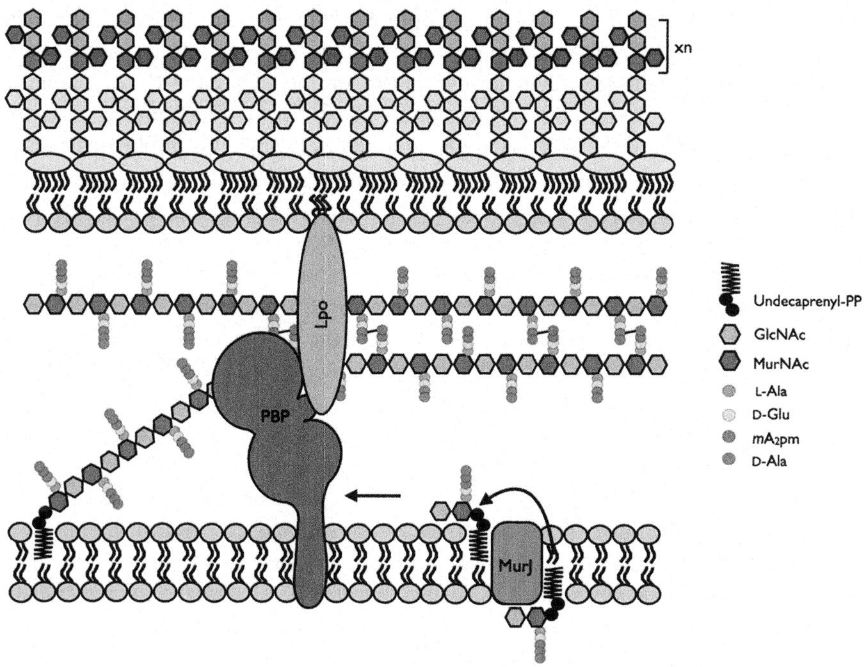

Fig. 3.6 Peptidoglycan assembly by the transglycosylation and transpeptidation activities of PBPs. The lipid II precursor for peptidoglycan assembly is composed of the disaccharide GlcNAc-MurNAc linked to the pentapeptide moiety and the lipid carrier undecaprenyl-P. In *B. subtilis* and *E. coli*, the pentapeptide is composed of L-Ala-D-Glu-mA_2pm-D-Ala-D-Ala. The flippase for the translocation of lipid II from the inner leaflet of the inner membrane to the outer leaflet has been proven to be MurJ, at least in *Proteobacteria* (see main text). In the envelope compartment, lipid II units are used as substrates by the PBPs that possess two distinct enzymatic domains for peptidoglycan assembly. In *Proteobacteria*, the bifunctional PBPs require activation by lipoprotein cofactors in the outer membrane (Lpo). Lipid II units are first polymerized in long glycan chains by the transglycosylation activity of PBPs. These nascent chains stay embedded in the inner membrane by the lipid carrier. They are added to the growing peptidoglycan network by the transpeptidase activity of PBPs that cleave the peptide bond between the terminal D-Ala-D-Ala residues of a stem peptide to form a peptide bound between the fourth residue of the donor stem peptide and the third residue of an adjacent acceptor stem peptide (D-Ala-mA_2pm crosslinks). Peptidoglycan assembly and maturation require the action of peptidoglycan hydrolases that cleave bonds in the peptidoglycan network, such as carboxypeptidases that remove the terminal D-Ala of unprocessed pentapeptide moieties

3.3.2 *Assembly of the Glycosidic Components of the Envelope: Peptidoglycan, LPS and Teichoic Acids*

3.3.2.1 Inner Membrane Flipping on the Undecaprenyl-P Lipid Carrier

The peptidoglycan precursor lipid II as well as the saccharide units for O-antigen and WTAs are translocated across the hydrophobic inner membrane by the lipid carrier undecaprenyl-P. This isoprenoid lipid is composed of a hydrophobic linear chain of 55 carbons embedded in a leaflet of the inner membrane and a polar phosphate group that can face the cytoplasm or the external side of the inner membrane (Bouhss et al. 2008; Manat et al. 2014). In the biochemical pathways for the biosynthesis of the glycosidic precursors for envelope assembly, hydrophilic saccharide-phosphate moieties are transferred from UDP nucleotide-activated sugars in the cytoplasm to the undecaprenyl-P lipid carrier in the inner leaflet of the inner membrane. This step performed by specific glycosyltransferase enzymes leads to the formation of undecaprenyl-pyrophosphate (undecaprenyl-PP)-linked glycosidic intermediates and to the release of UMP (Brown et al. 2013; Bouhss et al. 2008; Manat et al. 2014; Whitfield 1995). After the completion of the membrane step of the pathways for glycosidic precursor synthesis, the undecaprenyl-PP-linked glycosidic intermediates are translocated from the inner leaflet of the inner membrane to the outer leaflet by a specific ATP-independent isoprenoid lipid flippase that facilitates lipid diffusion or an ATP-dependent ABC transporter (Sanyal and Menon 2009). The undecaprenyl-PP-linked glycosidic intermediates in the outer leaflet of the inner membrane then act as activated donors in the envelope compartment. They are recognized as substrates by specific enzymes that use the energy of their sugar-phosphate intramolecular bonds to catalyze glycan transfer or polymerization on a specific acceptor molecule for assembly (Manat et al. 2014; Lovering et al. 2012). As the phosphodiester bonds between the lipid carrier and the glycosidic moiety of envelope precursors are cleaved, undecaprenyl-PP is released and further dephosphorylated into undecaprenyl-P by undecaprenyl-PP phosphatases for recycling (Manat et al. 2014; Lovering et al. 2012; Valvano 2008). However, the assembly of WTAs involves a different reaction that directly leads to the relase of undecaprenyl-P in the outer leaflet of the inner membrane (Brown et al. 2013; Kawai et al. 2011). We still do not understand how undecaprenyl-P is translocated back from the outer leaflet of the inner membrane to the inner leaflet (Manat et al. 2014; Valvano 2008).

3.3.2.2 Peptidoglycan Assembly

The peptidoglycan layer is the most important structural component of the bacterial envelope. It provides rigidity, flexibility and strength necessary for bacterial cells to grow and divide while maintaining their shape and withstanding their high internal osmotic pressure (Typas et al. 2012). The final lipid-linked precursor from the biochemical pathway for peptidoglycan biosynthesis is named lipid II. It is composed of the disaccharide GlcNAc-MurNAc, with the MurNAc sugar linked to both the pentapeptide and the lipid carrier. As the disaccharide GlcNAc-MurNAc is assembled on undecaprenyl-P from UDP-activated sugars, it results in an undecaprenyl-PP-precursor (see sect. 3.3.2.1) (Bouhss et al. 2008). In *B. subtilis* and *E. coli,* the pentapeptide is composed of the residues L-Ala-D-Glu-*m*A_2pm-D-Ala-D-Ala (Barreteau et al. 2008; Kouidmi et al. 2014). The mechanism of lipid II translocation across the inner membrane is not well understood and the candidates FtsW/RodA and MurJ (also called MviN) are at the center of a controversy on the identity of the lipid II flippase (Young 2014; Sham et al. 2014; Mohammadi et al. 2011). The conserved inner membrane proteins FtsW and RodA are members of the shape, elongation, division, and sporulation (SEDS) superfamily of proteins and are respectively required for cell division and elongation in rod-shaped bacteria like *B. subtilis* and *E. coli* (de Pedro et al. 2001; Henriques et al. 1998; Khattar et al. 1994). The inner membrane protein MurJ is a member of the multidrug/oligosaccharidyllipid/polysaccharide (MOP) exporter superfamily (Ruiz 2008; Hvorup et al. 2003) like the O-antigen flippase Wzx (Young 2014; Islam et al. 2013) (see Sect. 3.2.3 and Fig. 3.7 on LPS assembly). While FtsW has been shown to function as a lipid II flippase in an in vitro reconstituted system in proteoliposomes (Mohammadi et al. 2011, 2014), the data from *in vivo* assays support the function of MurJ as the lipid II flippase in *Proteobacteria* (Young 2014; Sham et al. 2014; Lara et al. 2005; Mohamed and Valvano 2014). MurJ is highly conserved in *Proteobacteria* but is not well conserved *Firmicutes* (Ruiz 2009). Anyhow, MurJ homologues from the species *B. subtilis* and *Streptococcus pneumoniae* from the phylum *Firmicutes* complement for the function of MurJ in *E. coli* (Ruiz 2009; Fay and Dworkin 2009), and the MurJ functional homologue is essential in *S. pneumoniae* (Ruiz 2009; Thanassi et al. 2002). However, the four putative MurJ homologues in *B. subtilis* can be inactivated without defects in growth, indicating that there must be another lipid II flippase in this species (Young 2014; Fay and Dworkin 2009).

Once lipid II units are translocated across the inner membrane, they are added to the growing peptidoglycan network by the high molecular weight Penicillin-Binding Proteins referred to as PBPs; the molecular targets of penicillin and other β-lactam antibiotics (Goffin and Ghuysen 1998; Spratt and Pardee 1975). These enzymes are embedded in the inner membrane and possess two distinct enzymatic domains on the external side of the inner membrane for their transglycosylation and transpeptidation activities (Sauvage et al. 2008). In *Proteobacteria* such as *E. coli* and *Vibrio cholerae*, the bifunctional PBPs require activation by lipoprotein cofactors in the outer membrane for proper activity (Paradis-Bleau et al. 2010; Typas et al. 2010; Lupoli et al. 2014; Dorr et al. 2014; Young 2010; Egan et al. 2014). For peptidoglycan assembly, the PBPs first use lipid II substrates to polymerize long glycan strands that stay embedded in the outer leaflet of the inner membrane by the lipid carrier (Fig. 3.6). In the first transglycosylation reaction, the phosphodiester-MurNAc bond of a lipid II unit is cleaved to energize the formation of a glycosidic bond between the MurNAc moiety of the donor substrate and the GlcNAc of the acceptor lipid II unit. Glycan chain polymerization then continues with the formation of new glycosidic bonds between MurNAc and GlcNAc residues concomitant with the release of an undecaprenyl-PP donor in each cycle (Lovering et al. 2012; Perlstein et al. 2007). The PBPs are the main lipid II polymerases and their transglycosylation activity is typically essential in bacteria (Lovering et al. 2012; Sauvage et al. 2008; Vollmer et al. 2008b; McPherson and Popham 2003). Many bacterial species from the phyla *Firmicutes* and *Proteobacteria* also have monofunctional transglycosylases in the inner membrane that perform a redundant transglycosylation reaction (Lovering et al. 2012; Di Berardino et al. 1996). Intriguingly, the transglycosylation activity of PBPs is not required in *B. subtilis* and no monofunctional transglycosylase-encoding gene has yet been identified in its genome. This indicates that the peptidoglycan glycan strands can be made by an unidentified novel type of enzyme in this species (McPherson and Popham 2003; Kunst et al. 1997). The last step of peptidoglycan assembly involves the attachment of nascent glycan strands to the growing peptidoglycan network. This is performed by the D,D-transpeptidase activity of the PBPs that attach peptide moieties from adjacent peptidoglycan glycan strands (Fig. 3.6). PBPs are serine transpeptidases that cleave the peptide bond between the terminal D-Ala-D-Ala residues of the peptidoglycan stem peptide to form a peptide bound between the fourth residue of the donor stem peptide and the third residue of an adjacent acceptor stem peptide (Fig. 3.6). The D-Ala-*m*A_2pm cross-link is predominant in the peptidoglycan of *Firmicutes* and *Proteobacteria* (Figs. 3.1 and 3.6) and the transpeptidase activity of PBPs is usually required for bacterial survival (Lovering et al. 2012; Sauvage et al. 2008; Vollmer et al. 2008b). Many species such as *B. subtilis* and *E. coli* also have L,D-transpeptidases that introduce peptide bonds between the third residues of the stem peptides (Magnet et al. 2007b, 2008; Mainardi et al. 2005). At the opposite of the D,D-transpeptidase activity of PBPs, the activity of L,D-transpeptidases is typically not inhibited by β-lactam antibiotics (Lecoq et al. 2012). The *m*A_2pm-*m*A_2pm cross-links are in low abundance in the peptidoglycan of *Firmicutes* and *Proteobacteria* (Vollmer et al. 2008b; Magnet et al. 2008; Lecoq et al. 2012; Glauner et al. 1988). However, their proportion increases in the stationary phase of growth or in cases of β-lactam resistance where bacterial species such as *Enterococcus faecium* from the phylum *Firmicutes* bypass the need for D,D-transpeptidation by PBPs by relying on L,D-transpeptidases for glycan strands cross-linking (Vollmer et al. 2008b; Lecoq et al. 2012; Pisabarro et al. 1985; Mainardi et al. 2000, 2008).

To assure proper peptidoglycan assembly, the addition of new material into the network by synthesizing enzymes must be tightly coordinated with the action of enzymes that cleave bonds in peptidoglycan (Typas et al. 2012; Lovering et al. 2012). Peptidoglycan hydrolases regroup different classes of enzymes that can collectively cleave almost any bond in the peptidoglycan network (Typas et al. 2012; Vollmer et al. 2008b). The collaborative work of peptidoglycan synthases and hydrolases assures proper peptidoglycan assembly and maturation for the maintenance of a functional structure (Typas et al. 2012). In rod-shaped bacteria like *B. subtilis* and *E. coli*, specific peptidoglycan-synthesizing complexes assemble for the purpose of cell elongation and division. These complexes are organized by the cytoskeletal elements MreB and FtsZ in the cytoplasm (Vollmer et al. 2008a). MreB forms cytoplasmic filaments that orient the elongation complex along the long axis of the bacterial cell while FtsZ makes the Z-ring at the center of the mother cell and directs the synthesizing complex involved in cell division (Vollmer et al. 2008a; Lovering et al. 2012; den Blaauwen et al. 2008; Margolin 2009).

3.3.2.3 Biogenesis of Lipopolysaccharides (LPS) and Transport by the Lpt System in *Proteobacteria*

The outer membrane of the diderm envelope of *Proteobacteria* is asymmetric; its inner leaflet is composed of phospholipids whereas its outer leaflet is composed mainly of LPS (Muhlradt and Golecki 1975; Kamio and Nikaido 1976) (Fig. 3.1b). These polyanionic glycolipids are typically made of a lipid A, a core oligosaccharide and a highly variable chain of repeating oligosaccharides named O-antigen. Lipid A anchors LPS in the outer leaflet of the outer membrane and consists of a glucosamine-based phospholipid. The core oligosaccharide is divided in an inner core linked to lipid A and an outer core attached to O-antigen. The inner core typically contains a few residues of the negatively charged sugar 3-deoxy-D-manno-oct-2-ulosonic acid (Kdo) followed by a few residues of L-glycero-D-manno heptose. The inner core is often decorated with other sugars, phosphate, phosphoethanolamine, pyrophosphorylethanolamine or phosphorylcholine residues, but only the structure of the main saccharidic backbone is well known. The outer core is more structurally diverse than the inner core and consists mostly of hexoses (Raetz and Whitfield 2002; Nikaido 2003). The O-antigen is made of oligosaccharide repeating units that are extremely variable with even a high degree of diversity between different strains of the same species (Raetz and Whitfield 2002; Bos et al. 2007). The strains of *E. coli* make about 170 different types of O-antigen that are defined as serotypes (Raetz and Whitfield 2002). The model system *E. coli* K12 acquired a mutation during its domestication in laboratory and only contains the essential portion of LPS consisting of the lipid A and Kdo portions, a phenotype referred to as "rough" due to the effect on colony morphology. Most wild-type strains and clinical isolates of *E. coli* contain complete LPS structures and form "smooth" colonies (Raetz and Whitfield 2002; Trent et al. 2006; Reeves et al. 1996). Anionic LPS are tightly packed to form the outer leaflet of the outer membrane and their negative charges are counterbalanced and stabilized by divalent cations such as magnesium and calcium that bridge LPS by strong lateral interactions (Nikaido 2003; Holst 2007).

In the pathway for biosynthesis of complete LPS, the lipid A-core and O-antigen units are made separately in the inner leaflet of the inner membrane and are further combined in the outer leaflet of the inner membrane (Raetz and Whitfield 2002) (Fig. 3.7). The lipid A-core is flipped from the inner leaflet to the outer leaflet of the inner membrane by the ABC transporter MsbA (Raetz and Whitfield 2002; King and Sharom 2012). As for lipid II units, O-antigen is synthesized on the lipid carrier undecaprenyl-P from nucleotide-activated sugar donor substrates. This leads to undecaprenyl-PP-linked O-antigen subunits or polysaccharides. Indeed, there are two main pathways for O-antigen assembly and transport. Individual O-antigen subunits can be made in the inner leaflet of the inner membrane and translocated to the outer leaflet by the flippase Wzx. O-antigen subunits are then assembled in mature O-antigen by the polymerase Wzy and the chain length regulator Wzz in the periplasm (Fig. 3.7). Alternatively, O-antigen can be assembled as a complete polymer in the inner leaflet of the inner membrane and further translocated to the

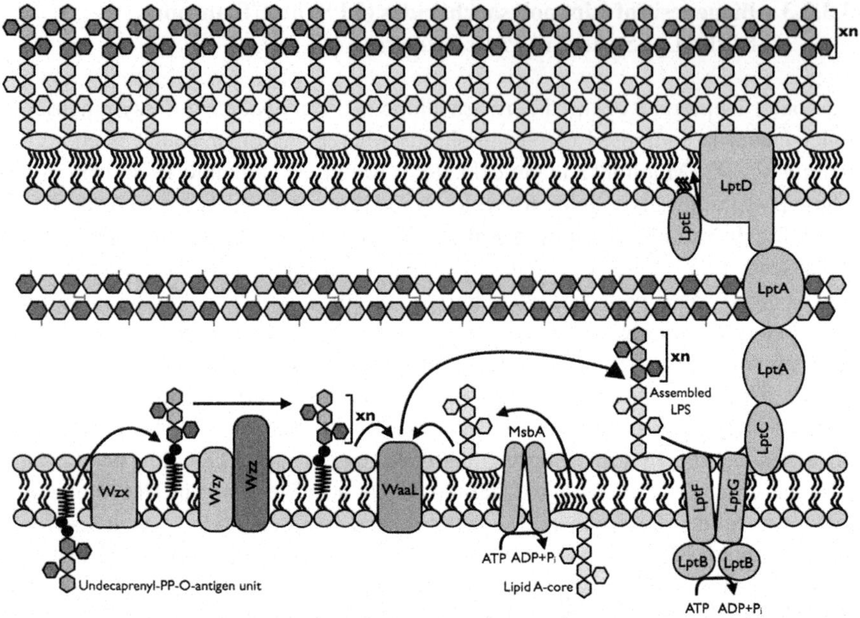

Fig. 3.7 Assembly of LPS and transport to the outer membrane by the Lpt system in *Proteobacteria*. The outer leaflet of the outer membrane of *Proteobacteria* is composed of LPS; anionic glycopolymers made of a lipid A anchor, a core oligosaccharide and a highly variable chain of repeating oligosaccharides named O-antigen. The lipid A-core and O-antigen units are made separately in the inner leaflet of the inner membrane and are combined in the outer leaflet. The lipid A-core is flipped from the inner leaflet of the inner membrane to the outer leaflet by the ABC transporter MsbA upon ATP hydrolysis. There are two main pathways for O-antigen biogenesis. As represented in the figure, O-antigen subunits can be linked to the lipid carrier undecaprenyl-P in the inner leaflet of the inner membrane and translocated to the outer leaflet by the flippase Wzx. O-antigen subunits are then assembled in mature O-antigen by the polymerase Wzy and the chain length regulator Wzz in the periplasm. O-antigen can alternatively be assembled as a complete polymer on the lipid carrier in the inner leaflet of the inner membrane and translocated to the outer leaflet by an ABC transporter (not represented in the figure). In both cases, mature O-antigens are transferred to the lipid A-core unit in the outer leaflet of the inner membrane by the ligase WaaL. Mature LPS are then transported to outer leaflet of the outer membrane by the Lpt system. The proteins LptA-G physically interact to span the entire envelope compartment. The LptBFG proteins form an ABC transporter in the inner membrane and associates with the LptC protein. The LptBCFG complex extracts LPS from the outer leaflet of the inner membrane and transfers it to LptA, a process requiring energy from ATP hydrolysis. The transport of LPS through the periplasm occurs inside the hydrophobic groove formed by the transenvelope protein complex. The β-barrel protein LptD and the lipoprotein LptE form a complex in the outer membrane for LPS incorporation and assembly of the outer leaflet. This figure is based on studies performed with the *E. coli* K12 model system that contains truncated LPS without O-antigens. It is not know if the transport of complete LPS requires additional components and/or mechanistic

outer leaflet by an ABC transporter (Raetz and Whitfield 2002; Whitfield 1995). In both cases, the mature O-antigen is transferred to the lipid A-core unit in the outer leaflet of the inner membrane by the ligase WaaL (Fig. 3.7). This enzyme utilizes the undecaprenyl-PP-linked O-antigen as a donor and the lipid A-core as an acceptor in the reaction, and consequently liberates undecaprenyl-PP (Raetz and Whitfield 2002; Han et al. 2012). The lipid A moiety of mature LPS can be modified by enzymes in the envelope compartment. However, the action of these enzymes is not required for proper bacterial physiology but rather modulates the virulence of some pathogens (Raetz et al. 2007).

Once LPS are assembled in the outer leaflet of the inner membrane, these amphipathic molecules are transported to the outer leaflet of the outer membrane by the LPS transport (Lpt) system (Ruiz et al. 2009; Sperandeo et al. 2009). The system includes the 7 proteins LptA-G that physically interact to span the entire envelope compartment (Chng et al. 2010; Freinkman et al. 2012; Sperandeo et al. 2011; Villa et al. 2013) (Fig. 3.7). The transenvelope complex mediates the extraction of LPS from the outer leaflet of the inner membrane, its transport across the periplasm and its insertion in the outer membrane for the construction of the outer leaflet (Polissi and Sperandeo 2014). The assembly of the Lpt system is quite complex and is tightly regulated to assure correct transenvelope protein complex formation and proper LPS transport (Polissi and Sperandeo 2014). The LptBFG proteins form an ABC transporter in the inner membrane. It uses the energy from cytoplasmic ATP hydrolysis to extract mature LPS from the outer leaflet of the inner membrane and transfer them to the LptC protein associated with the inner membrane and facing the periplasm (Fig. 3.7). The LptBFG complex then transfers LPS from LptC to the periplasmic protein LptA, a step that also requires energy from cytoplasmic ATP hydrolysis (Okuda et al. 2012). LptA connects the LptBCFG complex in the inner membrane to the complex in the outer membrane formed by the β-barrel protein LptD and the outer membrane lipoprotein LptE. It is still unclear how many LptA is required to bridge the transenvelope Lpt complex, but it is likely one or two molecules. The transport of LPS through the periplasm occurs inside the hydrophobic groove formed by the transenvelope Lpt protein complex (Polissi and Sperandeo 2014; Okuda et al. 2012; Sperandeo et al. 2008) (Fig. 3.7). The LptDE complex incorporates and assembles LPS in the outer membrane by an uncharacterized mechanism (Polissi and Sperandeo 2014). It should be noted that the study of the Lpt system has been performed on the *E. coli* K12 model system that contains only the lipid A and Kdo portions of LPS (Raetz and Whitfield 2002), and it is not know if the transport of complete LPS requires additional components and/or mechanistic (Polissi and Sperandeo 2014).

3.3.2.4 Biosynthesis of Wall Teichoic Acids (WTAs) and Attachment to Peptidoglycan in *Firmicutes*

The envelope of *Firmicutes* contains anionic glycopolymers named teichoic acids. They are made of a short saccharidic linkage unit attached to a long chain of

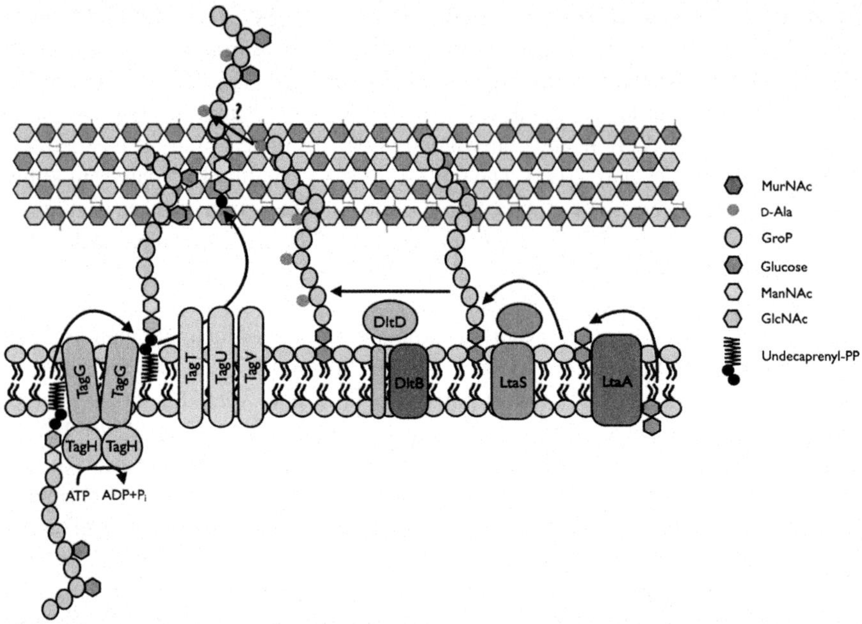

Fig. 3.8 Assembly of teichoic acids in *Firmicutes*. The envelope of *Firmicutes* contains teichoic acids made of a short saccharidic linkage unit and a long chain of phosphodiester-linked polyol repeat units. In the model system *B. subtilis* strains 168, WTAs contain the linkage unit GlcNAc-ManNAc-GroP and a polyol chain made of repeat units of GroP that is decorated with D-alanine and glucose moieties. WTAs are synthesized on the lipid carrier undecaprenyl-P in the inner leaflet of the membrane and glycosylated in the cytoplasm. They are translocated to the outer leaflet of the membrane by the ABC transporter TagGH upon ATP hydrolysis. The attachment of WTA to MurNac moieties of peptidoglycan is performed by the TagTUV enzymes. D-alanylation of WTAs occurs in the envelope compartment and is though to involve the transfer of D-alanine from LTAs. In *B. subtilis*, LTAs are embedded in the outer leaflet of the inner membrane by the glycolipid diacylglycerol-glucose-glucose. This anchor is attached to a polyol chain of GroP repeat units decorated with D-alanine residues. The glycolipid anchor is assembled on diacylglycerol in the inner leaflet of the inner membrane and is translocated to the outer leaflet by a flippase such as LtaA in *S. aureus* (the flippase of *B. subtilis* remains unidentified—see main text). After translocation, the glycolipid anchor serves as the acceptor for the assembly of the GroP polyol chain. GroP units are transferred from head groups of phosphatidylglycerol phospholipids in the outer leaflet of the inner membrane by LtaS. D-alanylation of LTAs occurs in the envelope compartment via the proteins DltB and DltD. The favored model for D-alanylation implies the translocation of undecaprenyl-P-linked activated D-alanine residues from the inner leaflet of the membrane to the outer leaflet for the attachment to LTAs. In most *B. subtilis* strains, the polyol chains of LTAs are decorated with GlcNAc moieties in the envelope compartment. This step is not understood and is not represented on the figure

negatively charged phosphodiester-linked polyol repeat units, a polyol being a compound with multiple hydroxyl functional groups. While LTAs are anchored in the inner membrane (see Sect. 3.2.5), WTAs are attached to peptidoglycan (Brown et al. 2013) (Figs. 3.1a and 3.8). The saccharidic linkage unit of WTAs is a disaccharide composed of GlcNAc and N-acetylmannosamine (ManNAc) further bound to one or two glycerol 3-phosphate (GroP) units, forming GlcNAc-ManNAc-GroP. While the disaccharide linkage unit of WTAs is highly conserved in *Firmicutes*, the composition of the repeat units varies and can differ within the same species. In *B. subtilis*, strains 168 and W23 contain repeat units of GroP and ribitol 5-phosphate, respectively (Brown et al. 2013). The polyol chain of WTAs is decorated with cationic D-alanyl esters and saccharides such as glucose or GlcNAc (Brown et al. 2013; Neuhaus and Baddiley 2003). The *B. subtilis* strains 168 and W23 have both D-alanyl esters and glucose moieties attached to their WTA polyol chains (Brown et al. 2013) (Figs. 3.1a and 3.8).

In the pathway for WTA biosynthesis, the disaccharide linkage unit is assembled on the undecaprenyl-P lipid carrier in the inner leaflet of the inner membrane from nucleotide-activated sugar donor substrates, yielding undecaprenyl-PP-GlcNAc-ManNAc-GroP. The phosphodiester-linked polyol repeat units are then polymerized at the GroP distal end of the disaccharide linkage unit (Brown et al. 2013; Xia and Peschel 2008). Glycosylation of the polyol chain of WTAs is performed by cytoplasmic glycosyltransferases that use specific UDP-activated sugar as donor substrates (Brown et al. 2013; Xia et al. 2010). Undecaprenyl-PP-linked glycosylated WTAs are translocated from the inner leaflet of the inner membrane to the outer leaflet by a two-component ABC transporter such as TagGH in *B. subtilis* 168 (Brown et al. 2013; Lazarevic and Karamata 1995; Schirner et al. 2011) (Fig. 3.8). In the final stage of WTA assembly, the glycopolymers are attached to MurNAc moieties of peptidoglycan by phosphodiester bonds. This implies a phosphotransfer reaction that binds phospho-WTAs from undecaprenyl-PP-linked WTAs to peptidoglycan and releases undecaprenyl-P (Brown et al. 2013; Kawai et al. 2011). The widespread LytR–Cps2A–Psr (LCP) protein family has been involved in WTA attachment to peptidoglycan in *B. subtilis*, and have been renamed TagTUV (Kawai et al. 2011). These three redundant enzymes have a phosphotransferase activity and are associated with the cytoskeleton element MreB that governs cellular elongation in rod-shaped bacteria (Kawai et al. 2011; Graumann 2009). This is consistent with the crucial role of WTAs for cell elongation (Schirner et al. 2009). Homologues of TagTUV have been involved in the ligation of WTAs to peptidoglycan in the species *Staphylococcus aureus* (Dengler et al. 2012; Over et al. 2011) and *S. pneumoniae* from the phylum *Firmicutes* (Eberhardt et al. 2012). The mecanistic of the phosphotransfer reaction is not understood and it is unclear whether WTAs are attached to nascent and/or mature peptidoglycan (Brown et al. 2013). In the model system *B. subtilis*, about 10 % of the peptidoglycan MurNAc moieties are covalently attached to a WTA disaccharide linkage unit that contains from 40 to 60 polyol repeats (Brown et al. 2013; Kojima et al. 1985).

The addition of cationic D-alanine residues to the anionic polyol chain of WTAs occurs in the envelope compartment and plays an important role in the modulation of the cell surface charge (Brown et al. 2013; Collins et al. 2002). Activated D-alanine residues from the cytoplasm are attached to the polyol chain of extracellular WTAs via the inner membrane proteins DltB and DltD through a mechanism that remains unclear (Brown et al. 2013; Kovacs et al. 2006; Perego et al. 1995). DltB is a member of the membrane-bound O-acetyltransferase and DltD is predicted to have esterase/thioesterase activity. The favored model implies the transfer of activated D-alanine residues from the cytoplasmic carrier protein DltC to undecaprenyl-P with the help of DltB. The lipid-linked donor substrates would then be translocated across the inner membrane for the attachment of D-alanine residues to LTAs, a process involving DltB and DltD (Brown et al. 2013; Percy and Grundling 2014; Perego et al. 1995). A subset of the D-alanyl esters on LTAs would further be transferred to WTAs by transesterification, but it is not clear if this step requires an enzyme (Brown et al. 2013; Reichmann and Grundling 2011; Schneewind and Missiakas 2014) (Fig. 3.8).

3.3.2.5 Assembly of Lipoteichoic Acids (LTAs) in *Firmicutes*

In addition to the WTAs described in Sect. 3.3.2.4, the envelope of *Firmicutes* includes LTAs. These amphipathic glycopolymers consist of long chains of negatively charged phosphodiester-linked polyol repeat units embedded in the outer leaflet of the inner membrane by a glycolipid anchor. There are five types of LTAs but most *Firmicutes*, including *B. subtilis*, synthesize type I LTAs (Percy and Grundling 2014). These LTAs are composed of a glycolipid anchor, typically containing a disaccharide linkage unit, attached to a polyol chain of GroP repeat units. While the composition of the repeat units of type I LTAs is conserved, the nature of their decorations and the number of repeats vary among bacterial species, so as the structure of the glycolipid anchor (Schneewind and Missiakas 2014). In *B. subtilis*, the glycolipid anchor is made of a diacylglycerol lipid bound to a disaccharide of glucose (Schneewind and Missiakas 2014) and the polyol chain contains between 15 and 50 GroP units (Fischer 1988). LTAs of *B. subtilis* are decorated with D-alanyl esters but their glycosylation varies among strains. Most *B. subtilis* strains have GlcNAc moieties on their LTA polyol chains and a subset of strains do not present any form of LTA glycosylation (Percy and Grundling 2014; Iwasaki et al. 1986, 1989).

The biosynthesis of LTAs does not involve the lipid carrier undecaprenyl-P. It rather directly involves the integral inner membrane phospholipid phosphatidylglycerol both as a source of lipid for glycolipid synthesis and membrane flipping, and of GroP for polyol chain elongation. In the inner leaflet of the inner membrane, diacylglycerol generated from phosphatidylglycerol turnover is used for the assembly of the LTA glycolipid anchor (Schneewind and Missiakas 2014; Ganfield and Pieringer 1980; Koch et al. 1984; Taron et al. 1983). The disaccharide linkage unit is made of two glucose units transferred from nucleotide-activated sugars in the cytoplasm, and the glycolipid diacylglycerol-glucose-glucose is translocated from the inner leaflet of the inner membrane to the outer leaflet by a flippase such as LtaA in *S. aureus* (Schneewind and Missiakas 2014; Grundling and Schneewind 2007a). While LtaA is conserved in many members of the *Firmicutes*, there is no apparent homologue of LtaA in *B. subtilis* and its diacylglycerol-glucose-glucose flippase remains unidentified (Reichmann and Grundling 2011; Schneewind and Missiakas 2014).

The LTA glycolipid anchor translocated in the outer leaflet of the inner membrane serves as the acceptor for the assembly of the GroP polyol chain in the envelope compartment. The head group of phosphatidylglycerol in the outer leaflet of the inner membrane is used as a GroP donor. The polymerization of the LTA polyol chain is made by the sequential addition of GroP units at the tip of the growing chain on the glycolipid anchor (Percy and Grundling 2014; Schneewind and Missiakas 2014; Koch et al. 1984; Fischer 1994). In *S. aureus*, this step is catalyzed by the LTA synthase (LtaS), an inner membrane protein with a large domain facing the cell exterior (Grundling and Schneewind 2007b; Lu et al. 2009; Karatsa-Dodgson et al. 2010). LtaS is conserved in most *Firmicutes*. *B. subtilis* has four functional LtaS homologues: LtaS (also named YflE), YqgS, YfnI and YvgJ (Reichmann and Grundling 2011; Schneewind and Missiakas 2014). *B. subtilis* LtaS seems to be the most important enzyme for LTA assembly while YqgS and YfnI appear to have overlapping roles for the adaptation to changing environmental conditions. Finally, *B. subtilis* YvgJ has a primase activity and adds the first GroP subunit to the glycolipid anchor. YvgJ is not required for LTA assembly in *B. subtilis* and it is suggested that the other LtaS homologues perform both the primase and polymerase activities as for the *S. aureus* LtaS enzyme (Reichmann and Grundling 2011; Sutcliffe 2011). The action of LtaS and its homologues liberates one diacylglycerol in the outer leaflet of the inner membrane for every GroP unit transferred from phosphatidylglycerol for the assembly of the LTA polyol chain. The diacylglycerol products return to the inner leaflet of the membrane and are further utilized for the synthesis of the LTA glycolipid or are recycled in phosphatidylglycerol (Percy and Grundling 2014; Koch et al. 1984).

LTAs are D-alanylated in the envelope compartment by the DltBD enzymes as described in the previous section on WTAs (see Sect. 3.2.4 and Fig. 3.8). While WTAs are glycosylated in the cytoplasm (see Sect. 3.2.4), saccharide moieties are added to the GroP units of LTAs in the envelope compartment. This step is not understood but seems to involve the lipid carrier undecaprenyl-P (Mancuso and Chiu 1982; Yokoyama et al. 1988). The current model implies the action of a cytoplasmic glycosyltransferase for the formation of the lipid-linked membrane precursor and a periplasmic glycosyltransferase for the glycosylation of LTAs from the translocated lipid donor (Percy and Grundling 2014).

To allow for glycopolymer polymerization, D-alanylation and glycosylation by the action of membrane proteins, LTAs must stay close to the inner membrane (Reichmann and Grundling 2011) (Fig. 3.8). This model of LTA localization is supported by LTA labeling experiments and observation by cryo-transmission electron microscopy (Matias and Beveridge 2008). It is also consistent with the crucial role of LTAs for cell division (Schirner et al. 2009). However, the presumed transfer of D-alanyl esters from LTAs to WTAs suggests that at least some LTAs extend in the peptidoglycan network (Reichmann and Grundling 2011) (Fig. 3.8).

3.4 Conclusion and Perspectives

The bacterial envelope is a functional structure of high complexity that fulfills many functions required for bacterial survival, morphogenesis, growth, adaptation and pathogenesis. The process of bacterial envelope assembly is extremely challenging and bacteria have evolved amazing molecular machines to accomplish this tremendous task while preserving cellular integrity and homeostasis. The study of the monoderm envelope of *Firmicutes* and the diderm envelope of *Proteobacteria* with the model systems *B. subtilis* and *E. coli*, respectively, have led to impressive breakthroughs over the last decades. We now have a broader understanding of how molecular machines assure the secretion, insertion and folding of the envelope proteins as well as the assembly of the glycosidic components of the envelope. This progress is very exciting and the availability of novel tools and expertise will surely lead to many more important discoveries in the field of envelope assembly in the upcoming years. This should provide answers and explain certain aspects of envelope assembly that remain unclear.

For example, we do not understand how α-helical transmembrane domains exit the gate of the Sec translocase for their insertion in the inner membrane. The mechanistic of YidC function is also not understood, so as the coordination of the action of SecA for the translocation of large protein domains in the periplasm with the process of membrane protein insertion (du Plessis et al. 2011). We do not know how protein translocation by the Tat system is energized, and how this system selectively allows the passage of its folded and/or cofactor-containing substrates while maintaining the impermeability of the inner membrane to other molecules (Palmer and Berks 2012). Even though the enzymes involved in peptidoglycan assembly are well studied, the mechanism of nascent glycan strand integration into the existing peptidoglycan layer is elusive (Holtje 1998; Sauvage et al. 2008; Perlstein et al. 2007) and the construction of the three-dimensional peptidoglycan meshwork is poorly understood (den Blaauwen et al. 2008). In *Proteobacteria*, the mechanism of lipoprotein incorporation in the inner leaflet of the outer membrane by LolB is undetermined (Okuda and Tokuda 2011). We also do not understand how phospholipids are translocated across the inner membrane and transported through the periplasm for the assembly of the inner leaflet of the outer membrane (Silhavy et al. 2010). The mechanism of LPS insertion in the outer membrane by LptDE is not elucidated and we do not know if the Lpt system requires additional components to accommodate complete LPS molecules with O-antigens (Polissi and Sperandeo 2014). The individual functions of the Bam proteins and the mechanistic of outer membrane protein insertion and folding remain largely unknown (Hagan et al. 2011). Regarding teichoic acid assembly in *Firmicutes*, the process of D-alanylation is not well characterized and the mechanism of WTA attachment to peptidoglycan is unclear (Brown et al. 2013). Finally, the glycosylation of LTAs in the envelope compartment remains a mystery (Percy and Grundling 2014). Looking at the bigger picture of envelope assembly, we are only starting to understand how the construction of the different envelope layers is regulated and coordinated for the maintenance of a functionally active structure (Silhavy et al. 2010; Ruiz et al. 2009; Weiner and Li 2008). As stated before, the novel tools and expertise for the study of the exiting field of bacterial envelope assembly will surely lead to important discoveries and provide some answers to these unresolved questions in the upcoming years.

References

Andersson H, von Heijne G (1993) Sec dependent and sec independent assembly of *E. coli* inner membrane proteins: the topological rules depend on chain length. EMBO J 12(2):683–691

Angelini S, Deitermann S, Koch HG (2005) FtsY, the bacterial signal-recognition particle receptor, interacts functionally and physically with the SecYEG translocon. EMBO Rep 6(5):476–481

Angelini S et al (2006) Membrane binding of the bacterial signal recognition particle receptor involves two distinct binding sites. J Cell Biol 174(5):715–724

Barreteau H et al (2008) Cytoplasmic steps of peptidoglycan biosynthesis. FEMS Microbiol Rev 32(2):168–207

Bartholomew JW, Mittwer T (1952) The gram stain. Bacteriol Rev 16(1):1–29

Beck K et al (2000) Discrimination between SRP- and SecA/SecB-dependent substrates involves selective recognition of nascent chains by SRP and trigger factor. EMBO J 19(1):134–143

Beveridge TJ (2001) Use of the gram stain in microbiology. Biotech Histochem 76(3):111–118

Beveridge TJ, Graham LL (1991) Surface layers of bacteria. Microbiol Rev 55(4):684–705

Boone DR, Castenholz RW, Garrity GM (2001) Bergey's manual of systematic bacteriology, 2nd edn. Springer, New York

Bos MP, Robert V, Tommassen J (2007) Biogenesis of the gram-negative bacterial outer membrane. Annu Rev Microbiol 61:191–214

Bouhss A et al (2008) The biosynthesis of peptidoglycan lipid-linked intermediates. FEMS Microbiol Rev 32(2):208–233

Braun V, Rehn K (1969) Chemical characterization, spatial distribution and function of a lipoprotein (murein-lipoprotein) of the *E. coli* cell wall. The specific effect of trypsin on the membrane structure. Eur J Biochem 10(3):426–438

Briggs MS et al (1986) Conformations of signal peptides induced by lipids suggest initial steps in protein export. Science 233(4760):206–208

Brown S, Santa Maria JP, Walker S Jr (2013) Wall teichoic acids of gram-positive bacteria. Annu Rev Microbiol 67:313–336

Buddelmeijer N, Young R (2010) The essential *Escherichia coli* apolipoprotein N-acyltransferase (Lnt) exists as an extracytoplasmic thioester acyl-enzyme intermediate. Biochemistry 49(2):341–346

Chevance FF, Hughes KT (2008) Coordinating assembly of a bacterial macromolecular machine. Nat Rev Microbiol 6(6):455–465

Chng SS, Gronenberg LS, Kahne D (2010) Proteins required for lipopolysaccharide assembly in *Escherichia coli* form a transenvelope complex. Biochemistry 49(22):4565–4567

Clark DP (2010) Molecular biology: academic cell update, vol xviii. Academic Press/Elsevier, Amsterdam/Boston, p 784

Collins LV et al (2002) *Staphylococcus aureus* strains lacking D-alanine modifications of teichoic acids are highly susceptible to human neutrophil killing and are virulence attenuated in mice. J Infect Dis 186(2):214–219

Comfort D, Clubb RT (2004) A comparative genome analysis identifies distinct sorting pathways in gram-positive bacteria. Infect Immun 72(5):2710–2722

Dalbey RE, Wang P, Kuhn A (2011) Assembly of bacterial inner membrane proteins. Annu Rev Biochem 80:161–187

de Pedro MA et al (2001) Constitutive septal murein synthesis in *Escherichia coli* with impaired activity of the morphogenetic proteins RodA and penicillin-binding protein 2. J Bacteriol 183(14):4115–4126
Delcour AH (2009) Outer membrane permeability and antibiotic resistance. Biochim Biophys Acta 1794(5):808–816
den Blaauwen T et al (2008) Morphogenesis of rod-shaped sacculi. FEMS Microbiol Rev 32(2):321–344
Dengler V et al (2012) Deletion of hypothetical wall teichoic acid ligases in *Staphylococcus aureus* activates the cell wall stress response. FEMS Microbiol Lett 333(2):109–120
Denoncin K et al (2012) Dissecting the *Escherichia coli* periplasmic chaperone network using differential proteomics. Proteomics 12(9):1391–1401
Depuydt M, Messens J, Collet JF (2011) How proteins form disulfide bonds. Antioxid Redox Signal 15(1):49–66
Di Berardino M et al (1996) The monofunctional glycosyltransferase of *Escherichia coli* is a member of a new class of peptidoglycan-synthesising enzymes. FEBS Lett 392(2):184–188
Dilks K et al (2003) Prokaryotic utilization of the twin-arginine translocation pathway: a genomic survey. J Bacteriol 185(4):1478–1483
Doerrler WT (2006) Lipid trafficking to the outer membrane of Gram-negative bacteria. Mol Microbiol 60(3):542–552
Dorr T et al (2014) Differential requirement for PBP1a and PBP1b in *in vivo* and *in vitro* fitness of *Vibrio cholerae*. Infect Immun 82(5):2115–2124
Dramsi S et al (2008) Covalent attachment of proteins to peptidoglycan. FEMS Microbiol Rev 32(2):307–320
Driessen AJ, Nouwen N (2008) Protein translocation across the bacterial cytoplasmic membrane. Annu Rev Biochem 77:643–667
du Plessis DJ, Nouwen N, Driessen AJ (2011) The Sec translocase. Biochim Biophys Acta 1808(3):851–865
Duguay AR, Silhavy TJ (2004) Quality control in the bacterial periplasm. Biochim Biophys Acta 1694(1–3):121–134
Dutton RJ et al (2008) Bacterial species exhibit diversity in their mechanisms and capacity for protein disulfide bond formation. Proc Natl Acad Sci U S A 105(33):11933–11938
Dziarski R, Gupta D (2005) Peptidoglycan recognition in innate immunity. J Endotoxin Res 11(5):304–310
Eberhardt A et al (2012) Attachment of capsular polysaccharide to the cell wall in *Streptococcus pneumoniae*. Microb Drug Resist 18(3):240–255
Egan AJ et al (2014) Outer-membrane lipoprotein LpoB spans the periplasm to stimulate the peptidoglycan synthase PBP1B. Proc Natl Acad Sci U S A 111(22):8197–8202
Egea PF et al (2004) Substrate twinning activates the signal recognition particle and its receptor. Nature 427(6971):215–221
Eisner G et al (2003) Ligand crowding at a nascent signal sequence. J Cell Biol 163(1):35–44
Facey SJ, Kuhn A (2010) Biogenesis of bacterial inner-membrane proteins. Cell Mol Life Sci 67(14):2343–2362
Fagan RP, Fairweather NF (2014) Biogenesis and functions of bacterial S-layers. Nat Rev Microbiol 12(3):211–222
Fay A, Dworkin J (2009) *Bacillus subtilis* homologs of MviN (MurJ), the putative *Escherichia coli* lipid II flippase, are not essential for growth. J Bacteriol 191(19):6020–6028
Ferbitz L et al (2004) Trigger factor in complex with the ribosome forms a molecular cradle for nascent proteins. Nature 431(7008):590–596
Fischer W (1988) Physiology of lipoteichoic acids in bacteria. Adv Microb Physiol 29:233–302
Fischer W (1994) Lipoteichoic acid and lipids in the membrane of *Staphylococcus aureus*. Med Microbiol Immunol 183(2):61–76
Focia PJ et al (2004) Heterodimeric GTPase core of the SRP targeting complex. Science 303(5656):373–377
Freinkman E et al (2012) Regulated assembly of the transenvelope protein complex required for lipopolysaccharide export. Biochemistry 51(24):4800–4806

Frobel J, Rose P, Muller M (2011) Early contacts between substrate proteins and TatA translocase component in twin-arginine translocation. J Biol Chem 286(51):43679–43689

Frobel J et al (2012) Transmembrane insertion of twin-arginine signal peptides is driven by TatC and regulated by TatB. Nat Commun 3:1311

Ganfield MC, Pieringer RA (1980) The biosynthesis of nascent membrane lipoteichoic acid of *Streptococcus faecium* (*S. faecalis* ATCC 9790) from phosphatidylkojibiosyl diacylglycerol and phosphatidylglycerol. J Biol Chem 255(11):5164–5169

Glauner B, Holtje JV, Schwarz U (1988) The composition of the murein of *Escherichia coli*. J Biol Chem 263(21):10088–10095

Goemans C, Denoncin K, Collet JF (2014) Folding mechanisms of periplasmic proteins. Biochim Biophys Acta 1843(8):1517–1528

Goffin C, Ghuysen JM (1998) Multimodular penicillin-binding proteins: an enigmatic family of orthologs and paralogs. Microbiol Mol Biol Rev 62(4):1079–1093

Goosens VJ, Monteferrante CG, van Dijl JM (2014) The Tat system of Gram-positive bacteria. Biochim Biophys Acta 1843(8):1698–1706

Graumann PL (2009) Dynamics of bacterial cytoskeletal elements. Cell Motil Cytoskeleton 66(11):909–914

Grundling A, Schneewind O (2007a) Genes required for glycolipid synthesis and lipoteichoic acid anchoring in *Staphylococcus aureus*. J Bacteriol 189(6):2521–2530

Grundling A, Schneewind O (2007b) Synthesis of glycerol phosphate lipoteichoic acid in *Staphylococcus aureus*. Proc Natl Acad Sci U S A 104(20):8478–8483

Hagan CL, Kim S, Kahne D (2010) Reconstitution of outer membrane protein assembly from purified components. Science 328(5980):890–892

Hagan CL, Silhavy TJ, Kahne D (2011) beta-Barrel membrane protein assembly by the Bam complex. Annu Rev Biochem 80:189–210

Haiko J, Westerlund-Wikstrom B (2013) The role of the bacterial flagellum in adhesion and virulence. Biology (Basel) 2(4):1242–1267

Han W et al (2012) Defining function of lipopolysaccharide O-antigen ligase WaaL using chemoenzymatically synthesized substrates. J Biol Chem 287(8):5357–5365

Hancock IC (1997) Bacterial cell surface carbohydrates: structure and assembly. Biochem Soc Trans 25(1):183–187

Harold FM (1972) Conservation and transformation of energy by bacterial membranes. Bacteriol Rev 36(2):172–230

Harshey RM (2003) Bacterial motility on a surface: many ways to a common goal. Annu Rev Microbiol 57:249–273

Hatahet F, Boyd D, Beckwith J (2014) Disulfide bond formation in prokaryotes: history, diversity and design. Biochim Biophys Acta 1844(8):1402–1414

Henriques AO et al (1998) Control of cell shape and elongation by the rodA gene in *Bacillus subtilis*. Mol Microbiol 28(2):235–247

Holst O (2007) The structures of core regions from enterobacterial lipopolysaccharides – an update. FEMS Microbiol Lett 271(1):3–11

Holtje JV (1998) Growth of the stress-bearing and shape-maintaining murein sacculus of *Escherichia coli*. Microbiol Mol Biol Rev 62(1):181–203

Hutchings MI et al (2009) Lipoprotein biogenesis in Gram-positive bacteria: knowing when to hold 'em, knowing when to fold 'em. Trends Microbiol 17(1):13–21

Hvorup RN et al (2003) The multidrug/oligosaccharidyl-lipid/polysaccharide (MOP) exporter superfamily. Eur J Biochem 270(5):799–813

Islam ST et al (2013) Proton-dependent gating and proton uptake by Wzx support O-antigen-subunit antiport across the bacterial inner membrane. MBio 4(5):e00678-13

Iwasaki H, Shimada A, Ito E (1986) Comparative studies of lipoteichoic acids from several Bacillus strains. J Bacteriol 167(2):508–516

Iwasaki H et al (1989) Structure and glycosylation of lipoteichoic acids in Bacillus strains. J Bacteriol 171(1):424–429

Jordan S, Hutchings MI, Mascher T (2008) Cell envelope stress response in Gram-positive bacteria. FEMS Microbiol Rev 32(1):107–146

Kadokura H, Beckwith J (2010) Mechanisms of oxidative protein folding in the bacterial cell envelope. Antioxid Redox Signal 13(8):1231–1246

Kamio Y, Nikaido H (1976) Outer membrane of *Salmonella typhimurium*: accessibility of phospholipid head groups to phospholipase c and cyanogen bromide activated dextran in the external medium. Biochemistry 15(12):2561–2570

Karatsa-Dodgson M, Wormann ME, Grundling A (2010) In vitro analysis of the *Staphylococcus aureus* lipoteichoic acid synthase enzyme using fluorescently labeled lipids. J Bacteriol 192(20):5341–5349

Kawai Y et al (2011) A widespread family of bacterial cell wall assembly proteins. EMBO J 30(24):4931–4941

Khattar MM, Begg KJ, Donachie WD (1994) Identification of FtsW and characterization of a new ftsW division mutant of *Escherichia coli*. J Bacteriol 176(23):7140–7147

Kim S et al (2007) Structure and function of an essential component of the outer membrane protein assembly machine. Science 317(5840):961–964

King G, Sharom FJ (2012) Proteins that bind and move lipids: MsbA and NPC1. Crit Rev Biochem Mol Biol 47(1):75–95

Koch HG, Muller M (2000) Dissecting the translocase and integrase functions of the *Escherichia coli* SecYEG translocon. J Cell Biol 150(3):689–694

Koch HU, Haas R, Fischer W (1984) The role of lipoteichoic acid biosynthesis in membrane lipid metabolism of growing *Staphylococcus aureus*. Eur J Biochem 138(2):357–363

Kojima N, Araki Y, Ito E (1985) Structure of the linkage units between ribitol teichoic acids and peptidoglycan. J Bacteriol 161(1):299–306

Kouidmi I, Levesque RC, Paradis-Bleau C (2014) The biology of Mur ligases as an antibacterial target. Mol Microbiol 94(2):242–253

Kovacs M et al (2006) A functional dlt operon, encoding proteins required for incorporation of d-alanine in teichoic acids in gram-positive bacteria, confers resistance to cationic antimicrobial peptides in *Streptococcus pneumoniae*. J Bacteriol 188(16):5797–5805

Kunst F et al (1997) The complete genome sequence of the gram-positive bacterium *Bacillus subtilis*. Nature 390(6657):249–256

Lara B et al (2005) Peptidoglycan precursor pools associated with MraY and FtsW deficiencies or antibiotic treatments. FEMS Microbiol Lett 250(2):195–200

Lazarevic V, Karamata D (1995) The tagGH operon of *Bacillus subtilis* 168 encodes a two-component ABC transporter involved in the metabolism of two wall teichoic acids. Mol Microbiol 16(2):345–355

Lecoq L et al (2012) Dynamics induced by beta-lactam antibiotics in the active site of *Bacillus subtilis* L,D-transpeptidase. Structure 20(5):850–861

Li J, Lee DS, Madrenas J (2013) Evolving bacterial envelopes and plasticity of TLR2-dependent responses: basic research and translational opportunities. Front Immunol 4:347

Lovering AL, Safadi SS, Strynadka NC (2012) Structural perspective of peptidoglycan biosynthesis and assembly. Annu Rev Biochem 81:451–478

Lu D et al (2009) Structure-based mechanism of lipoteichoic acid synthesis by *Staphylococcus aureus* LtaS. Proc Natl Acad Sci U S A 106(5):1584–1589

Lugtenberg EJ, Peters R (1976) Distribution of lipids in cytoplasmic and outer membranes of *Escherichia coli* K12. Biochim Biophys Acta 441(1):38–47

Lupoli TJ et al (2014) Lipoprotein activators stimulate *Escherichia coli* penicillin-binding proteins by different mechanisms. J Am Chem Soc 136(1):52–55

Lycklama ANJA, Driessen AJ (2012) The bacterial Sec-translocase: structure and mechanism. Philos Trans R Soc Lond B Biol Sci 367(1592):1016–1028

Magnet S et al (2007a) Identification of the L,D-transpeptidases responsible for attachment of the Braun lipoprotein to *Escherichia coli* peptidoglycan. J Bacteriol 189(10):3927–3931

Magnet S et al (2007b) Specificity of L,D-transpeptidases from gram-positive bacteria producing different peptidoglycan chemotypes. J Biol Chem 282(18):13151–13159

Magnet S et al (2008) Identification of the L,D-transpeptidases for peptidoglycan cross-linking in *Escherichia coli*. J Bacteriol 190(13):4782–4785

Mainardi JL et al (2000) Novel mechanism of beta-lactam resistance due to bypass of DD-transpeptidation in *Enterococcus faecium*. J Biol Chem 275(22):16490–16496

Mainardi JL et al (2005) A novel peptidoglycan cross-linking enzyme for a beta-lactam-resistant transpeptidation pathway. J Biol Chem 280(46):38146–38152

Mainardi JL et al (2008) Evolution of peptidoglycan biosynthesis under the selective pressure of antibiotics in Gram-positive bacteria. FEMS Microbiol Rev 32(2):386–408

Maloney PC, Kashket ER, Wilson TH (1974) A protonmotive force drives ATP synthesis in bacteria. Proc Natl Acad Sci U S A 71(10):3896–3900

Manat G et al (2014) Deciphering the metabolism of undecaprenyl-phosphate: the bacterial cell-wall unit carrier at the membrane frontier. Microb Drug Resist 20(3):199–214

Mancuso DJ, Chiu TH (1982) Biosynthesis of glucosyl monophosphoryl undecaprenol and its role in lipoteichoic acid biosynthesis. J Bacteriol 152(2):616–625

Margolin W (2009) Sculpting the bacterial cell. Curr Biol 19(17):R812–R822

Matias VR, Beveridge TJ (2008) Lipoteichoic acid is a major component of the *Bacillus subtilis* periplasm. J Bacteriol 190(22):7414–7418

McPherson DC, Popham DL (2003) Peptidoglycan synthesis in the absence of class A penicillin-binding proteins in *Bacillus subtilis*. J Bacteriol 185(4):1423–1431

Merdanovic M et al (2011) Protein quality control in the bacterial periplasm. Annu Rev Microbiol 65:149–168

Millman JS et al (2001) FtsY binds to the *Escherichia coli* inner membrane via interactions with phosphatidylethanolamine and membrane proteins. J Biol Chem 276(28):25982–25989

Mohamed YF, Valvano MA (2014) A Burkholderia cenocepacia MurJ (MviN) homolog is essential for cell wall peptidoglycan synthesis and bacterial viability. Glycobiology 24(6):564–576

Mohammadi T et al (2011) Identification of FtsW as a transporter of lipid-linked cell wall precursors across the membrane. EMBO J 30(8):1425–1432

Mohammadi T et al (2014) Specificity of the transport of lipid II by FtsW in *Escherichia coli*. J Biol Chem 289(21):14707–14718

Muhlradt PF, Golecki JR (1975) Asymmetrical distribution and artifactual reorientation of lipopolysaccharide in the outer membrane bilayer of *Salmonella typhimurium*. Eur J Biochem 51(2):343–352

Mullineaux CW et al (2006) Diffusion of green fluorescent protein in three cell environments in *Escherichia coli*. J Bacteriol 188(10):3442–3448

Nakayama H, Kurokawa K, Lee BL (2012) Lipoproteins in bacteria: structures and biosynthetic pathways. FEBS J 279(23):4247–4268

Narita S, Tokuda H (2010) Sorting of bacterial lipoproteins to the outer membrane by the Lol system. Methods Mol Biol 619:117–129

Natale P, Bruser T, Driessen AJ (2008) Sec- and Tat-mediated protein secretion across the bacterial cytoplasmic membrane—distinct translocases and mechanisms. Biochim Biophys Acta 1778(9):1735–1756

Nelson N (1994) Energizing porters by proton-motive force. J Exp Biol 196:7–13

Neuhaus FC, Baddiley J (2003) A continuum of anionic charge: structures and functions of D-alanyl-teichoic acids in gram-positive bacteria. Microbiol Mol Biol Rev 67(4):686–723

Nikaido H (1989) Outer membrane barrier as a mechanism of antimicrobial resistance. Antimicrob Agents Chemother 33(11):1831–1836

Nikaido H (2003) Molecular basis of bacterial outer membrane permeability revisited. Microbiol Mol Biol Rev 67(4):593–656

Okuda S, Tokuda H (2011) Lipoprotein sorting in bacteria. Annu Rev Microbiol 65:239–259

Okuda S, Freinkman E, Kahne D (2012) Cytoplasmic ATP hydrolysis powers transport of lipopolysaccharide across the periplasm in *E. coli*. Science 338(6111):1214–1217

Oliver DB (1996) Periplasm. ASM Press, Washington

Over B et al (2011) LytR-CpsA-Psr proteins in *Staphylococcus aureus* display partial functional redundancy and the deletion of all three severely impairs septum placement and cell separation. FEMS Microbiol Lett 320(2):142–151

Paetzel M et al (2002) Signal peptidases. Chem Rev 102(12):4549–4580
Pailler J et al (2012) Phosphatidylglycerol::prolipoprotein diacylglyceryl transferase (Lgt) of *Escherichia coli* has seven transmembrane segments, and its essential residues are embedded in the membrane. J Bacteriol 194(9):2142–2151
Pallen MJ, Chaudhuri RR, Henderson IR (2003) Genomic analysis of secretion systems. Curr Opin Microbiol 6(5):519–527
Palmer T, Berks BC (2012) The twin-arginine translocation (Tat) protein export pathway. Nat Rev Microbiol 10(7):483–496
Palmer T, Sargent F, Berks BC (2010) The Tat protein export pathway. In the *Escherichia coli* and *Salmonella*: cellular and molecular biology. ASM Press, Washington
Paradis-Bleau C et al (2010) Lipoprotein cofactors located in the outer membrane activate bacterial cell wall polymerases. Cell 143(7):1110–1120
Percy MG, Grundling A (2014) Lipoteichoic acid synthesis and function in gram-positive bacteria. Annu Rev Microbiol 68:81–100
Perego M et al (1995) Incorporation of D-alanine into lipoteichoic acid and wall teichoic acid in *Bacillus subtilis*. Identification of genes and regulation. J Biol Chem 270(26):15598–15606
Perlstein DL et al (2007) The direction of glycan chain elongation by peptidoglycan glycosyltransferases. J Am Chem Soc 129(42):12674–12675
Piddock LJ (2006) Multidrug-resistance efflux pumps – not just for resistance. Nat Rev Microbiol 4(8):629–636
Pisabarro AG, de Pedro MA, Vazquez D (1985) Structural modifications in the peptidoglycan of *Escherichia coli* associated with changes in the state of growth of the culture. J Bacteriol 161(1):238–242
Polissi A, Sperandeo P (2014) The lipopolysaccharide export pathway in *Escherichia coli*: structure, organization and regulated assembly of the Lpt machinery. Mar Drugs 12(2):1023–1042
Raetz CR (1978) Enzymology, genetics, and regulation of membrane phospholipid synthesis in *Escherichia coli*. Microbiol Rev 42(3):614–659
Raetz CR, Whitfield C (2002) Lipopolysaccharide endotoxins. Annu Rev Biochem 71:635–700
Raetz CR et al (2007) Lipid A modification systems in gram-negative bacteria. Annu Rev Biochem 76:295–329
Raivio TL (2005) Envelope stress responses and Gram-negative bacterial pathogenesis. Mol Microbiol 56(5):1119–1128
Ramos HC, Rumbo M, Sirard JC (2004) Bacterial flagellins: mediators of pathogenicity and host immune responses in mucosa. Trends Microbiol 12(11):509–517
Reeves PR et al (1996) Bacterial polysaccharide synthesis and gene nomenclature. Trends Microbiol 4(12):495–503
Reichmann NT, Grundling A (2011) Location, synthesis and function of glycolipids and polyglycerolphosphate lipoteichoic acid in Gram-positive bacteria of the phylum Firmicutes. FEMS Microbiol Lett 319(2):97–105
Ricci DP, Silhavy TJ (2012) The Bam machine: a molecular cooper. Biochim Biophys Acta 1818(4):1067–1084
Rizzitello AE, Harper JR, Silhavy TJ (2001) Genetic evidence for parallel pathways of chaperone activity in the periplasm of *Escherichia coli*. J Bacteriol 183(23):6794–6800
Robichon C, Vidal-Ingigliardi D, Pugsley AP (2005) Depletion of apolipoprotein N-acyltransferase causes mislocalization of outer membrane lipoproteins in *Escherichia coli*. J Biol Chem 280(2):974–983
Royet J, Dziarski R (2007) Peptidoglycan recognition proteins: pleiotropic sensors and effectors of antimicrobial defences. Nat Rev Microbiol 5(4):264–277
Ruiz N (2008) Bioinformatics identification of MurJ (MviN) as the peptidoglycan lipid II flippase in *Escherichia coli*. Proc Natl Acad Sci U S A 105(40):15553–15557
Ruiz N (2009) Streptococcus pyogenes YtgP (Spy_0390) complements *Escherichia coli* strains depleted of the putative peptidoglycan flippase MurJ. Antimicrob Agents Chemother 53(8):3604–3605

Ruiz N, Kahne D, Silhavy TJ (2009) Transport of lipopolysaccharide across the cell envelope: the long road of discovery. Nat Rev Microbiol 7(9):677–683

Samuelson JC et al (2000) YidC mediates membrane protein insertion in bacteria. Nature 406(6796):637–641

Sankaran K, Wu HC (1994) Lipid modification of bacterial prolipoprotein. Transfer of diacylglyceryl moiety from phosphatidylglycerol. J Biol Chem 269(31):19701–19706

Sanyal S, Menon AK (2009) Flipping lipids: why an' what's the reason for? ACS Chem Biol 4(11):895–909

Sarvas M et al (2004) Post-translocational folding of secretory proteins in Gram-positive bacteria. Biochim Biophys Acta 1694(1–3):311–327

Sauvage E et al (2008) The penicillin-binding proteins: structure and role in peptidoglycan biosynthesis. FEMS Microbiol Rev 32(2):234–258

Schirner K et al (2009) Distinct and essential morphogenic functions for wall- and lipo-teichoic acids in *Bacillus subtilis*. EMBO J 28(7):830–842

Schirner K, Stone LK, Walker S (2011) ABC transporters required for export of wall teichoic acids do not discriminate between different main chain polymers. ACS Chem Biol 6(5):407–412

Schneewind O, Missiakas DM (2012) Protein secretion and surface display in Gram-positive bacteria. Philos Trans R Soc Lond B Biol Sci 367(1592):1123–1139

Schneewind O, Missiakas D (2014) Lipoteichoic acids, phosphate-containing polymers in the envelope of gram-positive bacteria. J Bacteriol 196(6):1133–1142

Schneewind O, Model P, Fischetti VA (1992) Sorting of protein A to the staphylococcal cell wall. Cell 70(2):267–281

Sewell EW, Brown ED (2014) Taking aim at wall teichoic acid synthesis: new biology and new leads for antibiotics. J Antibiot (Tokyo) 67(1):43–51

Sham LT, Butler EK, Lebar MD, Kahne D, Bernhardt TG, Ruiz N (2014) Bacterial cell wall. MurJ is the flippase of lipid-linked precursors for peptidoglycan biogenesis. Science 345(6193):220–222

Silhavy TJ, Kahne D, Walker S (2010) The bacterial cell envelope. Cold Spring Harb Perspect Biol 2(5):a000414

Sklar JG et al (2007) Defining the roles of the periplasmic chaperones SurA, Skp, and DegP in *Escherichia coli*. Genes Dev 21(19):2473–2484

Sperandeo P et al (2008) Functional analysis of the protein machinery required for transport of lipopolysaccharide to the outer membrane of *Escherichia coli*. J Bacteriol 190(13):4460–4469

Sperandeo P, Deho G, Polissi A (2009) The lipopolysaccharide transport system of Gram-negative bacteria. Biochim Biophys Acta 1791(7):594–602

Sperandeo P et al (2011) New insights into the Lpt machinery for lipopolysaccharide transport to the cell surface: LptA–LptC interaction and LptA stability as sensors of a properly assembled transenvelope complex. J Bacteriol 193(5):1042–1053

Spirig T, Weiner EM, Clubb RT (2011) Sortase enzymes in Gram-positive bacteria. Mol Microbiol 82(5):1044–1059

Spratt BG, Pardee AB (1975) Penicillin-binding proteins and cell shape in *E. coli*. Nature 254(5500):516–517

Sutcliffe IC (2010) A phylum level perspective on bacterial cell envelope architecture. Trends Microbiol 18(10):464–470

Sutcliffe IC (2011) Priming and elongation: dissection of the lipoteichoic acid biosynthetic pathway in Gram-positive bacteria. Mol Microbiol 79(3):553–556

Swoboda JG et al (2010) Wall teichoic acid function, biosynthesis, and inhibition. Chembiochem 11(1):35–45

Taron DJ, Childs WC 3rd, Neuhaus FC (1983) Biosynthesis of D-alanyl-lipoteichoic acid: role of diglyceride kinase in the synthesis of phosphatidylglycerol for chain elongation. J Bacteriol 154(3):1110–1116

Taylor BL (1983) Role of proton motive force in sensory transduction in bacteria. Annu Rev Microbiol 37:551–573

Thanassi JA et al (2002) Identification of 113 conserved essential genes using a high-throughput gene disruption system in *Streptococcus pneumoniae*. Nucleic Acids Res 30(14):3152–3162

Thanassi DG, Bliska JB, Christie PJ (2012) Surface organelles assembled by secretion systems of Gram-negative bacteria: diversity in structure and function. FEMS Microbiol Rev 36(6):1046–1082

Trent MS et al (2006) Diversity of endotoxin and its impact on pathogenesis. J Endotoxin Res 12(4):205–223

Typas A et al (2010) Regulation of peptidoglycan synthesis by outer-membrane proteins. Cell 143(7):1097–1109

Typas A et al (2012) From the regulation of peptidoglycan synthesis to bacterial growth and morphology. Nat Rev Microbiol 10(2):123–136

Ulbrandt ND, Newitt JA, Bernstein HD (1997) The *E. coli* signal recognition particle is required for the insertion of a subset of inner membrane proteins. Cell 88(2):187–196

Utsumi R (2008) Bacterial signal transduction: networks and drug targets. Preface. Adv Exp Med Biol 631:v

Valvano MA (2008) Undecaprenyl phosphate recycling comes out of age. Mol Microbiol 67(2):232–235

van der Does C et al (2000) Non-bilayer lipids stimulate the activity of the reconstituted bacterial protein translocase. J Biol Chem 275(4):2472–2478

van der Sluis EO, Driessen AJ (2006) Stepwise evolution of the Sec machinery in Proteobacteria. Trends Microbiol 14(3):105–108

van Wely KH et al (2001) Translocation of proteins across the cell envelope of Gram-positive bacteria. FEMS Microbiol Rev 25(4):437–454

Villa R et al (2013) The *Escherichia coli* Lpt transenvelope protein complex for lipopolysaccharide export is assembled via conserved structurally homologous domains. J Bacteriol 195(5):1100–1108

Visweswaran GR, Dijkstra BW, Kok J (2011) Murein and pseudomurein cell wall binding domains of bacteria and archaea—a comparative view. Appl Microbiol Biotechnol 92(5):921–928

Vollmer W, Blanot D, de Pedro MA (2008a) Peptidoglycan structure and architecture. FEMS Microbiol Rev 32(2):149–167

Vollmer W et al (2008b) Bacterial peptidoglycan (murein) hydrolases. FEMS Microbiol Rev 32(2):259–286

Weiner JH, Li L (2008) Proteome of the *Escherichia coli* envelope and technological challenges in membrane proteome analysis. Biochim Biophys Acta 1778(9):1698–1713

Welte T et al (2012) Promiscuous targeting of polytopic membrane proteins to SecYEG or YidC by the *Escherichia coli* signal recognition particle. Mol Biol Cell 23(3):464–479

Whitfield C (1995) Biosynthesis of lipopolysaccharide O antigens. Trends Microbiol 3(5):178–185

Whitfield C (2006) Biosynthesis and assembly of capsular polysaccharides in *Escherichia coli*. Annu Rev Biochem 75:39–68

Wimley WC (2003) The versatile beta-barrel membrane protein. Curr Opin Struct Biol 13(4):404–411

Wooldridge K (2009) Bacterial secreted proteins: secretory mechanisms and role in pathogenesis, vol xii. Caister Academic Press, Wymondham, p 511

Xia G, Peschel A (2008) Toward the pathway of *S. aureus* WTA biosynthesis. Chem Biol 15(2):95–96

Xia G et al (2010) Glycosylation of wall teichoic acid in *Staphylococcus aureus* by TarM. J Biol Chem 285(18):13405–13415

Yakushi T et al (2000) A new ABC transporter mediating the detachment of lipid-modified proteins from membranes. Nat Cell Biol 2(4):212–218

Yokoyama K, Araki Y, Ito E (1988) The function of galactosyl phosphorylpolyprenol in biosynthesis of lipoteichoic acid in *Bacillus coagulans*. Eur J Biochem 173(2):453–458

Young KD (2010) New ways to make old walls: bacterial surprises. Cell 143(7):1042–1044

Young KD (2014) Microbiology. A flipping cell wall ferry. Science 345(6193):139–140

Chapter 4
Comparative Genomics and Evolutionary Modularity of Prokaryotes

Cedoljub Bundalovic-Torma and John Parkinson

Abstract The soaring number of high-quality genomic sequences has ushered in the era of post-genomic research where our understanding of organisms has dramatically shifted towards defining the function of genes within their larger biological contexts. As a result, novel high-throughput experimental technologies are being increasingly employed to uncover physical and functional associations of genes and proteins in complex biological processes. Through the construction and analysis of physical, genetic and metabolic networks generated for the model organisms, such as *Escherichia coli*, organizational principles of the genome have been deduced, such as modularity, which has important implications toward understanding prokaryotic evolution and adaptation to novel lifestyles.

Keywords Comparative genomics • Genomic-context • High-throughput interaction screening • Network biology • Modularity • Prokaryotic evolution

4.1 What Is Comparative Genomics?

Prokaryotes demonstrate a remarkable variety of lifestyles and strategies ranging from free-living in aquatic or terrestrial environments, to intimate associations (symbiosis) with other organisms with neutral, beneficial, or harmful, i.e. pathogenic, consequences for their hosts. Furthermore such host associations may occur either externally or internally, the latter involving either direct contact with host cytosol, or encasement within specialized vacuoles (Silva 2012). With recent advances in next-generation genome sequencing technologies resulting in the generation of thousands of high-quality prokaryotic genomes covering diverse taxa (Pagani et al. 2011),

C. Bundalovic-Torma
Department of Molecular Structure and Function, The Peter Gilgan Centre for Research and Learning, Hospital for Sick Children, 686 Bay St. Rm 21-9830, Toronto, ON, Canada, M5G 0A4
e-mail: ceda.bundalovic-torma@sickkids.ca

J. Parkinson (✉)
Department of Molecular Structure and Function, The Peter Gilgan Centre for Research and Learning, Hospital for Sick Children, 686 Bay St. Rm 20-9709, Toronto, ON, Canada, M5G 0A4
e-mail: jparkin@sickkids.ca

N.J. Krogan, M. Babu (eds.), *Prokaryotic Systems Biology*, Advances in Experimental Medicine and Biology 883, DOI 10.1007/978-3-319-23603-2_4

opportunities are emerging to understand the underlying genetic mechanisms that facilitate these diverse lifestyle strategies. For example, recent sequencing initiatives such as the Genomic Encyclopedia of Archaea and Bacteria (Wu et al. 2009) have uncovered a vast pool of novel uncharacterized prokaryotic genes from previously neglected prokaryotic phyla. Such genes offer enormous potential in driving the evolution of distinct lifestyle strategies. However, our ability to exploit these datasets is compromised by our limited understanding of prokaryotic biology, derived primarily from small-scale experiments of only a handful of prokaryotic model organisms, such as the Gram-negative and -positive model organisms *E. coli* and *B. subtilis*, respectively (Keseler et al. 2005; Barbe et al. 2009). Given the costs involved, it is unlikely in practice that such experimental investigations can be extended to cover even a fraction of currently uncharacterized genes. Consequently there has been much interest in the development and application of computational methods to functionally annotate novel genes and identify those responsible for driving innovations such as the ability of a pathogen to invade and cause disease.

Among the more widely adopted methods of predicting gene function are those that rely on sequence similarity searches that attempt to identify putative homologs of previously characterized genes. Such approaches range from the naive use of an established tool such as BLAST (Altschul et al. 1990), to more sophisticated tools that facilitate the concurrent detection of orthologous proteins across species (Kuzniar et al. 2008). Indeed, numerous pipelines now exist that facilitate automated functional annotation of novel genome sequences; two of the more notable being Rapid Annotation using Subsystem Technology RAST (Overbeek et al. 2014) and the NCBI Prokaryotic Genome Automatic Pipeline (Angiuoli et al. 2008). Further tools allow the prediction of specialized protein properties, such as cellular localization (Yu et al. 2010), and enzymatic function (Claudel-Renard et al. 2003; Hung et al. 2010; Karp et al. 2009). However, functional prediction based on sequence similarity can be compromised by the presence of gene-duplication events, where through sequence divergence one duplicate-copy may evolve a novel function (neofunctionalization), or the ancestral function may be divided between both of the duplicate-copies (subfunctionalization) (Taylor and Raes 2004). Also, through horizontal gene transfer (HGT) a functionally redundant duplicate may be present in a genome, known as a xenolog (Gabaldon and Koonin 2013). All of these aspects of prokaryotic evolution are widely recognized challenges towards automated functional annotation of novel prokaryotic genomes (Kuzniar et al. 2008), and numerous specialized methods have been devised towards resolving orthology relationships (Wall et al. 2003) across hundreds of genomes (COG, Tatsuov et al. 1997; InParanoid, Remm et al. 2001; ORTHOLUGE, Fulton et al. 2006; eggNOG, Powell et al. 2014; ORTHOMCL, Chen et al. 2006).

In addition to functional prediction, comparative genomics also facilitates the ability to identify core-conserved genes as well as lineage-specific innovations that are likely involved in environmental adaptations. With these defined orthologous groups, it is possible to examine the mechanisms that shaped the evolution of genes and generated distinct adaptations underlying prokaryotic lifestyle differences.

Determining the functions mediated by these genes requires an understanding of the biological context in which they operate, which can be elucidated through high-throughput interaction studies.

4.2 Generation of Biological Interaction Networks in Prokaryotes

Proteins do not function in isolation but typically form parts of integrated biological systems such as metabolic pathways, signaling networks, and protein complexes. Much of the current knowledge of biological systems are derived from experimentally tractable and well characterized model organisms such as *E. coli* and yeast (Keseler et al. 2005; Cherry et al. 1998). Allied to these investigations has been the establishment of reference resources providing curated information on biochemical pathways and complexes such as the Kyoto Encyclopaedia of Genes and Genomes (KEGG) (Kanehisa and Goto 2000), MetaCyc (Caspi et al. 2013) and MiPS (Mewes et al. 2002).

However, even for extensively studied prokaryotes such as *E. coli*, a large extent of genes lack annotation and require other methods of characterization. For example, in *E. coli* the largest proportion of uncharacterized genes are predicted to belong to the cellular membrane, a class of proteins refractory to traditional aqueous-based experimental approaches (Diaz-Mejia et al. 2008). However, the identification of such a gulf in our knowledge has resulted in the development of a variety of state-of-the-art experimental and computational-based approaches to begin to assign functional predictions for focused characterization.

Traditional low-throughput methods employed to predict gene function have typically relied on disrupting a gene in an organism of interest, either through directed knockout or random mutagenesis, and correlating its function to a change in phenotype (Smith et al. 1995; Wagner et al. 2002; Bernhardt and de Boer 2004; Buchanan et al. 2001). One disadvantage of such approaches is that they are limited to readily assayable phenotypes, and are thus challenging to implement. To overcome this challenge, high-throughput screens have been devised and can be used to examine gene function and represent this information as large-scale binary interaction datasets. These methods can be divided into two types: *physical interaction screens*, such as affinity-tag based Protein–Protein Interaction (PPI) purifications (Babu et al. 2009) and Two-Hybrid screens (Rajagopala et al. 2014); and *functional interaction screens* that identify genes involved in similar biological processes, which may not necessarily interact, such as gene expression microarrays which elucidate genes that are co-expressed in response to different physiological states (Richmond et al. 1999), and Genetic Interaction (GI) screens that explore epistatic buffering of genes with similar biological roles (Butland et al. 2008; Typas et al. 2008). Such approaches typically provide functional relationships between genes or proteins as a list of binary interactions, with an associated metric assessing the statistical significance of the interaction.

Affinity-tag based PPI purifications utilize genetically modified collections of bacterial strains, consisting of individual genes, known as *bait proteins*, modified with affinity-tag sequences that enable their transcribed protein sequences to effectively bind to an affinity column and be purified (Babu et al. 2009). An advantage to this approach over small-scale co-immunoprecipitation is that the development of specific antibodies for a bait of interest is not required (Monti et al. 2005). One important caveat, however, is that a bait protein will be identified with potentially multiple interactors, or *preys*, which may not all be valid due either to non-specific binding of proteins to the column, or over-expressed "background noise" proteins (i.e. the ribosome), which require additional filtering. The specificity of the identified bait-prey protein interactions are subject to a scoring metric (Pardo and Choudhary 2012; Armean et al. 2013) which indicates the statistical significance of the interaction, which can be represented in two ways: as a spoke-model where the bait is assumed to directly interact with all purified baits, or as a matrix-model where the baits are also assumed to interact with one another. The model of interaction is relevant as it may yield different sets of interactions depending on the PPI scoring metric utilized, possibly rejecting some that are genuine (Hakes et al. 2007).

Two-hybrid screens, initially developed in the yeast *S. cerevisaie* (Young 1998), utilize a different approach in detecting physically interacting proteins based on the *in vivo* reconstitution of a gene promoter system between directly interacting hybrid bait and prey proteins (Van Criekinge and Beyaert 1999; Uetz et al. 2000). This approach enables direct and transient protein interactions to be detected, which might otherwise be missed by affinity-tag purifications. In practice, the bait and prey fusion constructs are generated in separate yeast strains and conjugated, enabling high-throughput automated screening of interactions, although bacterial-based conjugation systems have been proposed (Joung et al. 2000; Clarke et al. 2005). The methodology is also prone to false-positive interactions resulting from auto-activation of promoter expression by certain protein baits or through non-specific interaction of promiscuously interacting preys, which require particular consideration (Van Criekinge and Beyaert 1999).

Gene expression microarrays enable prokaryotic biological processes to be examined both temporally and under various environmental conditions, and only recently is beginning to be replaced by microarray-independent methods such as RNA-Seq (Wang et al. 2009). The basis of the approach utilizes the binding of fluorescently labeled mRNA transcripts onto specially designed slides (microarrays), containing complementary transcripts of known genes for an organism of interest (Schena et al. 1995). The capacity for microarrays to assess the expression of thousands of genes in response to environmental change or under a disease state is a powerful tool for elucidating the roles of genes in biological pathways. This can be done either by examining the fold-expression change of individual genes from one condition to another, or by applying clustering algorithms to find groups of genes with correlated expression patterns across multiple conditions (Eisen et al. 1998). However this poses a challenge in data analysis, particularly in normalization of microarray expression values across multiple experiments, determining thresholds to select genes that are significantly altered in expression, or selecting the appropriate correlation metric when genes do not show a linear-coexpression pattern over time (Slonim and Yanai 2009; Song et al. 2012).

E. coli Synthetic Genetic Arrays (eSGA) is a novel high-throughput approach (Babu et al. 2011), previously devised for *S. cerevisiae* (Boone et al. 2007), that promises to shed further insight into the organization of prokaryotic biological networks beyond protein-interaction networks alone. The eSGA platform assesses pairwise Genetic Interactions (GIs) that occur when a double-gene knockout is observed to deviate in fitness from that expected when each knockout is considered in isolation. Such fitness deviations are termed *epistasis*, and can either be *alleviating* or *aggravating* in type and can be used to elucidate the biological relationships between genes. *Aggravating* GIs occur when, for a pair of genes serving a redundant biological role, deleting each separately does not impair fitness, yet deleting both together results in a significant decrease in fitness or death to the organism. Conversely, *alleviating* GIs occur when genes directly depend on each other to carry out a biological function, thus only disrupting one of the pair is necessary to generate a maximum decrease in fitness. The degree to which a GI is alleviating or aggravating can be quantified using the multiplicative model of epistasis (Boone et al. 2007). In this manner, functional predictions can be examined on the level of individual gene pairs, leading to focused experimental validation, or biological pathways can be deduced through the correlation and clustering of GI profiles (Butland et al. 2008). Setting cutoffs for significant GI scores in *E. coli* is currently a challenge, where only scores above an arbitrary significance-threshold are considered. Although it would be more appropriate to set biologically meaningful cutoffs from previously validated functional interactions, such a gold-standard is presently unavailable.

In addition to these experimental approaches, a number of computational methods, based on Genomic-Context (GC) information, have also been developed to predict physical or functional interactions of proteins in both model and understudied organisms. Unlike the experimental approaches previously described, GC methods infer functional associations for a given gene pair of interest (and their orthologs) based solely on informational features calculated from the genome sequence. The features typically examined include gene co-occurrence (phylogenetic profiles) (Pellegrini et al. 1999; Enault et al. 2003), gene-neighbourhood or conservation of gene-order (Dandekar et al. 1998; Overbeek et al. 1999; Korbel et al. 2004), chromosomal proximity (Yellaboina et al. 2007), and gene-fusion (Rosetta Stone) (Marcotte et al. 1999). These features are calculated for a given-gene pair of interest in a reference organism and their orthologs detected across multiple genomes; it is understood that the preservation or correlation of these features across different species indicates a co-evolutionary relationship between genes that are also likely to have related biological functions, or physically interact. Each method will be briefly summarized below.

Gene co-occurrence determines whether a pair of genes is likely to be functionally related based on their correlated patterns of conservation or absence in other organisms. Thus for each protein in a genome of interest a phylogenetic-profile is constructed, represented by a vector of the presence or absence of orthologs, across the proteomes of a set of compared genomes. The degree of functional interaction of a given pair of proteins is then determined by calculating the correlation between their phylogenetic profiles (Pearson Correlation, Jaccard Coefficient, or Mutual Information). It is important to note that species selection in the construction of phylogenetic profiles must be carefully considered (Yellaboina et al. 2007), gene-duplication (paralogy) may lead to false-positive predictions (Marcotte et al. 1999), and the approach tends not to be as effective for highly-conserved proteins.

The conservation of gene-neighbourhoods and their direction of transcription can also be utilized to discern groups of genes that are likely to be co-expressed, and thus functionally related. Importantly, gene-neighbourhood explicitly examines genes that are co-conserved, and can yield novel functional interactions missed by phylogenetic profiles. However, in a genome of interest not all functionally related genes may be located within the same neighbourhood, and chromosomal proximity can reveal these hidden relationships by detecting orthologs of gene pairs that are neighbours in the genome of a comparator species. In more extreme cases, a functionally related gene pair may have undergone a fusion event in another species, suggesting that the products of the pair physically interact. Such gene-fusion events occurring in distantly related species are termed *Rosetta-Stones*, as they provide a clue towards deciphering a functional relationship between uncharacterized genes of a genome of interest. The integration of GC methods with high-throughput screening, and small-scale experimental studies to aid in the functional prediction of uncharacterized proteins can be explored and retrieved from public online databases, such as STRING (Szklarczyk et al. 2011) (see Database Table 4.1).

4.3 Application of Large-Scale Biological Interaction Datasets in Prokaryotes

With the increasing availability of databases containing physical interactions and experimentally supported biological pathways for model organisms such as *E. coli*, *S. cerevisiae*, *D. melanogaster*, and *C. elegans* (see Database Table 4.1) (Kanehisa and Goto 2000; Caspi et al. 2013; Salwinski et al. 2004; Bader et al. 2003; Razick et al. 2008), it has been recently been possible examine how biological systems may evolve across diverse phylogenetic scales. In one of the first large-scale PPI networks in *E. coli*, Butland et al. (2004) evaluated the ability of GC based methods to recapitulate physically interacting protein pairs detected by using their novel affinity-tagging approach, finding only a small subset of interacting protein-pairs having a significant degree of phylogenetic co-occurrence (Pellegrini et al. 1999). Yellaboina et al. (2007) shortly after generated a large-scale functional network in *E. coli* using an integration of multiple GC approaches, and provided a comparison with two previously published PPI interaction datasets (Butland et al. 2004; Arifuzzaman et al. 2006), illustrating that proteins of core, widely-conserved biological processes tend to possess a high degree of physical interconnections.

Table 4.1 Publically available databases of prokaryotic biological networks

Database	Description	Types of biological interactions and data retrieval
KEGG (Kanehisa and Goto 2000)	Manually curated collection of databases of biological interactions based on literature evidence derived from reference organisms (e.g. *E. coli*, *B. subtilis*, *Homo sapiens*, etc.) that can also be utilized for network prediction in other fully sequenced genomes.	Defined biological pathways and physically interacting complexes; metabolic networks of enzymes and substrates. Although direct download of KEGG database networks requires a paid subscription, metabolic networks can be downloaded in flat-file format via an online web-service (http://www.kegg.jp/kegg/rest/), or as network maps via extensions of the biological network visualization tool, Cytoscape (Killcoyne et al. 2009).
MetaCyc/ EcoCyc (Caspi et al. 2013; Keseler et al. 2005)	MetaCyc represents a compendium of manually curated, experimentally defined metabolic pathways, presently derived from 2515 different organisms, of which EcoCyc represents those specific to the most-extensively studied model prokaryote, *E. coli* K12 MG1655.	Defined biological pathways, interacting protein complexes, metabolic pathways and enzymes, signaling pathways, transcriptional and post-transcriptional regulation is available for the gene content of *E. coli* K12 MG1655 (EcoCyc). EcoCyc, and lower-level curated organism-specific Pathway/Genome databases can be freely downloaded from the BioCyc collection of organism-specific databases, along with the PathwayTools (Karp et al. 2009) software enabling users to generate a Pathway/Genome database for a specific organism of interest (http://biocyc.org/download.shtml). A compendium of curated pathways from 2515 different organisms is also available through MetaCyc (Caspi et al. 2013) for large-scale analyses and can be downloaded via http://metacyc.org.
RegulonDB (Salgado et al. 2013)	Manually curated database of transcriptional regulation of the *E. coli* MG1655 K12 genome; both serving as a model of prokaryotic transcriptional regulation, or as a resource for comparative genomics studies. Genes are annotated by varying degrees of supporting evidence given to their transcriptional start-sites, promoter regions, and transcription factor binding sites.	Gene-regulatory interactions with transcriptional factors, transcriptional unit and operon organization. Flat-files of each type of interaction may be freely downloaded through http://regulondb.ccg.unam.mx.
Bacteriome.org (Su et al. 2008)	An *E. coli*-specific knowledge-base containing both high-quality experimental physical protein-interactions and theoretical interactions generated using an integration of high-throughput PPI and GC approaches. Users may search the networks via a web-interface, based on an *E. coli* gene of interest, or by BLAST-based search to a related protein of interest.	High quality physical interactions with experimental evidence and functional interactions derived from GC inferred interactions plus high-throughput physical interactions lacking direct literature support can be freely downloaded via http://www.compsysbio.org/bacteriome/download.php.

(continued)

Table 4.1 (continued)

Database	Description	Types of biological interactions and data retrieval
STRING (Franceschini et al. 2013)	Contains predicted protein–protein associations generated via the computational integration of a variety of evidence sources: gene co-expression, GC methods, experimentally determined associations from online databases, and text-mining of literature (see Szklarczyk et al. 2011 for more information). Using orthology mapping, predicted interactions are currently available for greater than 1100 organisms; users may also explore potential functional associations for a protein of interest using a BLAST-based search. Through a web-portal, users may browse functional interaction networks for a given protein of interest, as well as its supporting evidence.	Protein functional interactions and their contributing sources of evidence can be downloaded freely in flat-file format, for all available organisms, or a species of interest (http://string-db.org).
MINT (Licata et al. 2011)	Manually curated, experimentally verified protein–protein interactions detected using a variety of experimental approaches. Species coverage ranges from *Homo sapiens* to *Escherichia coli*. MINT employs several useful features to aid users in retrieving data relevant to their interests. These include: a unique scoring system that weights interaction reliability based on number of publications support, and the scale and reliability of the experimental approaches used; and, a detailed description of the interaction according to the Molecular Interaction Ontology of the Proteomics Standard Initiative (PSI-MI) (Kerrien et al. 2007).	Physical protein–protein interactions are freely downloadable (http://mint.bio.uniroma2.it/mint/download.do), providing varying levels of description. Note that prokaryote specific interactions are not separately provided and users must perform their own filtering.
IntAct (Kerrien et al. 2012)	Manually curated protein interactions covering over 200 organisms, derived from experimental literature (covering over 10 prokaryotic phyla). Users are provided with an easy to navigate web-based interface, allowing the selection of interactions based on taxonomic division, or through more specialized queries. Detailed description interactions are also provided using PSI-MI standard.	Physical and genetic interactions can be freely exported either through a customizable web based search tool, or the entire database can be downloaded via ftp, in a number of file formats (http://www.ebi.ac.uk/intact/pages/documentation/downloads.xhtml).
IRefWeb (Razick et al. 2008)	Provides protein interactions cross-referenced from over a dozen public repositories, derived from a number of experimental approaches in a select set of organisms (including *E. coli*, *Y. pestis* and *T. pallidum*). Unique to previous repositories, iRefWeb allows users the flexibility of generating custom interaction datasets for download through filters, i.e. by number of independent citations supporting an interaction, or the scale of experiment interactions were deduced (low- or high-throughput).	Physical and genetic interactions derived from literature curation, categorized by experimental approaches utilized. Tailored datasets may be downloaded through an online portal (http://wodaklab.org/iRefWeb/search/index), or can be constructed via a cytoscape extension, iRefScape 1.0 (iRefScape 2011).

Recently, two large-scale functional interaction datasets have been generated for *E. coli* (Hu et al. 2009; Peregrin-Alvarez et al. 2009a) using computational integration of high-quality physical interaction data, small-scale experiments, and GC methods to infer functional relationships between characterized and uncharacterized prokaryotic proteins. Clustering of the functional interactions enabled functional modules of genes to be identified in *E. coli*, consisting of both known physically interacting complexes, biochemical pathways, or complements of proteins enriched in similar annotated function. Importantly, such computational approaches for predicting gene-function can provide unique insight to biological organization of prokaryotes not provided by PPI studies alone, and are publically available (see Table 4.1 on Database Resources) for use as either for annotation purposes or examining the evolution of different protein functional classes across diverse prokaryotic phyla.

4.4 Network Biology and Biological Systems

One of the major aims of systems biology is to determine how the complex behaviours of a cell, i.e. a prokaryote, are carried out via underlying interactions of the multitude of genes and proteins encoded by a genome. In the previous sections we have discussed some of the main approaches in tackling this challenge through the methodologies and approaches in generating large-scale interaction datasets, which aim to place individual genes into their functional context. However, it is evident that these datasets can be further exploited to gain important insight on how genes and proteins evolve within the context of a biological system and enable adaptation of prokaryotes to distinct lifestyles. The integration, analysis, and visualization of such vast amounts of data require the extensive use of computational tools, and the concepts developed over the past decade of work in the field of network biology.

The elucidation of the organizational principles of complex biological systems and their evolution is the major goal of network biology. The analysis of biological networks and their evolution through comparative genomics can lead to an important understanding of how protein complexes and biochemical pathways function in redundant biological processes, and how changes/rewiring of these pathways may lead to the emergence of mutualistic, pathogenic or symbiotic associations. The primary mode of exploring and analyzing such networks involves network graphs. Although graphical analysis has long been applied to non-biological networks, recent study is revealing many common organizational features found in the physical interactome of *E. coli* to the Internet and even food-webs (Milo et al. 2002).

Large-scale physical, biochemical and regulatory relations occurring within a prokaryotic cell are typically represented by network graphs—abstractions of biological systems where nodes representing the components interest, e.g. genes, proteins, or metabolites, and edges connect those nodes with similar biological associations. The meaning of these relationships depends on the process described by the network. For example, nodes and edges in physical interaction networks represent proteins and their physical associations, respectively. In metabolic networks, nodes typically represent a metabolic enzyme, or reaction encoded by associated genes, with edges connecting them representing shared substrates. These networks demonstrate several interesting topological properties of biological importance. Nodes possess a varying number of interacting partners in order to carry out their roles in a cell; the number of these interactors is referred to as *degree*, and the number of edges that connect any two nodes in the network is referred to as a *path*. From these basic features topological properties of a network can be deduced and used to examine the organization principles of actual biological networks. Calculating the degree distribution of nodes illustrates that in biological networks, from eukaryotes (*S. cerevisiae*) to prokaryotes (*E. coli*), reveals a majority of nodes are sparsely connected, with the exception of a few nodes that are highly-connected, called *hubs*. This distribution follows a power-law and is termed scale-free, distinguishing them from random-networks (Butland et al. 2004; Arifuzzaman et al. 2006). The consequence is that the majority of biological processes in a cell are mediated by a subset genes or proteins, likely to enable robust responses to stimuli or environmental perturbation.

4.5 Deducing Modules from Biological Networks

Transitioning from *global* towards *local* topological properties of biological networks reveals that smaller sets of proteins appear to form regions enriched in functional connections with one another, called *clusters*, which often represent biological *modules*. Modules are groups of biological entities that can perform a distinct function in isolation; examples of biological modules range from prokaryotic membrane transporters involved in nutrient acquisition and antibiotic efflux, large molecular machines such as the bacterial flagellum, redundant iron-sulfur biogenesis clusters, and cell-envelope biogenesis pathways (Butland et al. 2008; Milo et al. 2002; Silhavy et al. 2010). Identifying modules in biological networks is not a trivial task and numerous clustering approaches have been developed for this end (Brohee and van Helden 2006). In regards to module identification in physical interaction networks, some commonly utilized computational algorithms include Molecular Complex Detection (MCODE) (Bader and Hogue 2003), Affinity Propagation (AP) (Frey and Dueck 2007), and Markov Clustering (MCL) (van Dongen and Abreu-Goodger 2012), of which MCL has been shown to perform with the greatest accuracy and with robustness against noisy data (Brohee and van Helden 2006; Vasblom and Wodak 2009) (see Fig. 4.1). Each method identifies modules based on different aspects of the underlying network topology and can be easily implemented in the popular network visualization tool, Cytoscape (Killcoyne et al. 2009; Saito et al. 2012).

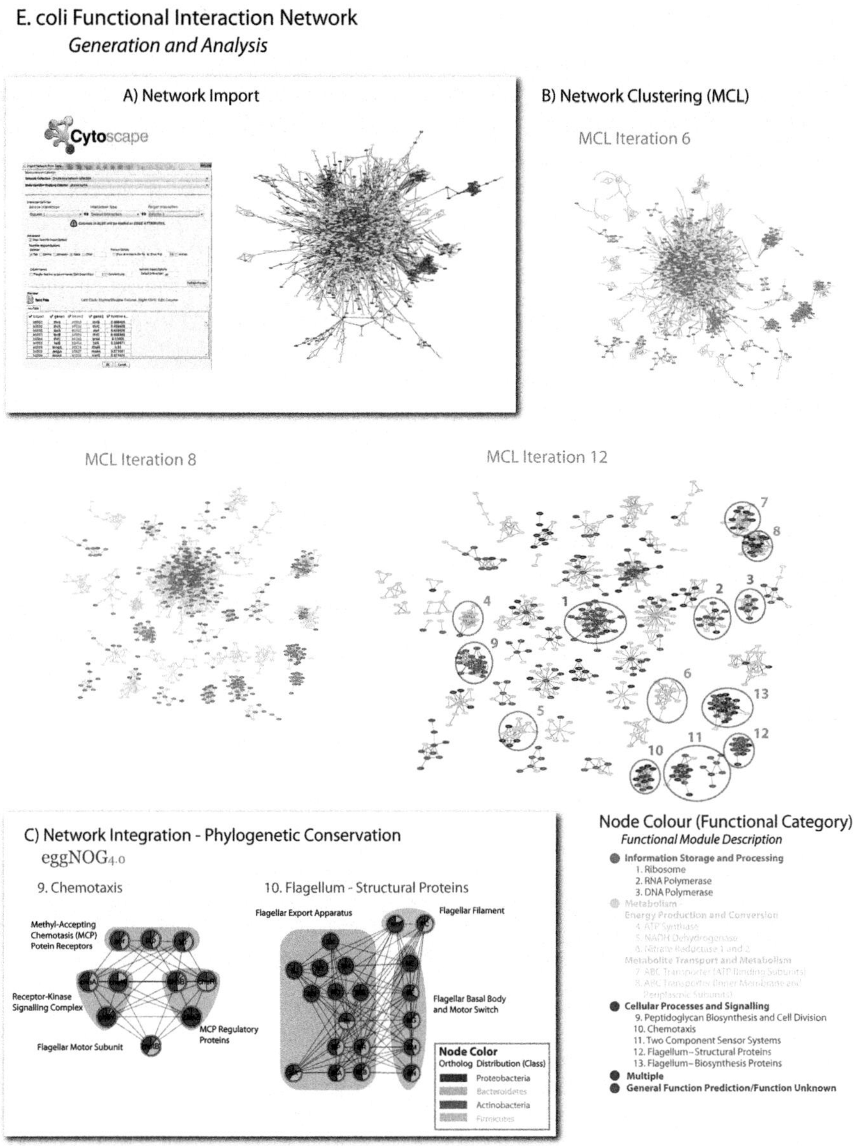

Fig. 4.1 (continued)

4.6 Evolution of Prokaryotic Biological Networks

Modularity is understood to be an organizational principle of biological systems that enables the development of complex behaviours, i.e. response to environmental change, from the combination and interchange of smaller functional units. Support of a modular organization in biological networks is exemplified by the successful recapitulation of growth and metabolite production by *in silico* metabolic reconstructions of well-studied prokaryotic species (Feist et al. 2007; Oh et al. 2007), and the application modular principles in synthetic biology to construct engineered bacterial strains through the combination of "biological parts" (Porcar et al. 2013). Furthermore, because prokaryotes, are capable of growth, reproduction, and survival under diverse environments, it follows that the genes or proteins that comprise biological modules underlying these processes should also demonstrate a significant degree of co-evolution. With high-quality physical interaction networks derived from small-scale experiments or high-throughput studies, it is now possible to examine the underlying mechanisms of prokaryotic evolution on a genomic scale.

It is well established that horizontal gene transfer is an important factor in the evolution of prokaryotes (Koonin et al. 2001), and is increasing our understanding of how modularity has influenced the evolution of prokaryotes; when gene(s) are transferred into a new genetic context, it is commonly understood that their

Fig. 4.1 Delimiting Functional Modules in *E. coli* using MCL and Applications for Comparative Genomics. (**a**) A previously published network of high-quality functional interactions in E. coli (http://www.compsysbio.org/bacteriome/download.php) was downloaded in tab-delimited format and imported into Cytoscape 3.1.1 Killcoyne et al. (2009). The left panel illustrates the organization of the data, where each line represents a binary interaction between a pair of proteins and their corresponding functional interaction likelihood score [see Peregrin-Alvarez et al. (2009a) for details]. After import, protein and interaction data are represented graphically as nodes and edges, respectively. The graphical representation of the overall functional network is shown after applying the Perfuse Force Directed layout algorithm (with default parameters. Node colors correspond to general COG functional categories Tatsuov et al. (1997) of each protein. (**b**) Increasing iterations of MCL clustering algorithm (inflation parameter = 2.5), provided by the ClusterMaker2 Cytoscape plugin Morris et al. (2011), are applied to illustrate how functional modules of different resolution can be extracted from a complex interaction network. Note that after 12 iterations of the MCL algorithm, functional modules are obtained which correspond to known E. coli complexes and pathways (as defined by EcoCyc Keseler et al. 2005). (**c**) Networks can also be used to integrate other datasets to aid in comparative genomics analyses. In this example, the relative phylogenetic distributions of Chemotaxis and Flagellum module proteins are shown among four major bacterial classes (Proteobacteria, Bacteroidetes, Firmicutes, and Actinobacteria, totalling 752 non-redundant species). In brief, orthologous groups containing of E. coli proteins were extracted from the EggNOG online database Powell et al. 2014 using in-house scripts, and mapped onto network nodes using the MultiColoredNodes Warsow et al. (2010) Cytoscape plugin. Note that not all components are uniformly conserved, and may represent specific innovations, i.e. components of the flagellar export apparatus in Proteobacteria Toft and Fares (2008), or may be replaced by a functionally equivalent but non-orthologous protein in distantly related bacteria. This serves as a valuable starting point for further investigation of how known biological modules have been adapted in prokaryotes of differing lifestyles

likelihood of being retained if the function provided is essentially modular, i.e. functioning in isolation without disruption of native biological processes. Such a notion of biological modularity has been demonstrated by the identification of "pathogenicity islands", or plasmids identified through comparative genomics studies of closely related strains of prokaryotes (Hacker and Kaper 2000). For example, numerous instances of horizontal transfer of partial to entire gene-neighbourhoods have been identified across phylogenetically diverse prokaryotes, from acquificales to proteobacteria, representing complexes and pathways as diverse as the ribosome, lipid biosynthesis and the NADH oxidoreductase (Omelchenko et al. 2003). Recently, the extensive genomic sequencing of Yersinia strains have has led to the discovery that the independent acquisition of plasmids and pathogenic determinants has resulted in the emergence of human pathogens Yersinia pseudotuberculosis and pestis from distinct environmental non-pathogenic lineages, contrary to the former notion of their divergence from a common ancestor (Reuter et al. 2014).

A study of the evolutionary conservation of metabolism across prokaryotic, archaean, and eukaryotic genomes (Peregrin-Alvarez et al. 2009b) identified a core set of widely conserved enzymes belonging to essential metabolic pathways, with a periphery of enzymes of limited conservation likely representing phyla-specific adaptations of using phylogenetic profiles of enzymes mapped to defined KEGG pathways, the evolutionary modularity of various pathways could be calculated by summing the jaccard coefficients of all enzyme pairs belonging to a pathway of interest, with higher jaccard coefficients indicating enzymes of a pathway are more frequently found together in a given genome. Given that many enzyme classes belong to highly conserved gene families, and that bias in species selection can bias phylogenetic profiling approaches, statistical significance of pathway modularity scores was assessed using distributions of shuffled enzyme phylogenetic profiles. From this approach, the authors were able to identify highly co-conserved submodules of enzymes within pathways, and also an appreciable extent of shared enzyme memberships across related pathways, indicating a degree of flexibility in metabolism across life.

One of the first large-scale PPI interaction networks generated by Butland et al. in *E. coli* (Butland et al. 2004) examined the network properties of both broadly conserved (across three domains of life) and *E. coli* specific protein baits (648 proteins in total). Proteins pairs with a high-degree of conservation, based on the number of genomes containing detectable orthologs, show an increased likelihood of physically interacting, and form a core network involved in essential bacterial processes. From the standpoint of prokaryotic evolution, the bias between ortholog conservation and physical interaction suggests that certain non-essential protein complexes detected in the *E. coli* PPI network may have evolved different interaction partners in other bacterial phyla, developing novel functional modules.

Utilizing a machine learning approach Hu et al. (2009) generated a large-scale functional network of *E. coli* using both experimental PPI and computational GC interactions. Clustering of binary interactions having overlapping support from these multiple datasets enabled 97 distinct *functional neighbourhoods* (modules) to be delimited, containing both uncharacterized proteins and those with consistent biological roles. The phylogenetic distribution of functional neighbourhood did not appear to be phyla-specific, suggesting that different biological processes in prokaryotes may consist of a core set of proteins with different extents of elaboration as phyla innovations. Among the predictions that were experimentally validated were novel components involved in several important prokaryotic biological processes, such as DNA replication, cell-envelope biogenesis, and antibiotic resistance. Immediately following this work, Peregrin-Alvarez et al. (2009a) constructed a high-quality functional interaction network encompassing over half of the *E. coli* proteome by utilizing a novel Bayesian-based approach to integrate an extensive number of high-throughput experimental and computationally generated interaction datasets. Based on the global topology of the functional network, genes predicted to have originated from horizontal transfer were less well connected, or *peripheral*, to the network. From the examination of the functional modules consisting of horizontally derived genes, many examples were found having roles in environmental adaptation and possessing interactions with native *E. coli* proteins with functional roles implicated in pathogenesis, i.e. iron-acquisition operons and iron-siderophore precursor biogenesis.

The study of modules is not limited to PPI networks alone, but can be applied to metabolic networks, networks of genetic interactions and networks of gene regulation. In a recent study, Babu et al. (2014) performed an unbiased genome-wide screen in of 163 genes representing diverse prokaryotic biological processes. Examination of GIs showed enrichment between the subunits of distinct complexes involved in related processes, such as DNA polymerase and DNA repair exonucleases, and iron–sulfur and ferric enterobactin biogenesis. A co-conservation analysis of the GI network based on mutual-information of phylogenetic-profiles of orthologous *E. coli* genes across 233 Gamma-Proteobacterial genomes, revealed that gene-pairs with highly correlated GI profiles also tended to show a greater degree of co-conservation. This trend was observed particularly for gene-pairs belonging to the same EcoCyc defined complex or pathway, suggesting that integrated biological processes revealed by GI interaction networks are built from the concerted action of distinct biological modules. However, when examining large complexes such as the flagellum, anti-correlation was observed mainly among subunits with low conservation (orthologs detected in <50 % of Gamma-proteobacterial genomes). One possibility is that the low-conserved subunits are lineage-specific acquisitions and may have specialized functional roles in flagellum assembly, illustrating how biological modules such as physically interacting complexes may evolve to elaborated function around an essential core of components.

Other studies have examined the role of modularity in the evolution of novel bacterial adaptations, with interesting insights. Broadly conserved prokaryotic stress responses, such as chemotaxis, spore formation can be strongly resolved into distinct submodules based on the biological functions of their components (structural proteins, environmental sensing, cell-signalling, pathway cross-talk), which strongly correlate with the lifestyles of different prokaryotic species (Singh et al. 2008); not surprisingly, components of these modules showing the greatest evolutionary divergence are involved either in direct environmental sensing or the last stage of internal cellular signaling cascades. Modularity has also been shown to play an important role in the evolution of phyla specific traits. For example, differences in the localization of stalk formation in the closely related Alpha-Proteobacteria *Caulobacter crecentus* and *Asticcacaulis* species was shown to be driven by the protein SpmX, which evolved from a cell-development regulator in *C. crescentus* into a stalk localization determinant in *Asticcacaulis*, recruiting peptidoglycan synthesis machinery commonly conserved across prokaryotes (Jiang et al. 2014). Prokaryotes also display the potential to adapt broadly conserved protein complexes to exploit novel lifestyle niches. For example, the twin-arginine export system (Tat) is one of two essential protein secretory pathways found across prokaryotes, and is involved in the transport of folded proteins across the bacterial membrane (Yuan et al. 2010). The Tat complex in the majority of prokaryotes is comprised of the three proteins TatA, TatB, and TatC, likely originating from the ancient duplication and sequence diversification of the ABC transporter family (Saurin et al. 1999). In a recent study, Jiang and Fares demonstrated that functional divergence of various subunits of the Tat complex was significantly increased in prokaryotic phyla containing pathogens (e.g. *Neisseria*, *Bartonella*, *Salmonella*) and species adapted to extreme environments (*Halobacteria*) (Jiang and Fares 2011). It was also found that many predicted Tat-dependent substrates in these species were enriched for functions that may serve as lifestyle adaptations, such as ribosomal proteins that may influence host immune response, and inorganic ion transport that may ensure ionic equilibrium in high-salt environments, respectively. These few examples illustrate that modularity can play an important role in prokaryotic evolution, through the combination of distinct pathways or processes, or evolution of components therein, generating novel environmental adaptations.

4.7 Conclusions and Future Directions

The study of prokaryotic evolution has been greatly aided by an unprecedented number of high-quality fully sequenced genomes, giving us a novel opportunity to study at a profound level the evolutionary relationships across the diverse prokaryotic world. However, genome sequencing initiatives have also revealed a vast expanse of uncharacterized genetic material, even among strains of the best characterized prokaryotic model organism, *E. coli*. The novel high-throughput screening approaches briefly mentioned here are aiding greatly in filling present gaps in our knowledge of gene function as well as characterizing the properties of complex biological systems beyond traditional reductionist approaches. It is apparent that computational approaches are becoming essential in resolving the wealth of information being generated, and may one day lead to the first fully integrated computational model of a known organism.

Despite existing gaps in our knowledge, model-organisms and databases that describe experimentally characterized biological systems are the key resources utilized for comparative genomic analyses. High-quality large-scale physical and genetic interaction networks are enabling functional predictions to be made for uncharacterized genes through guilt-by-association, that would otherwise be missed by smaller-scale studies. To do so, the use of graph theory to unravel the organization of these complex networks has been employed. With these insights, it has recently been possible to understand how functional associations between genes and proteins also manifest through genomic organization, leading to the generation of Genomic Context approaches and computationally inferred networks, which have also greatly aided in genome annotation. In this context we should note that theoretical concepts of interaction networks, particularly of modularity, has been invaluable towards the practical accomplishments of high-throughput experimental and computational efforts in genome annotation.

Conversely, the synergy of comparative genomics and large-scale interaction networks that describe the functional relationships of genes and proteins, has led to an expanded understanding of prokaryotic evolution. By applying the principle of modularity in the analysis of biological networks, it has been possible to identify subsets of genes that are involved in environmental adaptation across diverse prokaryotic taxa, and to understand the emergence of traits such as pathogenicity and antibiotic resistance. Looking toward the future, combining comparative genomics with high-throughput biological networks of greater scope and quality promises to only increase our understanding of the vast biological diversity of the prokaryotic world.

References

Altschul SF, Gish W, Miller W, Myers EW, Lipman DJ (1990) Basic local alignment search tool. J Mol Biol 215:403–410

Angiuoli SV, Gussman A, Klimke W, Cochrane G, Field D et al (2008) Toward an online repository of Standard Operating Procedures (SOPs) for (meta)genomic annotation. OMICS 12:137–141

Arifuzzaman M, Maeda M, Itoh A, Nishikata K, Takita C et al (2006) Large-scale identification of protein–protein interactions of *Escherichia coli* K-12. Genome Res 16:686–691

Armean IM, Lilley KS, Trotter MWB (2013) Popular computational methods to assess multiprotein complexes derived from label-free affinity purification and mass spectrometry (AP-MS) experiments. Mol Cell Proteomics 12:1–13

Babu M, Butland G, Pogoutse O, Li J, Greenblatt JF, Emili A (2009) Sequential peptide affinity purification system for the systematic isolation and identification of protein complexes from *Escherichia coli*. Methods Mol Biol 564:373–400

Babu M, Gagarinova A, Emili A (2011) Array-based synthetic genetic screens to map bacterial pathways and functional networks in *Escherichia coli*. Methods Mol Biol 781:99–126

Babu M, Arnold R, Bundalovic-Torma C, Gagarinova A, Wong KS et al (2014) Quantitative genome-wide genetic interaction screens reveal global epistatic relationships of protein complexes in *Escherichia coli*. PLoS Genet 10

Bader GD, Hogue CW (2003) An automated method for finding molecular complexes in large protein interaction networks. BMC Bioinformatics 4

Bader GD, Betel D, Hogue CWV (2003) BIND: the Biomolecular Interaction Network Database. Nucleic Acids Res 31:248–250

Barbe V, Cruveiller S, Kunst F, Lenoble P, Meurice G et al (2009) From a consortium sequence to a unified sequence: the *Bacillus subtilis* 168 reference genome a decade later. Microbiology 155:1758–1775

Bernhardt TG, de Boer PA (2004) Screening for synthetic lethal mutants in *Escherichia coli* and identification of EnvC (YibP) as a periplasmic septal ring factor with murein hydrolase activity. Mol Microbiol 52:1244–1269

Boone C, Bussey H, Andrews BJ (2007) Exploring genetic interactions and networks with yeast. Nat Rev Genet 8:437–449

Brohee S, van Helden J (2006) Evaluation of clustering algorithms for protein–protein interaction networks. BMC Bioinformatics 7:488–506

Buchanan G, Sargent F, Berks BC, Palmer T (2001) A genetic screen for suppressors of *Escherichia coli* Tat signal peptide mutations establishes a critical role for the second arginine within the twin-arginine motif. Arch Microbiol 177:107–112

Butland G, Peregrin-Alvarez JM, Li J, Yang W, Yang X et al (2004) Interaction network containing conserved and essential protein complexes in *Escherichia coli*. Nature 433:431–437

Butland G, Babu M, Diaz-Mejia JJ, Bohdana F, Phanse S et al (2008) eSGA: *E. coli* synthetic array analysis. Nat Methods 5:789–795

Caspi R, Altman T, Billington R, Dreher K, Foerster H et al (2013) The MetaCyc database of metabolic pathways and enzymes and the BioCyc collection of Pathway/Genome Databases. Nucleic Acids Res 42:D459–D471

Chen F, Mackey AJ, Stoeckert CJ, Roos DS (2006) OrthoMCL-DB: querying a comprehensive multi-species collection of ortholog groups. Nucleic Acids Res 34:D363–D368

Cherry JM, Adler C, Ball C, Chervitz SA, Dwight SS et al (1998) SGD: Saccharomyces Genome Database. Nucleic Acids Res 26:73–79

Clarke P, Vuiv PO, O'Connell M (2005) Novel mobilizable prokaryotic two-hybrid system vectors for high-throughput protein interaction mapping in *Escherichia coli* by bacterial conjugation. Nucleic Acids Res 33:e18

Claudel-Renard C, Chevalet C, Faraut T, Kahn D (2003) Enzyme-specific profiles for genome annotation: PRIAM. Nucleic Acids Res 15:6633–6639

Dandekar T, Snel B, Huynen M, Bork P (1998) Conservation of gene order: a fingerprint of proteins that physically interact. TIBS 23:325–328

Diaz-Mejia JJ, Babu M, Emili A (2008) Computational and experimental approaches to chart the *Escherichia coli* cell-envelope-associated proteome and interactome. FEMS Microbiol Rev 33:66–97

Eisen MB, Spellman PT, Brown PO, Botstein D (1998) Cluster analysis and display of genome-wide expression patterns. Proc Natl Acad Sci U S A 95:14863–14868

Enault F, Suhre K, Abergel C, Poirot O, Claverie J-M (2003) Annotation of bacterial genomes using improved phylogenomic profiles. Bioinformatics 19:i105–i107

Feist AM, Henry CS, Reed JL, Krummenacker M, Joyce AR, Karp PD, Broadbelt LJ, Hatzimanikatis V, Palsson BO (2007) A genome-scale metabolic reconstruction for *Escherichia coli* K-12 MG1655 that accounts for 1260 ORFs and thermodynamic information. Mol Syst Biol 3

Franceschini A, Szklarczyk D, Frankild S, Kuhn M, Simonovic M et al (2013) STRING v9.1: protein–protein interaction networks, with increased coverage and integration. Nucleic Acids Res 41:D808–D815

Frey BJ, Dueck D (2007) Clustering by passing messages between data points. Science 315:972–976

Fulton DL, Li YY, Laird MR, Horsman BG, Roche FM, Brinkman FS (2006) Improving the specificity of high-throughput ortholog prediction. BMC Bioinformatics 7:270–285

Gabaldon T, Koonin EV (2013) Functional and evolutionary implications of gene orthology. Nat Rev Genet 14:360–366

Hacker J, Kaper JB (2000) Pathogenicity islands and the evolution of microbes. Annu Rev Microbiol 54:641–679

Hakes L, Robertson DL, Oliver SG, Lovell SC (2007) Protein interactions from complexes: a structural perspective. Comp Funct Genomics 2007

Hu P, Janga SC, Babu M, Diaz-Mejia JJ, Butland G et al (2009) Global functional atlas of *Escherichia coli* encompassing previously uncharacterized proteins. PLoS Biol 7:e97

Hung SS, Wasmuth J, Sanford C, Parkinson J (2010) DETECT—a density estimation tool for enzyme classification and its application to *Plasmodium falciparum*. Bioinformatics 26:1690–1698

iRefScape (2011) A cytoscape plug-in for visualization and data mining of protein interaction data from iRefIndex. BMC Bioinformatics 12:388

Jiang X, Fares MA (2011) Functional diversification of the twin-arginine translocation pathway mediates the emergence of novel ecological adaptations. Mol Biol Evol 28:3183–3193

Jiang C, Brown PJ, Ducret A, Brun YV (2014) Sequential evolution of bacterial morphology by co-option of a developmental regulator. Nature 506:489–493

Joung JK, Ramm EI, Pabo CO (2000) A bacterial two-hybrid selection system for studying protein–DNA and protein–protein interactions. Proc Natl Acad Sci U S A 97:7382–7387

Kanehisa M, Goto S (2000) KEGG: Kyoto Encyclopedia of Genes and Genomes. Nucleic Acids Res 28:27–30

Karp PD, Paley SM, Krummenacker M, Latendresse M, Dale JM et al (2009) Pathway Tools version 13.0: integrated software for pathway/genome informatics and systems biology. Brief Bioinform 2:40–79

Kerrien S, Orchard S, Montecchi-Palazzi L, Aranda B, Quinn AF et al (2007) Broadening the horizon – level 2.5 of the HUPO-PSI format for molecular interactions. BMC Biol 5:44

Kerrien S, Aranda B, Breuza L, Bridge A, Broackes-Carter F et al (2012) The IntAct molecular interaction database in 2012. Nucleic Acids Res 40:D841–D846

Keseler IM, Collado-Vides J, Gama-Castro S, Ingraham J, Paley S, Paulsen IT, Peralta-Gil M, Karp PD (2005) EcoCyc: a comprehensive database resource for *Escherichia coli*. Nucleic Acids Res 33:D334–D337

Killcoyne S, Carter GW, Smith J, Boyle J (2009) Cytoscape: a community-based framework for network modeling. Methods Mol Biol 563:219–239

Koonin EV, Makarova KS, Aravind L (2001) Horizontal gene transfer in prokaryotes: quantification and classification. Annu Rev Microbiol 55:709–742

Korbel JO, Jensen LJ, von Mering C, Bork P (2004) Analysis of genomic context: prediction of functional associations from conserved bidirectionally transcribed gene pairs. Nat Biotechnol 22:911–917

Kuzniar A, van Ham RC, Pongor S, Leunissen JA (2008) The quest for orthologs: finding the corresponding gene across genomes. Trends Genet 24:539–551

Licata L, Briganti L, Peluso D, Perfetto L, Iannuccelli M et al (2011) MINT, the molecular interaction database: 2012 update. Nucleic Acids Res 40:D857–D861

Marcotte EM, Pellegrini M, Ng H-L, Rice DW, Yeates TO, Eisenberg D (1999) Detecting protein function and protein–protein interactions from genome sequences. Science 285:751–753

Mewes HW, Frishman D, Guldener U, Mannhaupt G, Mayer K et al (2002) MIPS: a database for genomes and protein sequences. Nucleic Acids Res 30:31–34

Milo R, Shen-Orr S, Itzkovitz S, Kashtan N, Chklovskii D, Alon U (2002) Network motifs: simple building blocks of complex networks. Science 298:824–827

Monti M, Orru S, Pagnozzi D, Picci P (2005) Interaction proteomics. Biosci Rep 25:45–56

Morris JH, Apeltsin L, Newman AM, Baumbach J, Wittkop T, Bader GD, Ferrin TE (2011) clusterMaker: a multi-algorithm clustering plugin for Cytoscape. BMC Bioinformatics 12: 436–449

Oh YK, Palsson BO, Park SM, Schilling CH, Mahadevan R (2007) Genome-scale reconstruction of metabolic network in *Bacillus subtilis* based on high-throughput phenotyping and gene essentiality data. J Biol Chem 282:28791–28799

Omelchenko MV, Makarova KS, Wolf YI, Rogozin IB, Koonin EV (2003) Evolution of mosaic operons by horizontal gene transfer and gene displacement *in situ*. Genome Biol 4

Overbeek R, Fonstein M, D'Souza M, Pusch GD, Maltsev N (1999) The use of gene clusters to infer functional coupling. Proc Natl Acad Sci U S A 96:2896–2901

Overbeek R, Olson R, Pusch GD, Olsen GJ, Davis JJ et al (2014) The SEED and the Rapid Annotation of microbial genomes using Subsystems Technology (RAST). Nucleic Acids Res 42:D206–D214

Pagani I, Liolios K, Jansson J, Chen I-MA, Smirnova T, Nosrat B, Markowitz M, Kyrpides NC (2011) The Genomes OnLine Database (GOLD) v.4: status of genomic and metagenomic projects and their associated metadata. Nucleic Acids Res 40:D571–D579

Pardo M, Choudhary JS (2012) Assignment of protein interactions from affinity purification/mass spectrometry data. J Proteome Res 11:1462–1474

Pellegrini M, Marcotte EM, Thompson MJ, Eisenberg MJ, Yeates TO (1999) Assigning protein functions by comparative genome analysis: protein phylogenetic profiles. Proc Natl Acad Sci U S A 96:4285–4288

Peregrin-Alvarez JM, Xiong X, Su C, Parkinson J (2009a) The modular organization of protein interactions in *Escherichia coli*. PLoS Comp Biol 5

Peregrin-Alvarez JM, Sanford C, Parkinson J (2009b) The conservation and evolutionary modularity of metabolism. Genome Biol 10

Porcar M, Latorre A, Moya A (2013) What symbionts teach us about modularity. Front Bioeng Biotechnol 1

Powell S, Forslund K, Szklarczyk D, Trachana K, Roth A et al (2014) eggNOG v4.0: nested orthology inference across 3686 organisms. Nucleic Acids Res 42:D231–D239

Rajagopala SV, Sikorski P, Kumar A, Mosca R, Vasblom J et al (2014) The binary protein–protein interaction landscape of *Escherichia coli*. Nat Biotechnol 32:285–293

Razick S, Magklaras G, Donaldson IM (2008) IRefIndex: a consolidated protein interaction database with provenance. BMC Bioinformatics 9

Remm M, Storm CE, Sonnhammer EL (2001) Automatic clustering of orthologs and in-paralogs from pairwise species comparisons. J Mol Biol 314:1041–1052

Reuter S, Connor TR, Barquist L, Walker D, Feltwell T et al (2014) Parallel independent evolution of pathogenicity within the genus *Yersinia*. Proc Natl Acad Sci U S A 111:6768–6773

Richmond CS, Glasner JD, Mau R, Jin H, Blattner FR (1999) Genome-wide expression profiling in *Escherichia coli* K-12. Nucleic Acids Res 19:3821–3835

Saito R, Smoot ME, Ono K, Ruscheinski J, Wang PL, Lotia S, Pico AR, Bader GD, Ideker T (2012) A travel guide to Cytoscape plugins. Nat Methods 9:1069–1076

Salgado H, Peralta-Gil M, Gama-Castro S, Santos-Zavaleta A, Muniz-Rascado L et al (2013) RegulonDB v8.0: omics data sets, evolutionary conservation, regulatory phrases, cross-validated gold standards and more. Nucleic Acids Res 41:D203–D213

Salwinski L, Miller CS, Smith AJ, Pettit FK, Bowie JU, Eisenberg D (2004) The database of interacting proteins: 2004 update. Nucleic Acids Res 32:D449–D451

Saurin W, Hofnung M, Dassa E (1999) Getting in or out: early segregation between importers and exporters in the evolution of ATP-binding cassette (ABC) transporters. J Mol Evol 48:22–41

Schena M, Shalon D, Davis RW, Brown PO (1995) Quantitative monitoring of gene expression with a complementary DNA microarray. Science 270:467–470

Silhavy TJ, Kahne D, Walker S (2010) The bacterial cell envelope. Cold Spring Harb Perspect Biol 2

Silva MT (2012) Classical labeling of bacterial pathogens according to their lifestyle in the host: inconsistencies and alternatives. Front Microbiol 3:71

Singh AH, Wolf DM, Wang P, Arkin AP (2008) Modularity of stress response evolution. Proc Natl Acad Sci U S A 105:7500–7505

Slonim DK, Yanai I (2009) Getting started in gene expression microarray analysis. PLoS Comput Biol 5

Smith V, Botsteinm D, Brown PO (1995) Genetic footprinting: a genomic strategy for determining a gene's function given its sequence. Proc Natl Acad Sci U S A 92:6479–6483

Song L, Langfelder P, Horvath S (2012) Comparison of co-expression measures: mutual information, correlation, and model based indices. BMC Bioinformatics 13:328–348

Su C, Peregrin-Alvarez JM, Butland G, Panse S, Fong V, Emili A, Parkinson J (2008) Bacteriome.org—an integrated protein interaction database for *E. coli*. Nucleic Acids Res 36:D632–D636

Szklarczyk D, Franceschini A, Kuhn M, Simonovic M, Roth A et al (2011) The STRING database in 2011: functional interaction networks of proteins, globally integrated and scored. Nucleic Acids Res 39:D561–D568

Tatsuov RL, Koonin EV, Lipman DJ (1997) A genomic perspective on protein families. Science 278:631–637

Taylor JS, Raes J (2004) Duplication and divergence: the evolution of new genes and old ideas. Annu Rev Genet 38:615–643

Toft C, Fares MA (2008) The evolution of the flagellar assembly pathway in endosymbiotic bacterial genomes. Mol Biol Evol 25:2069–2076

Typas A, Nichols RJ, Siegele DA, Shales M, Collins S et al (2008) A tool-kit for high-throughput, quantitative analyses of genetic interactions in *E. coli*. Nat Methods 5:781–787

Uetz P, Giot L, Cagney G, Mansfield TA, Judson RS et al (2000) A comprehensive analysis of protein–protein interactions in *Saccharomyces cerevisiae*. Nature 403:623–627

Van Criekinge W, Beyaert R (1999) Yeast two-hybrid: state of the art. Biol Proced Online 2:1–38

van Dongen S, Abreu-Goodger C (2012) Using MCL to extract clusters from networks. Methods Mol Biol 804:281–295

Vasblom J, Wodak SJ (2009) Markov clustering versus affinity propagation for the partitioning of protein interaction graphs. BMC Bioinformatics 10

Wagner C, de Saizieu A, Schonfeld H-J, Kamber M, Lange R et al (2002) Genetic analysis and functional characterization of the *Streptococcus pneumoniae vic* operon. Infect Immun 70:6121–6128

Wall DP, Fraser HB, Hirsh AE (2003) Detecting putative orthologs. Bioinformatics 19:1710–1711

Wang Z, Gerstein M, Snyder M (2009) RNA-Seq: a revolutionary tool for transcriptomics. Nat Rev Genet 10:57–63

Warsow G, Greber B, Falk SS, Harder C, Siatkowski M et al (2010) ExprEssence-revealing the essence of differential experimental data in the context of an interaction/regulation net-work. BMC Syst Bil 4:164–191

Wu D, Hugenholtz P, Mavromatis K, Pukall R, Dalin E et al (2009) A phylogeny-driven genomic encyclopaedia of Bacteria and Archaea. Nature 462:1056–1060

Yellaboina S, Goyal K, Mande SC (2007) Inferring genome-wide functional linkages in *E. coli* by combining improved genome context methods: comparison with high-throughput experimental data. Genome Res 17:527–535

Young KH (1998) Yeast two-hybrid: so many interactions, (in) so little time.... Biol Reprod 58:302–311

Yu NY, Wagner JR, Liard MR, Melli G, Rey S et al (2010) PSORTb 3.0: improved protein subcellular localization prediction with refined localization subcategories and predictive capabilities for all prokaryotes. Bioinformatics 26:1608–1615

Yuan J, Zweers JC, van Dijl JM, Dalbey RE (2010) Protein transport across and into cell membranes in bacteria and archaea. Cell Mol Life Sci 67:179–199

Chapter 5
Predicting Functional Interactions Among Genes in Prokaryotes by Genomic Context

G. Moreno-Hagelsieb and G. Santoyo

Abstract Genomic context methods for finding functions of unannotated genes were implemented very early after the publication of the first few prokaryotic genomes. The ideas behind these methods include gene fusions, conservation of gene adjacency, and the patters of co-occurrence of genes across available genomes. A later addition was the prediction of features related to functional organization, such as operons, stretches of genes co-transcribed into a single messenger RNA. The ideas behind these methods tend to be easy to understand, while the strategies for transforming those basic ideas into predictions can vary in complexity, mostly because genes whose products are known to functionally interact vary in the way they relate to those basic ideas. We present here a view of genomic context methods for predicting functional interactions, with simple examples of their implementation as compared and evaluated using genes whose products are known to functionally interact.

Keywords Operons • Interactome • Conservation of gene order • Phylogenetic profiles • Gene fusion • Operon rearrangement • Comparative genomics • Prokaryotes

5.1 Introduction

One of the problems noticed when the first genomes were sequenced was the existence of a large number of genes with no known function. Researchers wasted no time before proposing computational methods to try and predict the missing

G. Moreno-Hagelsieb (✉)
Department of Biology, Wilfrid Laurier University, 75 University Ave. W., Waterloo, ON, N2L 3C5, Canada
e-mail: gmoreno@wlu.ca

G. Santoyo
Instituto de Investigaciones Químico Biológicas, Universidad Michoacana de San Nicolás de Hidalgo, Ciudad Universitaria, Edificio A1', 58030 Morelia, Michoacán, Mexico

N.J. Krogan, M. Babu (eds.), *Prokaryotic Systems Biology*, Advances in Experimental Medicine and Biology 883, DOI 10.1007/978-3-319-23603-2_5

functions. The most direct approach to predicting functions is by transference of functions from a homologous, functionally characterized, gene into the genes without a known function (Galperin and Koonin 2000; Stormo and Tan 2002). Transference by homology continues to be a main source for putative functions, and is in itself a dynamic field worth of a review. However, here we will concentrate on methods building on top of homology. Methods that infer functions by associations among gene products. That is, predictions by genomic context, so called because they take into account genomic organization and evolutionary expectations of genes whose products functionally interact. The three main genomic contexts proposed as evidence for functional interactions were: (a) gene fusions (Enright et al. 1999; Marcotte et al. 1999), (b) conservation of gene order (Overbeek et al. 1999; Dandekar et al. 1998), and (c) phyletic patterns or phylogenetic profiles (Gaasterland and Ragan 1998; Pellegrini et al. 1999), with a fourth one being developed later on: the rearrangement of predicted operons (Rogozin et al. 2002; Snel et al. 2002; Yanai et al. 2002; Janga et al. 2005).

While the ideas behind genomic context are easy to understand, their implementation can vary due to possible confounding factors, such as the evolutionary plasticity of functional interactions. Here we will explore further predictions based on each genomic context, and evaluate an example each of their implementations using genes whose products are known to interact in *Escherichia coli* K12 MG1655, as presented in the EcoCyc database (Keseler et al. 2011) and RegulonDB (Gama-Castro et al. 2011). Readers can explore genomic context results using the STRING database (Szklarczyk et al. 2015), which also includes predictions based on high-throughput experiments, and which remain today as one of the best web-based tools for navigating predictions by genomic context.

5.2 Data and Methods

5.2.1 Gold Standards

To evaluate predictions, for positive gold standards (true positives), we used genes whose encoded proteins work in the same biochemical pathways, or are part of the same protein complex, as available in EcoCyc (Keseler et al. 2011), a curated database containing information on experimentally confirmed pathway, regulatory, and other interactions. We also use experimentally confirmed operons as present in RegulonDB (Gama-Castro et al. 2011). As true negatives we used genes present in different biochemical pathways found using the data in EcoCyc as described previously (Moreno-Hagelsieb and Jokic 2012).

5.2.2 Genomes and Orthologs

Using a web-based tool (Moreno-Hagelsieb et al. 2013), we selected a non-redundant genome dataset filtered using a genomic similarity score (Moreno-Hagelsieb and Jokic 2012; Moreno-Hagelsieb et al. 2013; Moreno-Hagelsieb and Janga 2008) chosen to keep the equivalent of one genome per represented species ($GSSa = 0.90$) out of the 2733 prokaryotic genomes available at the RefSeq database (Pruitt et al. 2007) (ftp://ftp.ncbi.nih.gov/genomes/Bacteria/) by the end of December 2013. We further filtered this non-redundant genome dataset to keep genomes longer than 2.5 Mbp.

For phylogenetic profiles, we produced a second reduced genome dataset filtered at a *GSSa* threshold of 0.75, a threshold previously shown to produce phylogenetic profiles with good discrimination between genes coding for functionally interacting proteins and genes coding for non-interacting proteins (Moreno-Hagelsieb and Janga 2008). Presence of an ortholog (see below) was represented with 1, absence with 0. We used mutual information (*MI*), measured in bits, to compare the similarity of phylogenetic profiles (Moreno-Hagelsieb and Janga 2008; Huynen et al. 2000):

$$MI = \sum_{i=0}^{1}\sum_{j=0}^{1} P_{ij} log_2 \frac{P_{ij}}{P_i P_j} \tag{5.1}$$

We used NCBI's *blastp* (Camacho et al. 2009) to determine orthologs as reciprocal best hits (RBHs) as described previously (Moreno-Hagelsieb and Latimer 2008; Ward and Moreno-Hagelsieb 2014).

5.3 Gene Fusions

The idea behind exploring gene fusions for inferring functional interactions is the easiest to understand and implement. Essentially, if separate genes in a target genome (the genome we want to functionally annotate), are found as a fused gene in an informative genome, we can infer that the two genes in the target genome functionally interact (Enright et al. 1999; Marcotte et al. 1999).

Implementations might take care of making sure that an apparent fusion is not the result of a sequencing error, and measured their confidence in the prediction by counting the number of genomes where the fusion is found (von Mering et al. 2003). The functional inference is the most direct of all.

Here we predicted functional associations by gene fusions without taking care of potential sequencing errors. Given that gene fusions can be thought of as a special case for conservation of gene order, we mix predictions based on gene fusions with predictions based on operon rearrangements (see below).

5.4 Conservation of Adjacency

The inspiration for looking for conservation of gene order goes beyond a naïve expectation that functionally associated genes would be found closer together in a chromosome. Since the first genomes available for comparative genomics were prokaryotic, the expectation was based on the knowledge of operons, stretches of adjacent genes in the same strand that are transcribed into a single messenger RNA. Operons tend to contain genes that work together [reference to Salgado], and it was expected that they should be conserved across genomes. While some work challenged the idea of conservation of gene order in general (Mushegian and Koonin 1996), and of operon structures in particular Itoh et al. (1999), other works found that even if there is no complete conservation, a signal is still detectable (Moreno-Hagelsieb et al. 2001). No only that, methods based on the idea of conservation of adjacency were already fruitful (Overbeek et al. 1999; Huynen et al. 2000).

Several implementations exist. For example, Overbeek et al. (1999) scored the conservation of adjacency by a measure of the evolutionary distance between the genomes compared. Others calculated scores by the number of genomes where they found the genes next to each other. Here we will use an implementation that estimates the confidence (*CV*) of meaningful conservation of adjacency by comparing the conservation of genes in the same strand (P_{same}), as compared with the conservation of genes in different strands ($P_{opposite}$) (Ermolaeva et al. 2001; Moreno-Hagelsieb 2006):

$$CV = 1 - 0.5\frac{P_{opposite}}{P_{same}} \quad (5.2)$$

Predictions based on conservation of gene order tend to have very good quality (Fig. 5.1). For example, we have used these predictions to test if genes in operons in prokaryotes have the same tendencies to have short distances between genes as experimentally determined operons from *Escherichia coli* (Moreno-Hagelsieb et al. 2001; Moreno-Hagelsieb and Collado-Vides 2002; Moreno-Hagelsieb 2006). We found that conserved pairs tend to have such intergenic distances, which gave us confidence that we could predict operons by intergenic distances (see below) in most prokaryotes.

5.5 Phylogenetic Profiles or Phyletic Patterns (PP)

The genomic contexts discussed above already assume the co-occurrence of the genes within the genomes explored. Phyletic patterns or phylogenetic profiles (PP) are about the co-occurrences themselves. The idea is that if the products of two genes functionally interact in a target genome, then whenever one of the genes is present in another genome, the other gene should be expected to also be present.

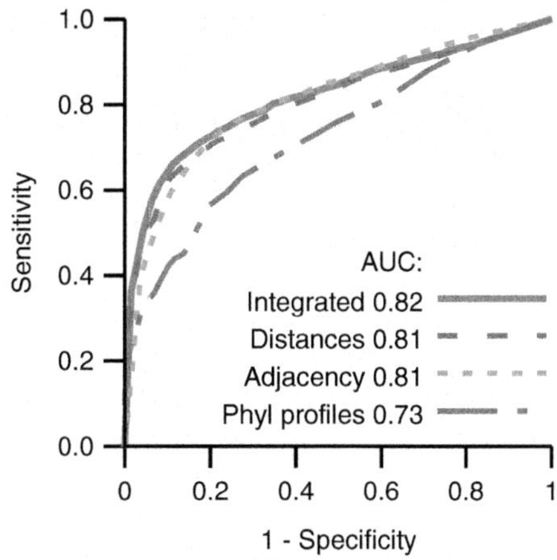

Fig. 5.1 Quality of predicted functional associations. Predicted associations were evaluated against positive gold standards based on experimentally-determined interactions, such as genes whose products work in the same biochemical pathways (Keseler et al. 2011; Moreno-Hagelsieb and Jokic 2012), or belong to the same operon (Gama-Castro et al. 2011; Moreno-Hagelsieb and Jokic 2012), while gold negatives were genes in different biochemical pathways or present in different, but adjacent, transcription units (Moreno-Hagelsieb and Jokic 2012). The intergenic distance-based predictions, and the ones based on conservation of gene order include predictions based on operon rearrangements

Absent one, the other should also be absent. The genes whose products functionally interact should co-occur (Gaasterland and Ragan 1998; Pellegrini et al. 1999). Co-occurrence would be visible as a pattern of co-presence, co-absence, of the genes in question across a series of informative genomes.

Here we used mutual information as a measure of co-occurrence (Huynen et al. 2000; Moreno-Hagelsieb and Janga 2008). PP can be calculated for any gene in a genome, and therefore, PP would be expected to produce the largest amount of predictions. When filtering for high-quality predictions, however, PP do not seem to produce a number of predictions obviously above those produced by other methods (Fig. 5.2). Their quality, using this implementation, does not seem better than the quality of predictions by other methods either (Fig. 5.1). Other implementations try and work on potential problems using PP, for example, the problem that co-occurrence should be expected for reasons other than functional interactions. For example, evolutionarily close organisms would be expected to share more genes, while evolutionarily distance ones should be expected to share less. Therefore, a phylogenetic signal might be mistaken for functionally-related co-occurrence. In our own work, we filter out redundant genomes (Moreno-Hagelsieb and Janga 2008; Moreno-Hagelsieb et al. 2013), which seems to improve the functionally-related signal (Moreno-Hagelsieb and Janga 2008). It remains a challenge to improve results.

5.6 Operon Predictions by Intergenic Distances

As explained above, operons were part of the inspiration for exploring conservation of gene order (Overbeek et al. 1999), and conservation itself has been used for predicting operons (Ermolaeva et al. 2001; Moreno-Hagelsieb 2006; Hu et al. 2009). Another method for predicting functional association would be the prediction of operons regardless of their conservation. To this end, the first successful method explored was based on the expectation that operons would have short distances between their genes, since genes inside the operon would not require such signals as those required for transcriptional regulation (Salgado et al. 2000).

Here we used the method for predicting operons based on calculating log-likelihoods for adjacent genes to be in the same operon given a distance. The method is trained on known operons and adjacent genes in the same strand known to be in different transcription units (transcription unit boundaries) (Salgado et al. 2000; Moreno-Hagelsieb and Collado-Vides 2002). Predictions based on internecine distances have qualities comparable to those produced by conservation of gene order (Fig. 5.1). Intergenic distances have been the most informative feature for predicting operons for a good while (Stormo and Tan 2002; Price et al. 2005; Ferrer et al. 2010; Chuang et al. 2012). However, our current results suggest that, with the increasing number of available genomes, conservation of gene order might have come close enough to intergenic distances to compete as the most informative feature (Fig. 5.2).

5.7 Operon Rearrangements

Two methods above can predict operons, one by conservation of gene order, the other by the distances between adjacent genes. While predicted operons might

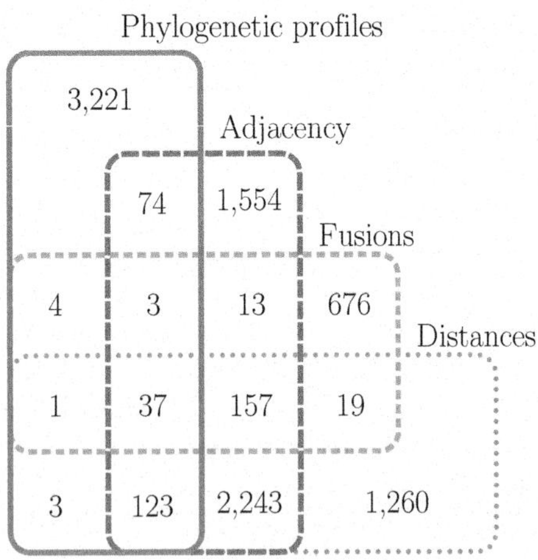

Fig. 5.2 Venn diagram comparing the number of predictions by each genomic context methodwithout operon rearrangements

contain mostly genes with related functions, the predicted associations are limited to neighbouring genes. However, even though operons are more conserved than neighbouring genes in different transcription units, their conservation is still small (Moreno-Hagelsieb et al. 2001). Galperin and Koonin (2000) suggested that there would be an advantage to the variable conservation of gene order should a method for predicting operons be developed. The idea was that if two genes were predicted to be in the same operon in an informative genome, then the case for a functional interaction between the corresponding genes in a target genome could be made. That would increase the potential of operons for predicting functional interactions among genes that are not necessarily in close proximity in the target genome.

With that idea in mind, some authors managed to project predicted functional interactions by conservation of gene order, onto their non-adjacent counterpart genes in other genomes (Rogozin et al. 2002; Snel et al. 2002; Yanai et al. 2002). Later on, another group explored the same idea, except with operons predicted by internecine distances (Janga et al. 2005).

Here we explore the prediction of functional interactions by transference of predicted interactions from informative operons as predicted by conservation of gene order (Moreno-Hagelsieb 2006; Hu et al. 2009), and by intergenic distances (Salgado et al. 2000; Moreno-Hagelsieb and Collado-Vides 2002; Janga et al. 2005; Hu et al. 2009). The latter corresponding to the method presented previously as Nebulon (Janga et al. 2005). The expansion in the number of predicted interactions is quite notable (Fig. 5.3).

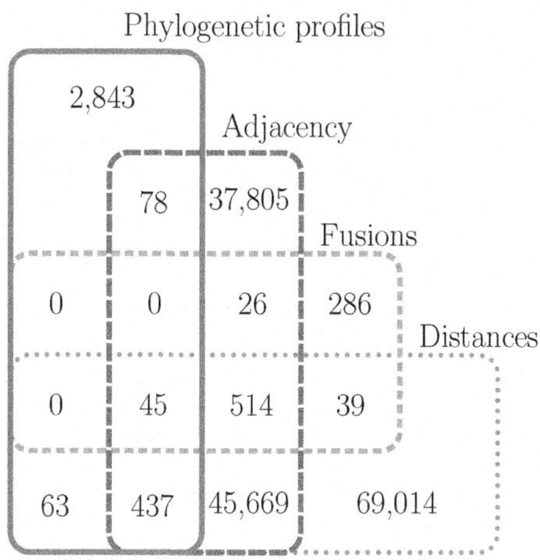

Fig. 5.3 Venn diagram comparing the number of predictions by each genomic context method when operon rearrangements are also present. The number of predictions increases enormously compared those in Fig. 5.2

5.8 Concluding Remarks

In this work we presented some example predictions based on genomic context. The predictions were evaluated with experimentally confirmed functional interaction databases. The results are encouraging in terms of helping us annotate unknown genes with potential functions based on their annotated predicted partners (Hu et al. 2009).

5.9 Competing Interests

The authors declare that they have no competing interests.

Acknowledgements We thank The Shared Hierarchical Academic Research Computing Network (SHARCNET) for computing facilities. Work supported with a Discovery Grant to GM-H from the Natural Sciences and Engineering Research Council of Canada (NSERC).

References

Camacho C, Coulouris G, Avagyan V, Ma N, Papadopoulos J, Bealer K, Madden TL (2009) BLAST+: architecture and applications. BMC Bioinf 10:421

Chuang L-Y, Chang H-W, Tsai J-H, Yang C-H (2012) Features for computational operon prediction in prokaryotes. Brief Funct Genomics 11(4):291–299

Dandekar T, Snel B, Huynen M, Bork P (1998) Conservation of gene order: a fingerprint of proteins that physically interact. Trends Biochem Sci 23(9):324–328

Enright AJ, Iliopoulos I, Kyrpides NC, Ouzounis CA (1999) Protein interaction maps for complete genomes based on gene fusion events. Nature 402(6757):86–90

Ermolaeva MD, White O, Salzberg SL (2001) Prediction of operons in microbial genomes. Nucleic Acids Res 29(5):1216–1221

Ferrer L, Dale JM, Karp PD (2010) A systematic study of genome context methods: calibration, normalization and combination. BMC Bioinf 11:493

Gaasterland T, Ragan MA (1998) Microbial genescapes: phyletic and functional patterns of ORF distribution among prokaryotes. Microb Comp Genomics 3(4):199–217

Galperin MY, Koonin EV (2000) Who's your neighbor? New computational approaches for functional genomics. Nat Biotechnol 18(6):609–613

Gama-Castro S, Salgado H, Peralta-Gil M, Santos-Zavaleta A, Muniz-Rascado L, Solano-Lira H, Jimenez-Jacinto V, Weiss V, Garcia-Sotelo JS, Lopez-Fuentes A, Porron-Sotelo L, Alquicira-Hernandez S, Medina-Rivera A, Martinez-Flores I, Alquicira-Hernandez K, Martinez-Adame R, Bonavides-Martinez C, Miranda-Rios J, Huerta AM, Mendoza-Vargas A, Collado-Torres L, Taboada B, Vega-Alvarado L, Olvera M, Olvera L, Grande R, Morett E, Collado-Vides J (2011) Regulondb version 7.0: transcriptional regulation of escherichia coli k-12 integrated within genetic sensory response units (gensor units). Nucleic Acids Res 39(Database issue):98–105

Hu P, Janga SC, Babu M, Diaz-Mejia JJ, Butland G, Yang W, Pogoutse O, Guo X, Phanse S, Wong P, Chandran S, Christopoulos C, Nazarians-Armavil A, Nasseri NK, Musso G, Ali M, Nazemof N, Eroukova V, Golshani A, Paccanaro A, Greenblatt JF, Moreno-Hagelsieb G, Emili A (2009) Global functional atlas of escherichia coli encompassing previously uncharacterized proteins. PLoS Biol 7(4):96

Huynen M, Snel B, Lathe W, Bork P (2000) Predicting protein function by genomic context: quantitative evaluation and qualitative inferences. Genome Res 10(8):1204–1210

Itoh T, Takemoto K, Mori H, Gojobori T (1999) Evolutionary instability of operon structures disclosed by sequence comparisons of complete microbial genomes. Mol Biol Evol 16(3): 332–346

Janga SC, Collado-Vides J, Moreno-Hagelsieb G (2005) Nebulon: a system for the inference of functional relationships of gene products from the rearrangement of predicted operons. Nucleic Acids Res 33(8):2521–2530

Keseler IM, Collado-Vides J, Santos-Zavaleta A, Peralta-Gil M, Gama-Castro S, Muniz-Rascado L, Bonavides-Martinez C, Paley S, Krummenacker M, Altman T, Kaipa P, Spaulding A, Pacheco J, Latendresse M, Fulcher C, Sarker M, Shearer AG, Mackie A, Paulsen I, Gunsalus RP, Karp PD (2011) Ecocyc: a comprehensive database of escherichia coli biology. Nucleic Acids Res 39(Database issue):583–590

Marcotte EM, Pellegrini M, Ng H-L, Rice DW, Yeates TO, Eisenberg D (1999) Detecting protein function and protein-protein interactions from genome sequences. Science (New York, NY) 285(5428):751–753

von Mering C, Huynen M, Jaeggi D (2003) STRING: a database of predicted functional associations between proteins. Nucleic Acids Res 31:258-261

Moreno-Hagelsieb G (2006) Operons across prokaryotes: Genomic analyses and predictions 300+ genomes later. Curr Genet 7:163–170

Moreno-Hagelsieb G, Collado-Vides J (2002) A powerful non-homology method for the prediction of operons in prokaryotes. Bioinformatics 18(Suppl 1):329–336

Moreno-Hagelsieb G, Janga SC (2008) Operons and the effect of genome redundancy in deciphering functional relationships using phylogenetic profiles. Proteins 70(2):344–352

Moreno-Hagelsieb G, Jokic P (2012) The evolutionary dynamics of functional modules and the extraordinary plasticity of regulons: the escherichia coli perspective. Nucleic Acids Res 40(15):7104–7112

Moreno-Hagelsieb G, Latimer K (2008) Choosing BLAST options for better detection of orthologs as reciprocal best hits. Bioinformatics 24(3):319–324

Moreno-Hagelsieb G, Trevino V, Perez-Rueda E, Smith TF, Collado-Vides J (2001) Transcription unit conservation in the three domains of life: a perspective from escherichia coli. Trends Genet 17(4):175–177

Moreno-Hagelsieb G, Wang Z, Walsh S, ElSherbiny A (2013) Phylogenomic clustering for selecting non-redundant genomes for comparative genomics. Bioinformatics 29(7):947–949

Mushegian AR, Koonin EV (1996) Gene order is not conserved in bacterial evolution. Trends Genet 12(8):289–290

Overbeek R, Fonstein M, D'Souza M, Pusch GD, Maltsev N (1999) The use of gene clusters to infer functional coupling. Proc Natl Acad Sci USA 96(6):2896–2901

Pellegrini M, Marcotte EM, Thompson MJ, Eisenberg D, Yeates TO (1999) Assigning protein functions by comparative genome analysis: protein phylogenetic profiles. Proc Natl Acad Sci USA 96(8):4285–4288

Price MN, Huang KH, Alm EJ, Arkin AP (2005) A novel method for accurate operon predictions in all sequenced prokaryotes. Nucleic Acids Res 33(3):880–892

Pruitt KD, Tatusova T, Maglott DR (2007) NCBI reference sequences (RefSeq): a curated non-redundant sequence database of genomes, transcripts and proteins. Nucleic Acids Res 35(Database issue):61–65

Rogozin IB, Makarova KS, Murvai J, Czabarka E, Wolf YI, Tatusov RL, Szekely LA, Koonin EV (2002) Connected gene neighborhoods in prokaryotic genomes. Nucleic Acids Res 30(10):2212–2223

Salgado H, Moreno-Hagelsieb G, Smith TF, Collado-Vides J (2000) Operons in escherichia coli: genomic analyses and predictions. Proc Natl Acad Sci USA 97(12):6652–6657

Snel B, Bork P, Huynen MA (2002) The identification of functional modules from the genomic association of genes. Proc Natl Acad Sci USA 99(9):5890–5895

Stormo GD, Tan K (2002) Mining genome databases to identify and understand new gene regulatory systems. Curr Opin Microbiol 5(2):149–153

Szklarczyk D, Franceschini A, Wyder S, Forslund K, Heller D, Huerta-Cepas J, Simonovic M, Roth A, Santos A, Tsafou KP, Kuhn M, Bork P, Jensen LJ, von Mering C (2015) STRING v10: protein-protein interaction networks, integrated over the tree of life. Nucleic Acids Res 43(Database issue):447–452

Ward N, Moreno-Hagelsieb G (2014) Quickly finding Orthologs as reciprocal best hits with BLAT, LAST, and UBLAST: how much do we miss? PLoS ONE 9(7):101850

Yanai I, Mellor JC, DeLisi C (2002) Identifying functional links between genes using conserved chromosomal proximity. Trends Genet 18(4):176–179

Chapter 6
Functional Implications of Domain Organization Within Prokaryotic Rhomboid Proteases

Rashmi Panigrahi and M. Joanne Lemieux

Abstract Intramembrane proteases are membrane embedded enzymes that cleave transmembrane substrates. This interesting class of enzyme and its water mediated substrate cleavage mechanism occurring within the hydrophobic lipid bilayer has drawn the attention of researchers. Rhomboids are a family of ubiquitous serine intramembrane proteases. Bacterial forms of rhomboid proteases are mainly composed of six transmembrane helices that are preceded by a soluble N-terminal domain. Several crystal structures of the membrane domain of the *E. coli* rhomboid protease ecGlpG have been solved. Independently, the ecGlpG N-terminal cytoplasmic domain structure was solved using both NMR and protein crystallography. Despite these structures, we still do not know the structure of the full-length protein, nor do we know the functional role of these domains in the cell. This chapter will review the structural and functional roles of the different domains associated with prokaryotic rhomboid proteases. Lastly, we will address questions remaining in the field.

Keywords Domain organization • Prokaryotic rhomboid proteases • Serine intramembrane proteases

6.1 Intramembrane Proteolysis

Proteases are essential components that regulate cellular processes in all organisms, and their dysregulation can be the major causative factor of human disease. Decades of study has provided detailed insight into their catalytic mechanism advancing our understanding of their catalytic mechanisms and development of both high-affinity inhibitors and drug therapies (Drag and Salvesen 2010). More recently, attention

R. Panigrahi • M.J. Lemieux (✉)
Department of Biochemistry, Faculty of Medicine & Dentistry, Membrane Protein Disease Research Group, University of Alberta, Edmonton, AB, Canada
e-mail: panigrah@ualberta.ca; joanne.lemieux@ualberta.ca

N.J. Krogan, M. Babu (eds.), *Prokaryotic Systems Biology*, Advances in Experimental Medicine and Biology 883, DOI 10.1007/978-3-319-23603-2_6

has been given to intramembrane proteases, which also play a major role in several disease states. These peptidases cleave transmembrane substrates to facilitate roles in several biological pathways (Strisovsky 2013).

Four classes of intramembrane proteases exist, all of which are shown to play key roles in human health. The serine protease rhomboid is linked to Parkinson's disease (Shi et al. 2011; Whitworth et al. 2008) and cancer (Etheridge et al. 2013; Abba et al. 2009; Blaydon et al. 2012), while the aspartyl protease presenilin is involved in Alzheimer's disease (De Strooper et al. 1998; Scheuner et al. 1996; Sherrington et al. 1995). The metalloprotease site-2-protease plays key roles in cholesterol metabolism (Rawson et al. 1997). Glutamic intramembrane proteases have also been identified, yet their physiologic role remains to be determined (Manolaridis et al. 2013). Bacterial intramembrane proteases also contribute to human diseases in pathogenic organisms including *Mycobacterium tuberculosis* (Schneider et al. 2014; Sklar et al. 2010), and others (Schneider and Glickman 2013; Rather 2013; Urban 2009; Ye 2013).

Compared to canonical proteases, less is known about intramembrane protease family. Over the past 15 years, however, there have been major advances in structure–function studies with the rhomboid family of proteases, which has become a strong model for enhancing our understanding of the mechanism of intramembrane proteolysis (Strisovsky 2013).

Proteolysis within the membrane is an important feature for bacterial systems. This process, within the confines of the two-dimensional space of the lipid bilayer, provides a means for precise regulation, akin to fidelity, which occurs in other signaling pathways (Tian et al. 2007). The lipid bilayer, an essential barrier to these single celled organisms, also provides a mechanism for communication with the outside environment. While most classes of intramembrane proteases have their catalytic active site located near the cytoplasmic face of membrane lipid bilayer, only rhomboids have their site of action at the extracellular leaflet, suggesting a common role in secretion (Urban and Freeman 2002). Indeed, in eukaryotic systems, rhomboid proteases play roles in secreting epidermal growth factor (Urban et al. 2002).

6.2 Rhomboid Family

Rhomboids have been identified in every kingdom of life, implying a strong evolutionary relationship between prokaryotic and eukaryotic enzymes (Urban 2009; Lemberg and Freeman 2007; Koonin et al. 2003). While not found in viruses and proteobacteria, they have been identified in acidobacteria and higher species with several isoforms identified in humans. In thirty-two bacteria, at least one or more rhomboids can be found. Due to the conservation of key polar amino acids including histidine, it is postulated that their initial role was in peptide transport and these polytopic membrane proteins evolved to cut peptides (Koonin et al. 2003). Rhomboid proteases bear no structural homology to soluble serine proteases, suggesting independent evolutionary origins (Wang et al. 2006). Phylogenetic analysis suggests rhomboids evolved in bacteria and through horizontal gene transfer were acquired by the eukaryotic systems. Further phylogenetic analysis is ongoing to provide insight into its evolutionary and functional relationship (Kinch and Grishin 2013).

Despite the fact that rhomboids belong to the same family, there is little sequence similarity between the family members. Thus topological classification was required to assign rhomboids into various classes (Lemberg and Freeman 2007; Bergbold and Lemberg 2013). Rhomboids can be classified into four distinct classes based on their sequence and topological arrangement (Fig. 6.1) (Lemberg and Freeman 2007). The first type is the basic secretase that consists of 6 transmembrane segments (TM). An 100 amino acid cytoplasmic domain is located at the N-terminus. The 1+6 secretase-type has an identical predicted topological arrangement as the basic secretase type except an extra transmembrane segment is appended at the C-terminus. The iRhom class of rhomboid proteases has similar predicted topology as the 6+1 secretase type, except there is a large domain found between helix 1 and 2. Furthermore, iRhoms also have a cytoplasmic domain at the N-terminus. These are non-catalytic forms of rhomboid proteases that have evolved to play regulatory roles in metazoans (Bergbold and Lemberg 2013; Lemberg 2013). Lastly, the presenilin-associated rhomboid like class, which has a 1+6 TM topology, and are located in mitochondria of the eukaryotic cells.

6.3 Domain Organization for Prokaryotic Rhomboids

In the bacterial systems, only basic secretase type and the 6+1 secretase type are found. Bacterial rhomboid proteases consist of two main conserved domains. At the N-terminus there is a conserved 90 amino acid soluble cytoplasmic domain. The C-terminus consists of a hydrophobic domain that is embedded within the lipid bilayer. Only the *Haemophilus influenzae* and *Providencia stuartii* rhomboid proteases (hiGlpG and AarA) lack this domain. Because of their ease of expression, structural analysis of several prokaryotic rhomboid domains has contributed to our knowledge of intramembrane proteolysis. The remainder of this chapter will focus on the different domains of the rhomboid protease and summarize our current knowledge of their structure and function.

6.3.1 Soluble Domain Structure

The soluble N-terminal cytoplasmic domain of prokaryotic rhomboids is highly conserved and consists of approximately 90 residues (Lazareno-Saez et al. 2013). Our first structural glimpse came from an NMR study of the *Pseudomonas aeruginosa*

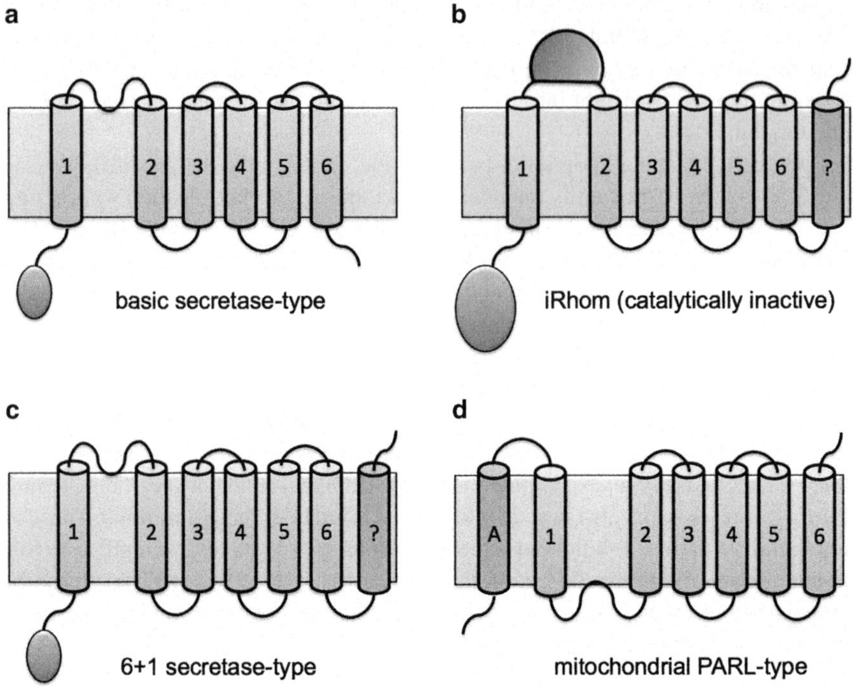

Fig. 6.1 The four topological forms of the rhomboid protease are represented. (**a**) The basic secretase class of rhomboid protease has a six transmembrane topology with an N-terminal cytoplasmic domain. The majority of prokaryotic rhomboid proteases have this topology. (**b**) The 6+1 topological form has an extra transmembrane helix appended at the C-terminus. (**c**) The iRhoms are pseudoproteases with catalytically inactive membrane domains. Some have an extra helix at the C-terminus. A large soluble domain exists between transmembrane helix 1 and 2. (**d**) The mitochondrial PARL type rhomboid has an extra transmembrane helix at the N-terminus

N-terminal cytoplasmic domain structure to reveal a compact fold of both α-helices and β-sheet (Ghasriani et al. 2014). Subsequently, a similar NMR structure was determined from the *E. coli* rhomboid (ecGlpG) N-terminal cytoplasmic domain (Sherratt et al. 2012). In contrast to these globular structures, a crystal structure and subsequent NMR structure of the N-terminal cytoplasmic domain of ecGlpG revealed that there is an extended conformation with extensive domain swapping for this domain (Fig. 6.2) (Lazareno-Saez et al. 2013; Ghasriani et al. 2014). When purified without the membrane domain, the N-terminal cytoplasmic domain of ecGlpG is monomeric with a small proportion being dimeric (Lazareno-Saez et al. 2013). It is known that domain swapping is an energetically demanding process. In vitro, the presence of certain detergents was shown to facilitate the switch from monomeric to dimeric forms. Furthermore, domain swapping could be induced with elevated temperatures (Lazareno-Saez et al. 2013; Ghasriani et al. 2014). More recently, the oligomeric state of the N-terminal cytoplasmic domain of ecGlpG was analyzed upon shedding from full-length ecGlpG in native lipid bilayers (Arutyunova et al. 2014). In this case, all N-terminal cytoplasmic domain of ecGlpG was dimeric suggesting that *in vivo* both membrane domain and cytoplasmic domains oligomerize, confirming that this is an intrinsic feature of the protein molecule and not a crystallization artifact. At this stage it is uncertain whether the soluble domain is globular *in vivo* before forming domain swapped dimers.

6.4 Membrane Domain Structure

The six-transmembrane topology was confirmed from the first crystal structures of ecGlpG membrane domain (Fig. 6.2) (Wang et al. 2006; Wu et al. 2006; Ben-Shem et al. 2007) and hiGlpG (Lemieux et al. 2007). The six alpha-helical bundle is organized into a catalytic and structural core (Urban and Shi 2008). A GxxxG motif brings together the catalytic dyad consisting of a nucleophilic serine on helix 4 and the general base histidine on helix 6 (Urban 2010; Ha et al. 2013). This dyad is obscured from the hydrophobic lipid bilayer by helices 2 and 5. From the aqueous periplasmic region, loop 5 reduces accessibility to the active site. The mechanism for access to the buried active site has been debated however recent reports suggest that loop 5 and the top of helix 5 move to allow for substrate access to the buried active site (Zoll et al. 2014). Between helix 1 and 2 is a conserved loop of 20 amino acids that has been shown to play a role in substrate recognition.

Structural stability of the membrane domain is conferred by a series of sequence motifs. Aside from these motifs, there is little sequence similarity in the rhomboid family (Lemberg and Freeman 2007). Several signature sequence motifs have been identified including a GxSx motif surrounding the catalytic serine in the fourth helix, and a conserved histidine in the sixth helix. This histidine is in a dimerization motif AHxxGxxxG that interacts with second dimerization motif, GxxxExxxG, located on helix 2. In the hydrophilic loop separating helices one and two, there is a conserved ExWRxxT motif that also confers structural stability to the rhomboid protease. Typically membrane proteins have weak stability. However rhomboids can be denatured with SDS and refolded, suggesting inherent structural features are present in rhomboids to confer stability (Baker and Urban 2012). Indeed, a study of over 150 mutants of the core of ecGlpG-membrane domain revealed several regions are critical for the maintenance of rhomboid protease stability and function.

6.5 Dimeric Nature of Rhomboid Proteases

Although we have structures of the ecGlpG membrane domain and its cytoplasmic domain, there are no full-length structures of the *E. coli* rhomboid protease (Fig. 6.2). Prokaryotic rhomboids behave as dimers in detergent solution and early studies suggest this dimeric nature is important for function (Arutyunova

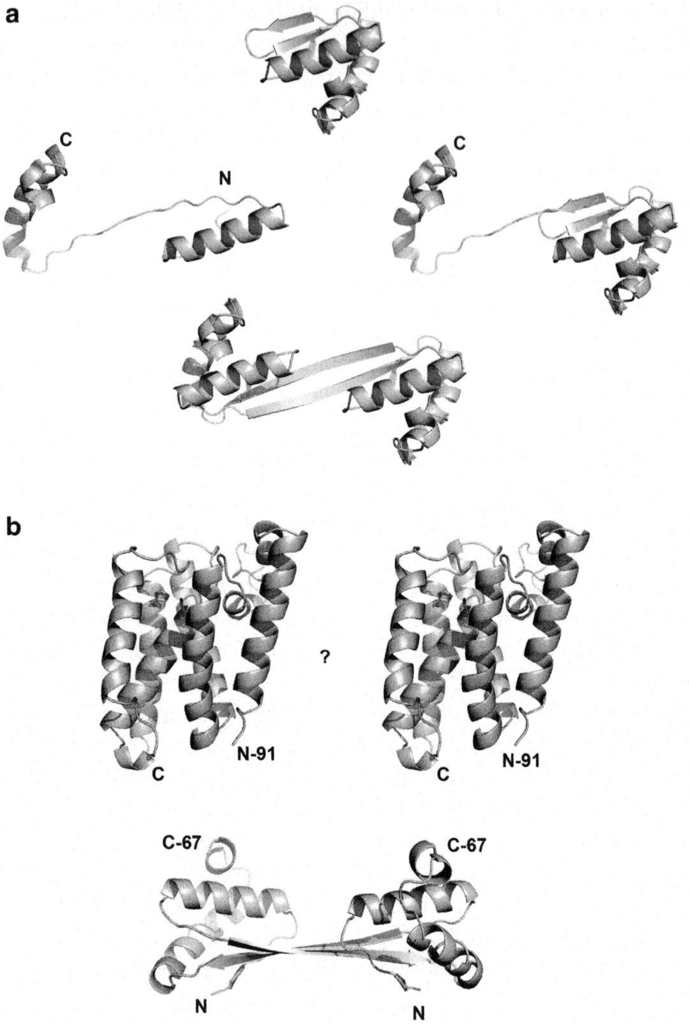

Fig. 6.2 Structural representation of the *E. coli* GlpG (ecGlpG) domains. (**a**) In solution, the N-terminal cytoplasmic domain of ecGlpG can be in either a globular form (*pink*) or an extended form (*green*). The extended form dimerizes to form a domain swapped dimer (*bottom*). *in vivo* we know the N-terminal cytoplasmic domain of ecGlpG is dimeric. It remains to be determined whether the cytoplasmic domain is globular *in vivo*. (**b**) The structure of the ecGlpG rhomboid protease membrane domain has revealed a monomer (*gold*, residues 91–272, 2IC8.pdb), and the interface for the dimerization is unknown. The N-terminal cytoplasmic domain of ecGlpG is also dimeric (4HDD.pdb). Residues 2–67 are shown in cartoon representation in *pale green*. The full-length ecGlpG structure, revealing how the membrane domain associates with the cytoplasmic domain, remains to be determined

et al. 2014). Initial studies focused on prokaryotic rhomboids in detergent solution (Sampathkumar et al. 2012). Analytical ultracentrifugation revealed three different prokaryotic rhomboids, hiGlpG, ecGlpG and YqgP, were primarily dimeric in these conditions. A further examination of hiGlpG, which inherently lacks an N-terminal soluble domain, with co-immunoprecipitation and gel filtration studies confirmed that membrane domain alone is sufficient for dimerization. Thus, although the cytoplasmic domain is dimeric, the membrane domain can form dimers independently. The membrane domain alone is capable of catalyzing the peptide bond cleavage and the presence of cytoplasmic domain seems to be unnecessary for activity with model a substrate. However, with the physiological substrate the cytoplasmic domain could play some role in rhomboid function, for example substrate recognition and tethering (Lazareno-Saez et al. 2013). It is possible domain swapping in the cytoplasmic domain is a means for facilitating the dimerization of membrane domain of rhomboid protease, a feature shown to enhance cleavage rate of substrates (Arutyunova et al. 2014). In fact, early NMR studies suggest that this domain interacts with the membrane domain (Sherratt et al. 2009). Further structural analysis is needed to determine the structure of the full-length dimeric rhomboid, whether the cytoplasmic domain interacts with the membrane domain, and how dimerization facilitates a structural change to enhance substrate cleavage. It would also be interesting to determine if the dimeric nature is transient and whether this is a means for regulation of rhomboid activity.

6.6 Cytoplasmic Domain Function

Given the high conservation of this domain in bacterial rhomboids no functional role has been established for the N-terminal cytoplasmic domain. Initial studies suggested that the cytoplasmic domain enhanced the catalytic rate of ecGlpG, which could be explained by the physical interaction between two domains. Conserved residues, predicted to be involved in the interaction with membrane domain were mutated, however no change in activity was observed (Lazareno-Saez et al. 2013). The membrane domain alone is enough for catalysis, however at present due to a lack of known substrates, it is unknown whether the cytoplasmic domain has a regulatory role in rhomboid protease activity. Little is known about the regulation of these apparently promiscuous enzymes (Dickey et al. 2013). With the eukaryotic rhomboid RHBDL4, a soluble domain was shown to play a role in substrate recognition (Fleig et al. 2012). Ironically AarA, the only rhomboid protease with a known physiological substrate, is one of the few rhomboid proteases lacking a N-terminal cytosolic domain. Domain swapping could be a regulatory signal for substrate recognition, since upon swapping the new surface is exposed which could be the recognition site for the substrate.

At the time of publication, no homologous structures were identified for the crystal structure of the ecGlpG N-terminal soluble domain, with weak homology attributed to the Type III secretion EscJ protein from *E. coli* (Lazareno-Saez et al.

2013). A more recent search revealed structural similarities with PrgK, a component of the Type III secretion system (Bergeron et al. 2015). The type III secretion system is a multi-protein pore that assembles to facilitate transport of effector proteins into the cytoplasm of the host cells (Ghosh 2004). There is currently no evidence that the *E. coli* N-terminal cytoplasmic domain plays a similar role in a Type III secretion system.

6.7 Membrane Domain

Although rhomboid proteases have been identified in bacteria, we have limited knowledge of their substrates and hence function (Rather 2013). Genetic knockout of rhomboid proteases had led to a characterizable phenotype in only three bacteria. The rhomboid protease GlpG from *E. coli* is so far the best-studied rhomboid protein, yet its function remains to be determined. *E. coli* strains lacking GlpG have no apparent phenotype, however, under laboratory conditions these strains are slightly more susceptible to the β-lactam antibiotics cefotaxime (Clemmer et al. 2006). In *Bacillus subtilis*, a knockout of the rhomboid protease YqgP resulted in defects including glucose uptake and cell division. It must be noted that in these studies effects of downstream genes in the operons were not considered (Rather 2013).

Most of our functional knowledge of prokaryotic rhomboids comes from the rhomboid protease AarA of the pathogenic gram-negative bacteria, *Providencia stuartii*. AarA was identified in a screen to characterize mutations leading to quorum sensing defect. A loss of AarA was found to increase transcription of aac(2′), an enzyme possessing O-acetyltransferase activity (Rather et al. 1993). This enzyme is known to o-acetylate peptidoglycan. The acc″(2′) levels were also affected by loss of AarA in high density cultures, suggesting the rhomboid protease plays a role in quorum sensing (Stevenson et al. 2007).

AarA has been shown to cleave the TatA protein, the only *bone fide* physiological prokaryotic rhomboid substrate. TatA is a single-pass transmembrane protein that along with TatB and TatC forms the pore for the twin-arginine secretion pathway; a secretion system that plays a role in the secretion of fully folded proteins (Porcelli et al. 2002). In *P. stuartii*, the AarA rhomboid protease was reported to cleave the psTatA protein (Stevenson et al. 2007). This had an effect on growth of *E. coli* and the possible secretion of a quorum factor in the bacteria (Rather and Orosz 1994). Cleavage of TatA enhances the oligomerization of the pore and contribute to its widening to facilitate the transport of fully folded proteins (Stevenson et al. 2007). The TatA secreted proteins play various physiological roles in the cell including respiratory and photosynthetic energy metabolism, iron and phosphate acquisition, cell division, cell motility, quorum sensing, organophosphate metabolism, resistance to heavy metals and antimicrobial peptides, and symbiotic nitrogen fixation (reviewed in Palmer and Berks 2012). While the TatA pathway exists in *E. coli* and other bacteria, they are not involved in the generation of the Tat translocon (Rather 2013). The lack of substrates for the *E. coli* rhomboid remains one of the biggest questions in the rhomboid field. An *E. coli* rhomboid substrate would provide not only a physiological role for this well-studied enzyme, but also a substrate to analyse the catalytic parameters of intramembrane proteolysis.

6.8 Summary

The rhomboid protease family, although representing a model for intramembrane proteolysis in general, also has complexity due to the various topological forms. Despite the fact that prokaryotic rhomboids have a simple topology and we know the individual structure of the membrane and cytoplasmic domain of *E. coli* rhomboid, we do not know how the dimer assembles in the membrane, nor do we know the conformational changes that take place upon dimerization. To date there is little information on prokaryotic substrates and the function for these enzymes. In particular, for the well-studied *E. coli* rhomboid protease ecGlpG, a lack of physiological substrate has hampered true measurement of catalytic parameters and specificity determinants. These questions are at the forefront of the field and given the interest in rhomboid protease, will hopefully be answered in the coming years.

Acknowledgements Operating support by the Canadian Institute for Health Research. MJL is an Alberta Innovates Health Solutions Scientist and RP is supported by a graduate studentship from the Natural Sciences and Engineering Research Council-supported International Research Training Group in Membrane Biology. We thank Dr. Elena Arytyunova for critical review of the manuscript.

References

Abba MC et al (2009) Rhomboid domain containing 2 (RHBDD2): a novel cancer-related gene over-expressed in breast cancer. Biochim Biophys Acta 1792(10):988–997

Arutyunova E et al (2014) Allosteric regulation of rhomboid intramembrane proteolysis. EMBO J 33(17):1869–1881

Baker RP, Urban S (2012) Architectural and thermodynamic principles underlying intramembrane protease function. Nat Chem Biol 8(9):759–768

Ben-Shem A, Fass D, Bibi E (2007) Structural basis for intramembrane proteolysis by rhomboid serine proteases. Proc Natl Acad Sci U S A 104(2):462–466

Bergbold N, Lemberg MK (2013) Emerging role of rhomboid family proteins in mammalian biology and disease. Biochim Biophys Acta 1828(12):2840–2848

Bergeron JR et al (2015) The modular structure of the inner-membrane ring component PrgK facilitates assembly of the type III secretion system basal body. Structure 23(1):161–172

Blaydon DC et al (2012) RHBDF2 mutations are associated with tylosis, a familial esophageal cancer syndrome. Am J Hum Genet 90(2):340–346

Clemmer KM et al (2006) Functional characterization of *Escherichia coli* GlpG and additional rhomboid proteins using an aarA mutant of *Providencia stuartii*. J Bacteriol 188(9):3415–3419

De Strooper B et al (1998) Deficiency of presenilin-1 inhibits the normal cleavage of amyloid precursor protein. Nature 391(6665):387–390

Dickey SW et al (2013) Proteolysis inside the membrane is a rate-governed reaction not driven by substrate affinity. Cell 155(6):1270–1281
Drag M, Salvesen GS (2010) Emerging principles in protease-based drug discovery. Nat Rev Drug Discov 9(9):690–701
Etheridge SL et al (2013) Rhomboid proteins: a role in keratinocyte proliferation and cancer. Cell Tissue Res 351(2):301–307
Fleig L et al (2012) Ubiquitin-dependent intramembrane rhomboid protease promotes ERAD of membrane proteins. Mol Cell 47(4):558–569
Ghasriani H et al (2014) Micelle-catalyzed domain swapping in the GlpG rhomboid protease cytoplasmic domain. Biochemistry 53(37):5907–5915
Ghosh P (2004) Process of protein transport by the type III secretion system. Microbiol Mol Biol Rev 68(4):771–795
Ha Y, Akiyama Y, Xue Y (2013) Structure and mechanism of rhomboid protease. J Biol Chem 288(22):15430–15436
Kinch LN, Grishin NV (2013) Bioinformatics perspective on rhomboid intramembrane protease evolution and function. Biochim Biophys Acta 1828(12):2937–2943
Koonin EV et al (2003) The rhomboids: a nearly ubiquitous family of intramembrane serine proteases that probably evolved by multiple ancient horizontal gene transfers. Genome Biol 4(3):R19
Lazareno-Saez C et al (2013) Domain swapping in the cytoplasmic domain of the *Escherichia coli* rhomboid protease. J Mol Biol 425(7):1127–1142
Lemberg MK (2013) Sampling the membrane: function of rhomboid-family proteins. Trends Cell Biol 23(5):210–217
Lemberg MK, Freeman M (2007) Functional and evolutionary implications of enhanced genomic analysis of rhomboid intramembrane proteases. Genome Res 17(11):1634–1646
Lemieux MJ et al (2007) The crystal structure of the rhomboid peptidase from *Haemophilus influenzae* provides insight into intramembrane proteolysis. Proc Natl Acad Sci U S A 104(3):750–754
Manolaridis I et al (2013) Mechanism of farnesylated CAAX protein processing by the intramembrane protease Rce1. Nature 504(7479):301–305
Palmer T, Berks BC (2012) The twin-arginine translocation (Tat) protein export pathway. Nat Rev Microbiol 10(7):483–496
Porcelli I et al (2002) Characterization and membrane assembly of the TatA component of the *Escherichia coli* twin-arginine protein transport system. Biochemistry 41(46):13690–13697
Rather P (2013) Role of rhomboid proteases in bacteria. Biochim Biophys Acta 1828(12):2849–2854
Rather PN, Orosz E (1994) Characterization of aarA, a pleiotrophic negative regulator of the 2′-N-acetyltransferase in *Providencia stuartii*. J Bacteriol 176(16):5140–5144
Rather PN et al (1993) Characterization and transcriptional regulation of the 2′-N-acetyltransferase gene from *Providencia stuartii*. J Bacteriol 175(20):6492–6498
Rawson RB et al (1997) Complementation cloning of S2P, a gene encoding a putative metalloprotease required for intramembrane cleavage of SREBPs. Mol Cell 1(1):47–57
Sampathkumar P et al (2012) Oligomeric state study of prokaryotic rhomboid proteases. Biochim Biophys Acta 1818(12):3090–3097
Scheuner D et al (1996) Secreted amyloid beta-protein similar to that in the senile plaques of Alzheimer's disease is increased *in vivo* by the presenilin 1 and 2 and APP mutations linked to familial Alzheimer's disease. Nat Med 2(8):864–870
Schneider JS, Glickman MS (2013) Function of site-2 proteases in bacteria and bacterial pathogens. Biochim Biophys Acta 1828(12):2808–2814
Schneider JS, Sklar JG, Glickman MS (2014) The Rip1 protease of *Mycobacterium tuberculosis* controls the SigD regulon. J Bacteriol 196(14):2638–2645
Sherratt AR et al (2009) Insights into the effect of detergents on the full-length rhomboid protease from *Pseudomonas aeruginosa* and its cytosolic domain. Biochim Biophys Acta 1788(11):2444–2453

Sherratt AR et al (2012) Activity-based protein profiling of the *Escherichia coli* GlpG rhomboid protein delineates the catalytic core. Biochemistry 51(39):7794–7803

Sherrington R et al (1995) Cloning of a gene bearing missense mutations in early-onset familial Alzheimer's disease. Nature 375(6534):754–760

Shi G et al (2011) Functional alteration of PARL contributes to mitochondrial dysregulation in Parkinson's disease. Hum Mol Genet 20(10):1966–1974

Sklar JG et al (2010) *M. tuberculosis* intramembrane protease Rip1 controls transcription through three anti-sigma factor substrates. Mol Microbiol 77(3):605–617

Stevenson LG et al (2007) Rhomboid protease AarA mediates quorum-sensing in *Providencia stuartii* by activating TatA of the twin-arginine translocase. Proc Natl Acad Sci U S A 104(3):1003–1008

Strisovsky K (2013) Structural and mechanistic principles of intramembrane proteolysis—lessons from rhomboids. FEBS J 280(7):1579–1603

Tian T et al (2007) Plasma membrane nanoswitches generate high-fidelity Ras signal transduction. Nat Cell Biol 9(8):905–914

Urban S (2009) Making the cut: central roles of intramembrane proteolysis in pathogenic microorganisms. Nat Rev Microbiol 7(6):411–423

Urban S (2010) Taking the plunge: integrating structural, enzymatic and computational insights into a unified model for membrane-immersed rhomboid proteolysis. Biochem J 425(3):501–512

Urban S, Freeman M (2002) Intramembrane proteolysis controls diverse signalling pathways throughout evolution. Curr Opin Genet Dev 12(5):512–518

Urban S, Shi Y (2008) Core principles of intramembrane proteolysis: comparison of rhomboid and site-2 family proteases. Curr Opin Struct Biol 18(4):432–441

Urban S, Schlieper D, Freeman M (2002) Conservation of intramembrane proteolytic activity and substrate specificity in prokaryotic and eukaryotic rhomboids. Curr Biol 12(17):1507–1512

Wang Y, Zhang Y, Ha Y (2006) Crystal structure of a rhomboid family intramembrane protease. Nature 444(7116):179–180

Whitworth AJ et al (2008) Rhomboid-7 and HtrA2/Omi act in a common pathway with the Parkinson's disease factors Pink1 and Parkin. Dis Model Mech 1(2–3):168–174, discussion 173

Wu Z et al (2006) Structural analysis of a rhomboid family intramembrane protease reveals a gating mechanism for substrate entry. Nat Struct Mol Biol 13(12):1084–1091

Ye J (2013) Roles of regulated intramembrane proteolysis in virus infection and antiviral immunity. Biochim Biophys Acta 1828(12):2926–2932

Zoll S et al (2014) Substrate binding and specificity of rhomboid intramembrane protease revealed by substrate-peptide complex structures. EMBO J 33(20):2408–2421

Chapter 7
Mapping Transcription Regulatory Networks with ChIP-seq and RNA-seq

Joseph T. Wade

Abstract Bacterial genomes encode numerous transcription factors, DNA-binding proteins that regulate transcription initiation. Identifying the regulatory targets of transcription factors is a major challenge of systems biology. Here I describe the use of two genome-scale approaches, ChIP-seq and RNA-seq, that are used to map transcription factor regulons. ChIP-seq maps the association of transcription factors with DNA, and RNA-seq determines changes in RNA levels associated with transcription factor perturbation. I discuss the strengths and weaknesses of these and related approaches, and I describe how ChIP-seq and RNA-seq can be combined to map individual transcription factor regulons and entire regulatory networks.

Keywords Regulatory network • Transcription factor • Microarray • Transcription profiling • RNA-seq • Chromatin immunoprecipitation • ChIP-chip • ChIP-seq

7.1 Transcription Regulatory Networks

A key goal of systems biology is to understand, on a global scale, the mechanisms by which bacteria modulate their gene expression in response to environmental signals. Gene regulation has been studied for many decades, but until recently, studies have been limited to regulation of individual genes. These focused studies have led to a deep understanding of all aspects of bacterial gene expression, in particular transcription. However, they have provided little insight into the regulation of transcription on a genomic scale. The advent of genomic approaches, in particular microarrays and next-generation sequencing, has revolutionized our understanding of transcription regulation, and revealed unexpected phenomena that remain to be fully understood. Here I discuss the genome-scale approaches that are most widely

J.T. Wade (✉)
New York State Department of Health, Wadsworth Center, Albany, NY12208, USA

Department of Biomedical Sciences, University at Albany, Albany, NY12201, USA
e-mail: joseph.wade@health.ny.gov

N.J. Krogan, M. Babu (eds.), *Prokaryotic Systems Biology*, Advances in Experimental Medicine and Biology 883, DOI 10.1007/978-3-319-23603-2_7

used to study bacterial transcription. I describe the specific methods, their strengths and weaknesses, and how a combination of techniques provides a powerful approach for mapping entire regulatory networks.

The primary point at which gene expression is regulated is transcription initiation. To initiate transcription, RNA polymerase (RNAP) must associate with an accessory σ factor that recognizes specific DNA sequences in the promoter region. RNAP:σ then forms an open complex in which the DNA is melted around the transcription start site (TSS), followed by the transition to productive transcription elongation. The affinity of RNAP:σ for promoter DNA, and the rate at which RNAP transitions to elongation, are controlled by a wide array of sequence-specific DNA-binding proteins called transcription factors (TFs). TFs typically bind close to the promoter region and either positively regulate transcription through protein–protein interactions with RNAP:σ, or negatively regulate transcription by occluding the promoter elements such that RNAP:σ cannot bind DNA. It is the TFs that coordinate transcription initiation with the growth conditions. TF activity or expression levels are coupled to specific environmental signals through a wide range of mechanisms.

The set of genes directly regulated by a TF is referred to as a "regulon". The sum of all regulons in a single bacterium is referred to as the "transcription regulatory network" (TRN). In most bacterial species there are >100 TFs. A given TF regulon can include as few as one transcript, and as many as several hundred transcripts. Thus, the TRN is highly complex. Many studies have sought to identify the regulons of bacterial TFs. Comprehensive identification of a TF regulon is important because it provides insight into how gene expression is coupled to growth conditions, and it can reveal fundamental principles of TF function. TF regulons can be mapped on a genomic scale in two ways. First, the binding of the TF can be mapped using a chromatin immunoprecipitation (ChIP) coupled with either microarrays (ChIP-chip) or next-generation sequencing (ChIP-seq). Second, the effect on global RNA levels of deleting or overexpressing a TF-encoding gene can be measured using microarrays or RNA-seq. Here, I discuss these methods, their strengths and weaknesses, and how they can be effectively combined to identify TF regulons. I also discuss unexpected complexities of bacterial TRNs that have been revealed using these genome-scale methods. The ultimate goal of these approaches is to identify the complete TRN, i.e. all TF regulons. This will allow for accurate modeling of transcription without the need for experimental data, and will be a valuable resource for synthetic biology applications.

7.2 Mapping TF Binding Genome-Wide Using Chromatin Immunoprecipitation

To map a TF regulon, two pieces of information are required: (1) the location of TF binding genome-wide, and (2) the impact of the TF on genome-wide RNA levels. There are many methods to map the position and/or strength of TF-DNA

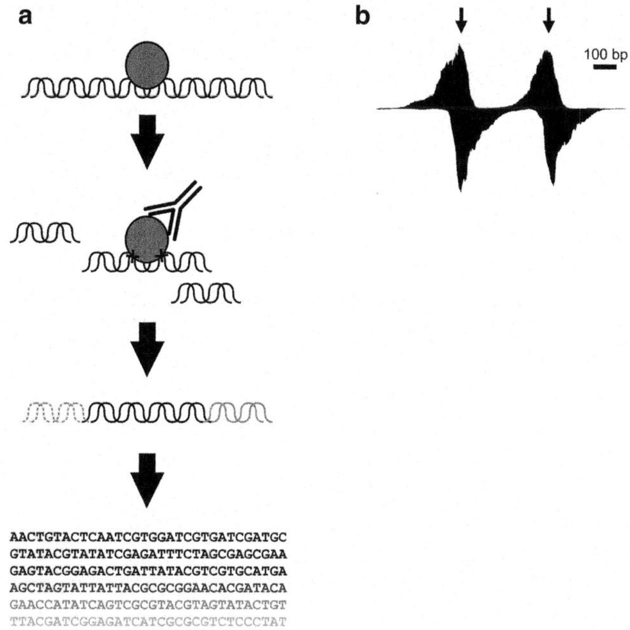

Fig. 7.1 ChIP-seq. (**a**) Schematic showing ChIP-seq method. ChIP involves cross-linking of DNA-binding proteins to DNA, cell lysis, sonication (to fragment DNA), and immunoprecipitation with an antibody against the protein of interest. For ChIP-seq, ChIP-enriched DNA is converted into a library for next-generation sequencing. Sequence reads are aligned to a reference genome sequence. (**b**) Example of peaks in ChIP-seq data for a transcription factor from *Salmonella enterica*. The graph shows piled up sequence reads from the ChIP-seq experiment. Reads mapping to the plus strand are plotted above the horizontal line, and reads mapping to the minus strand are plotted below the horizontal line. Two binding sites for the transcription factor are indicated with *vertical arrows*. Note the characteristic stagger between the peaks on the plus and minus strands

interactions in vitro, e.g. electromobility shift assay, DNase I footprinting. However, one method in particular has become the standard approach for genome-scale mapping of TF binding. Chromatin Immunoprecipitation (ChIP) is a method to measure the strength and location of protein–DNA interactions in living cells. Cells are cross-linked using formaldehyde, lysed, sonicated to fragment DNA, and the DNA-binding protein of interest is immunoprecipitated using a specific antibody (Fig. 7.1a). Enrichment of bound genomic regions can then be determined relative to a control region using quantitative PCR (Aparicio et al. 2005). Advantages of ChIP over other methods used to map protein–DNA interactions are (1) ChIP measures binding *in vivo*, under physiological conditions, (2) ChIP can detect indirect interactions between protein and DNA, and (3) ChIP can detect protein–DNA interactions for proteins with specific post-translational modifications. ChIP was originally developed to study individual protein–DNA interactions (Solomon et al. 1988), and has been widely used to measure the strength of TF binding to specific target sites. However, this requires that the genomic location of a TF binding site is already known. To determine the position and strength of previously unknown TF-DNA interactions using ChIP, it must be combined with either microarrays (ChIP-chip/ChIP-on-chip), or sequencing (ChIP-seq).

ChIP-chip was developed shortly after microarrays became a popular tool for studying gene expression on a genomic scale (Blat and Kleckner 1999; Iyer et al. 2001; Lieb et al. 2001; Reid et al. 2000; Ren et al. 2000). It has been used extensively to map TF-DNA interactions in eukaryotes (Cawley et al. 2004; ENCODE 2007; Harbison et al. 2004; Lee et al. 2002), but less so for bacteria (Grainger et al. 2004; Laub et al. 2002; Wade et al. 2007). Nonetheless, ChIP-chip is an effective method for mapping TF binding in bacteria. Microarray design is critical for the success of a ChIP-chip experiment. Some ChIP-chip studies of bacterial proteins have used microarrays containing probes only within genes, i.e. no intergenic sequence. Higher density microarrays, including "tiling" arrays in which every genome position is covered by at least one probe, reduce bias and increase resolution. The relatively small size of bacterial genomes is ideal for the use of high density microarrays. As discussed below, ChIP-chip studies have provided surprising new insights into the global pattern of TF binding.

With the advent of next-generation sequencing, ChIP-chip has been largely replaced by ChIP-seq (Park 2009; Robertson et al. 2007). For ChIP-seq, ChIP-enriched DNA is converted into a library that is suitable for next-generation sequencing. Advantages of ChIP-seq over ChIP-chip are (1) higher resolution (a typical ChIP-seq experiment gives 10 bp resolution), (2) no limit on dynamic range, and (3) no bias with respect to genome position. A typical ChIP-seq experiment generates between 1 and 100 million sequence reads that are aligned to a reference genome. Although the resolution of ChIP-seq is high, it is still not possible to identify the binding site for a DNA-binding protein with single nucleotide resolution using ChIP-seq data alone. ChIP-exo is a modified ChIP-seq approach that incorporates an exonuclease treatment of the protein-bound DNA during immunoprecipitation to achieve single nucleotide resolution (Rhee and Pugh 2011). Although ChIP-exo has proven effective for mapping TF binding in eukaryotes, it has only recently applied to bacterial systems.

7.3 Analysis of ChIP-seq Data

There are many available programs for analysis of ChIP-seq data. In each case, the primary goal is to identify enriched genomic regions, "peaks" (Fig. 7.1b), from sequence reads that have been aligned to a reference genome. There are currently not enough bacterial ChIP-seq studies to evaluate the effectiveness of each of the available methods. Indeed, the most effective approach may be combine multiple programs (Schweikert et al. 2012). Hence, I will not highlight any particular program. Rather, I will describe the key features of these programs for accurately calling peaks. All programs use the density of sequence reads as the primary parameter for peak calling. Another critical parameter is the "shape" of the peak. The expected shape of a ChIP-seq peak is a symmetric distribution of reads on each strand, with a characteristic stagger between the plus and minus strands (Fig. 7.1b) (Valouev et al. 2008). Genomic regions with a high density of sequence reads that do not show this shape are unlikely to represent genuine sites of enrichment. A third component of many peak calling programs is comparison of experimental data to a control (see below). Finally, it is often helpful to eliminate repetitive sequence since it is impossible to determine which repeat a given sequence read corresponds to (Bonocora et al. 2013).

Bacterial TFs often bind to multiple, closely-spaced sites. The size of DNA fragments generated in a typical ChIP experiment is 200–400 bp. Hence, TF binding sites located <200 bp apart are expected to be found in many of the same DNA fragments. Nonetheless, it is possible to bioinformatically infer the presence of multiple closely-spaced binding sites using ChIP-seq data alone. The CSDeconv program accurately deconvolutes contiguous regions of high sequence density that contain multiple binding sites (Lun et al. 2009). Importantly, CSDeconv has been trained and tested on bacterial ChIP-seq datasets. The importance of this approach was recently demonstrated by a study of *E. coli* ArcA (Park et al. 2013b). This study demonstrated, using CSDeconv, that ArcA preferentially binds to multimers of a previously described motif, typically spaced only 11 bp apart.

7.4 ChIP-seq Artifacts and Controls

Several artifacts have been described for ChIP-seq datasets. The two major artifacts are associated with regions of differential nucleosome density (Auerbach et al. 2009; Vega et al. 2009), and highly transcribed regions (Park et al. 2013a; Teytelman et al. 2013). The former is irrelevant for bacteria, since they lack histones. The second artifact has, thus far, only been described for yeast samples. However, we have observed high ChIP-chip and ChIP-seq signal at highly transcribed regions for control samples from *E. coli* and *Salmonella enterica*. This can result in erroneous identification of ChIP-seq peaks in highly transcribed regions (Cho et al. 2014). The cause of the bias associated with highly transcribed regions is unclear. Given that there are known artifacts in ChIP-seq data, use of a control dataset is important (Bonocora et al. 2013; Galagan et al. 2013). There are a variety of options for controls. The most straightforward option is “input” DNA. Input DNA is the DNA from lysed, sonicated cell extracts, before immunoprecipitation. Thus, input DNA represents the starting pool from which ChIP-enriched DNA is isolated. The main advantage of input DNA is that it is plentiful and does not require additional samples to be generated. However, it fails to control for the known artifacts associated with ChIP-seq. A better control is DNA generated by a mock ChIP. This can be a ChIP performed without antibody, ChIP from a strain in which the relevant TF-encoding gene has been deleted, or in the case of TFs that are ChIPped using an antibody against a fused epitope tag, ChIP from an untagged strain. Each of these will control for artifacts associated with highly transcribed regions. A potential drawback of these controls is that the amount of DNA generated is expected to be far less than for an experimental sample, which may lead to limiting library complexity. This would be reflected by multiple sequence reads corresponding to a single DNA fragment.

7.5 Barcodes and Multiplexing

Bacterial genomes are relatively small. Therefore, it is often possible to combine multiple samples in a single ChIP-seq experiment. This mitigates the high cost of next-generation sequencing. Multiplexing of samples can be achieved using barcodes, equivalent to those used for sequencing of genomic DNA libraries. The degree of multiplexing is dependent upon the proteins being studied. Proteins that bind few genomic locations require fewer sequence reads in order to identify their targets. In contrast, proteins such as RNAP subunits require many sequence reads since they bind to many regions. In principle, it is also possible to combine ChIP-seq samples from different organisms without the need for barcoding, assuming that the genome sequences are sufficiently divergent.

7.6 Identifying TF-Binding Motifs Using Genome-Scale ChIP Data

ChIP-seq maps TF-DNA interactions with high resolution. However, this is insufficient to precisely map TF binding sites using ChIP-seq data alone. Most TFs bind DNA in a sequence-specific manner, and their binding to DNA is associated with a specific DNA sequence motif. Hence, it is possible to infer the precise location of binding sites using a combination of ChIP-seq data and analysis of DNA sequence. There are several programs that identify overrepresented DNA motifs in ChIP-enriched sequences. The most commonly used program is MEME (Bailey and Elkan 1994). Such programs can facilitate identification of TF binding sites at nucleotide resolution even with relatively low resolution data. The reliability of MEME can be increased further by identifying overrepresented motifs that are located centrally within ChIP-enriched DNA sequences (Bailey and Machanick 2012). Although programs such as MEME are valuable tools for identifying TF binding sites, it is important to note that not all TF binding sites are associated with the expected motif (see below), and some TFs bind with little or no discernible sequence specificity.

7.7 Surprising Aspects of TF Binding Revealed by Genome-Scale ChIP

ChIP-chip and ChIP-seq of bacterial TFs have revealed three surprising features of TF binding profiles. First, TF binding upstream of many genes is not associated with detectable regulation of those genes. Second, TF binding often correlates poorly with the presence of the expected DNA sequence motif. Third, many TF binding sites are located within genes, far from the start of an annotated gene. The extent of each of these phenomena was completely unexpected based on previous studies of individual TF targets, and highlights the importance of genome-scale approaches such as ChIP-chip and ChIP-seq.

7.7.1 TF Binding Sites Not Associated with Detectable Regulation of the Downstream Gene

Many TF binding sites identified by ChIP-chip and ChIP-seq are not associated with detectable regulation of nearby gene(s). Although there are many examples of TFs that function in a condition-specific manner, the extent to which seemingly non-functional TF binding sites have been discovered is unexpected. There are likely to be several reasons for this phenomenon, although it remains to be comprehensively investigated. First, some TF binding sites might simply be non-functional. This is unlikely to be the case since the relevant TF binding sites are in intergenic regions; although there is extensive TF binding inside genes (see below), only 10 % of a typical bacterial genome is intergenic, and most TFs bind more intergenic sites than expected by chance. Second, some TFs are only active under specific growth conditions. However, most TFs mapped using ChIP-chip or ChIP-seq are sufficiently well studied that the relevant growth conditions are known. Hence, failure to detect regulation of nearby genes is not likely to be a consequence of using the "wrong" growth conditions. Third, and most likely, TFs rarely work alone. Combinatorial regulation by groups of two or more TFs is likely to be the norm, and dependency/redundancy between TFs could result in condition-specific requirements for individual TFs that cannot easily be predicted without knowing all the TFs that regulate a given gene. A recent study identified binding sites of FNR in *E. coli* using ChIP-seq (Myers et al. 2013). Many of the FNR binding sites identified in this study are upstream of genes that were not detectably regulated by FNR under the conditions tested. However, the authors analyzed other transcriptomic datasets for cells grown under different growth conditions. This analysis strongly suggested that FNR regulation of certain target genes is modulated by the activity of at least four other TFs that are each specific to a different growth condition.

7.7.2 Poor Correlation Between TF Binding and Motif Score

Most TFs bind DNA with sequence specificity. Hence, a motif can be identified that corresponds to the preferred binding site. Motifs can be represented as position weight matrices (PWMs), which can be used to computationally predict TF binding based on DNA sequence alone. In some cases, DNA sequence is highly predictive of binding. For example, *Escherichia coli* LexA binds to essentially all occurrences of its binding site (Wade et al. 2005). However, in most cases, DNA sequence is a poor predictor of *in vivo* binding (Galagan et al. 2013). TFs often fail to bind to seemingly excellent binding sites. Conversely, many TFs bind some sequences that bear little or no resemblance to the expected motif. For example, although LexA binding can be predicted accurately by the presence of an appropriate sequence motif, LexA also binds many sites across the *E. coli* genome that differ greatly from the motif (Wade et al. 2005). There are a variety of explanations for the poor predictive value of

DNA sequence for many TFs: (1) bioinformatic prediction of binding site strength based on DNA sequence may be inaccurate. For example, the interdependence of pairs of positions within the site is usually ignored (Salama and Stekel 2010), as is cooperativity between adjacent sites (Courey 2001). Nonetheless, for some TFs, the difference between predicted and actual binding is so great that other factors are likely to be important (Galagan et al. 2013); (2) TF binding may be affected by the binding of other TFs to nearby sites. This effect may be cooperative, i.e. binding of one TF may require binding of a second TF to an adjacent site (Kallipolitis et al. 1997; Wade et al. 2001), or competitive, i.e. two TFs bind overlapping sites but cannot bind simultaneously. In the latter case, nucleoid-associated proteins (NAPs), TFs that bind with low sequence specificity to extensive regions of the genome, are likely to be important players. H-NS is a NAP in *E. coli* and has been shown to prevent binding of FNR, a sequence-specific TF, to many of its predicted target sites (Myers et al. 2013); (3) TF binding could be influenced by other factors that control DNA structure, such as DNA methylation, supercoiling, and bending. The degree of DNA supercoiling has been shown to affect the binding of OmpR to its target sites in *Salmonella enterica* (Cameron and Dorman 2012).

7.7.3 Extensive TF Binding Inside Genes and Far from Annotated Gene Starts

Decades of studies of bacterial TFs have focused almost exclusively on TF binding sites located a short distance upstream of a gene. The majority of TFs whose binding has been mapped by ChIP-chip or ChIP-seq have been observed to bind many sites inside genes, far from the nearest annotated gene start. The frequency of intragenic binding sites varies for each TF. For example, *E. coli* AraC binds only one intragenic site (Stringer et al. 2014) while 70 % of *E. coli* RutR binding sites are intragenic (Shimada et al. 2008). Only 10–20 % of bacterial genomes consist of intergenic sequence. Hence, almost all TFs mapped by ChIP-chip and ChIP-seq have an enrichment of intergenic sites relative to the genome content. However, a comprehensive ChIP-seq study of over 50 TFs in *Mycobacterium tuberculosis* identified some TFs that bind even more intragenic sites than expected by chance based on the genome composition (Galagan et al. 2013). Despite the enormous number of intragenic TF binding sites, the function of very few intragenic sites has been determined. Possible functions include regulation of RNAs that initiate from intragenic promoters (Singh et al. 2014), repression of transcription elongation by forming a roadblock for RNAP (Belitsky and Sonenshein 2013), and a role in maintaining higher order chromosome structures (Qian et al. 2012).

7.8 Identifying TF-Regulated Genes by Measuring Genome-Wide Changes in RNA Levels

Identifying the binding sites of a TF is insufficient to reconstruct the regulon. As discussed above, many TF binding sites are not associated with regulation of the associated gene. Furthermore, TFs have both direct and indirect effects on transcription. To determine a TF regulon it is necessary to combine binding site locations with information on the impact of the TF on RNA levels. There are a variety of methods used to identify and quantify changes in RNA levels between strains and/or conditions, e.g. rtPCR, Northern blot. However, most of these methods are targeted, i.e. they measure regulation of a single gene. Genome-scale analysis of changes in RNA levels, commonly referred to as "transcription profiling", was first possible when microarrays were developed. More recently, RNA-seq has begun to replace microarrays for this application.

Transcription profiling can be used to identify all direct and indirect regulatory targets for a given TF by either comparing RNA levels between a wild-type and a mutant strain (i.e. strain with/without the TF-encoding gene), or between conditions known to influence TF abundance/activity. While ChIP methods and transcription profiling provide overlapping information, there are several key differences. First, ChIP methods detect binding but do not provide any information on the impact of the TF on RNA levels. Hence, it is impossible to determine from ChIP data alone whether a given TF binding site is associated with regulation of transcription under the conditions tested. Second, transcription profiling identifies entire regulated transcripts. ChIP-based methods typically identify a binding site near a transcript but the transcript itself cannot be directly inferred to be regulated. Transcription profiling is particularly important in cases where a ChIP-identified binding site falls between divergently transcribed genes, and/or cases in which the TF regulates several genes in an operon. Third, transcription profiling identifies both direct and indirect regulatory targets and it is not possible to distinguish the two. Nonetheless, indirect effects can be limited by transiently expressing the TF and comparing to cells in which the TF was not expressed, as has been shown for regulatory RNAs (Zhang et al. 1998). Fourth, transcription profiling cannot distinguish between changes in transcription and changes in RNA stability. Given the strengths and weaknesses of the two approaches, combining ChIP-chip/ChIP-seq and transcription profiling data is the most effective way to identify a TF regulon (discussed below).

The first microarrays used for transcription profiling were based on PCR products amplified from cDNA libraries (Schena and Yamamoto 1988). Thus, the microarray design was limited in its scope and its resolution. Improvements in microarray fabrication permitted the use of microarrays that had higher resolution and covered whole genomes. This permitted true genome-scale analysis of RNA levels (DeRisi et al. 1997; Wodicka et al. 1997). Using microarrays, transcription profiling was used to identify regulatory targets of TFs by comparing RNA levels in wild-type and TF mutant cells, or between conditions known to affect a specific TF (Courcelle et al. 2001). Microarray density is often a limiting factor for analysis of RNA levels since there are many unannotated RNAs and RNAs can occur on either strand. High density tiling microarrays have permitted identification of novel RNAs (Reppas et al. 2006), but the resolution and sensitivity of even the highest density microarray is less than that of RNA-seq.

RNA-seq involves next-generation sequencing of a cDNA library. Advantages of RNA-seq over microarrays for transcription profiling are (1) higher resolution (RNA-seq can give single bp resolution), (2) no limit on dynamic range, and (3) no bias with respect to genome position or strand. Thus, RNA-seq is more sensitive, and can detect unannotated RNAs. One drawback of RNA-seq is that ribosomal RNA (rRNA) represents a large proportion of all cellular RNA. Without removal of rRNA, most of the sequencing reads will correspond to rRNA and hence will provide little or no useful information. However, there are several efficient kits for removal of rRNA from total RNA samples. As for ChIP-seq samples, RNA-seq samples can be barcoded and multiplexed to increase efficiency. For a typical experiment, 10 million sequencing reads per sample is sufficient for an average sized bacterial genome. Early RNA-seq experiments used cDNA libraries that lacked strand information. There are now a variety of methods and kits that can be used to generate strand-specific libraries. Nevertheless, there is always a small amount of "bleed-through" onto the opposite strand. It is important to take this bleed-through into account when analyzing RNA-seq data.

7.9 Analysis of RNA-seq Data

Analysis of RNA-seq data is more straightforward than that of ChIP-seq data. The genomic coordinates of specific RNA species are selected and the number of sequence reads mapping to those regions is determined for the control and experimental samples. There is a wide variety of analysis programs, the most widely used being Cufflinks/CuffDiff (Trapnell et al. 2010). The specific RNA species to be analyzed is an important parameter since the list must be determined prior to analysis. For some RNA-seq analysis programs, there is a first step that identifies RNAs from the data rather than from a predefined list. This is possible due to the high resolution of RNA-seq data, and permits inclusion of 5′ and 3′ UTRs, analysis of operonic transcripts rather than single genes, and permits inclusion of unannotated transcripts (Fig. 7.2). Thus, RNA discovery is combined with analysis of RNA levels. As discussed below, recent studies using modified RNA-seq methods indicate the existence of hundreds, possibly thousands of unannotated RNAs in any given bacterial genome. The most effective program for identifying UTRs, operons, and unannotated transcripts from bacterial RNA-seq datasets and including them in the analysis of differential RNA levels is Rockhopper (McClure et al. 2013), which was trained on bacterial datasets.

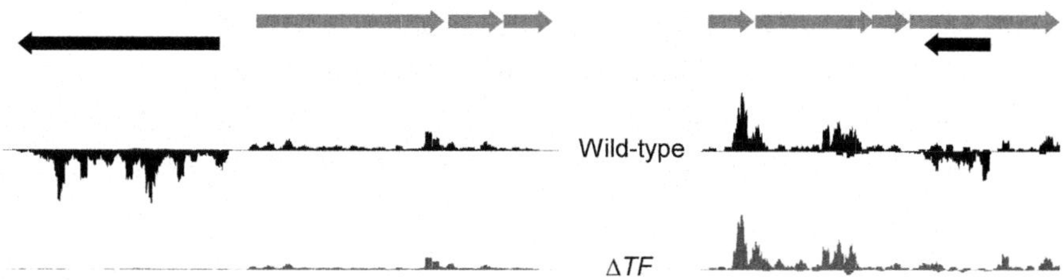

Fig. 7.2 RNA-seq. Example of RNA-seq data for wild-type *Escherichia coli* and cells containing a deletion of a transcription factor-encoding gene. Two representative genomic regions are shown. The graph shows piled up sequence reads from the RNA-seq experiment. Reads mapping to the plus strand are plotted above the horizontal line, and reads mapping to the minus strand are plotted below the horizontal line. *Gray horizontal arrows* indicate non-regulated genes. *Black horizontal arrows* indicate significantly regulated genes. Note that the *black arrow* in the right panel represents an unannotated, non-coding gene

7.10 Modified RNA-seq Approaches

Several modifications to standard RNA-seq methods have been described. In particular, RNA-seq can be adapted to map RNA 5′ or 3′ ends. Two methods have been described to map TSSs. The first method, Differential RNA-seq (dRNA-seq), combines standard RNA-seq with an exonuclease that specifically degrades RNAs with monophosphorylated 5′ ends (Sharma et al. 2010). Libraries are made with and without exonuclease treatment and sequenced separately. Thus, triphosphorylated RNA 5′ ends (corresponding to TSSs) can be distinguished from monophosphorylated RNA ends (corresponding to RNA processing sites). The second method also distinguishes between mono- and triphosphorylated RNA 5′ ends. Libraries are made that include only RNA 5′ ends, rather than all portions of each RNA. Libraries are generated from samples treated with or without a phosphorylase that converts tri- to monophosphorylated 5′ ends (Cho et al. 2014; Kim et al. 2012; Singh and Wade 2014; Singh et al. 2014). TSSs are identified by comparing RNA 5′ end abundance between the phosphorylase-treated and untreated libraries. TSS mapping using either of these methods is more sensitive than standard RNA-seq, and can identify RNAs that initiate inside genes in the sense orientation, which is usually not possible using standard RNA-seq. TSSs mapping has identified large numbers of unannotated transcripts in a variety of bacterial species, including thousands of RNAs that initiate inside genes (Wade and Grainger 2014).

A second method that is particularly effective at identifying unannotated RNAs is Native Elongating Transcript (NET)-seq (Churchman and Weissman 2011). NET-seq involves isolation of elongating RNAP complexes and library construction from the 3′ ends of the associated nascent transcripts. Thus, NET-seq measures the level of nascent rather than mature RNA, and is not impacted by differences in RNA stability. NET-seq is a relatively new approach and hence has rarely been applied to bacterial systems (Qi et al. 2013).

7.11 Inferring Regulons and Regulatory Networks by Combining ChIP-seq and RNA-seq Data

Inferring a TF regulon is straightforward using a combination of ChIP-seq and RNA-seq data for that TF, and relies on the assumption that TFs regulate nearby genes. Binding sites identified by ChIP-seq are paired with adjacent transcripts (single gene or multi-gene operon) that are significantly regulated. As discussed below, TF binding sites can regulate distally encoded genes, and TF binding sites inside genes can be associated with regulation of the overlapping gene. Incorporating these effects into TF regulon identification is problematic since these phenomena are not well understood.

With the exception of one study in *Mycobacterium tuberculosis* (see below), ChIP-seq and RNA-seq have been used to identify TF regulons for only a handful of TFs (Myers et al. 2013; Park et al. 2013b; Stringer et al. 2014). In each case, many significantly regulated genes were identified that were not associated with nearby TF binding. These indirectly regulated genes supplement the TF regulon and their regulation is due either to a chain of TF-TF regulation (e.g. the gene is regulated by TF#2 which is itself regulated by TF#1, the focus of the experiment), or to physiological changes brought about by mutation of the TF-encoding gene.

Inferring entire regulatory networks requires ChIP-seq and RNA-seq data for multiple TFs. To infer the regulatory network for an entire organism, every TF must be analyzed. Only one study has approached this level of complexity (Galagan et al. 2013). The binding location of 50 TFs in *M. tuberculosis* was determined using ChIP-seq. In parallel, RNA-seq was used to determine the effect of overexpressing each of the TFs. 25 % of all TF binding sites were associated with regulation of a gene <1 kb away, and the degree of regulation correlated with the strength of TF binding. However, for TFs >1 kb from a regulated gene, there was a significant association of TF binding with regulation of genes up to 4 kb away. Furthermore, for TF binding sites within genes there was a significant association of binding with regulation of the overlapping gene. Hence, simply combining TF binding site information with regulation of the adjacent gene may be insufficient to identify all direct regulatory interactions. However, until our mechanistic understanding of these processes improves, it is not possible to confidently associate binding sites inside genes with regulation of the overlapping gene, or to confidently associate binding sites and regulated genes that are spatially separated.

The study of TFs in *M. tuberculosis* is the only one to date that used ChIP-seq and RNA-seq data to infer an entire regulatory network. Although this study did not include all *M. tuberculosis* TFs, it included most that are known to be associated with regulation during hypoxia and regulation of lipid metabolism. Comparing regulons for different TFs identified extensive cross-talk. A few TFs were identified as "hubs", based on their regulation of many genes. Most TFs regulated themselves, consistent with similar observations for TFs in *E. coli* (Shen-Orr et al. 2002). Other well-characterized network topologies, such as feed-forward loops (Shen-Orr et al. 2002), were also identified. By combining regulon information for all TFs in the study, the authors were able to reconstruct a regulatory network. This network was used to model expression of all genes during hypoxic growth, and was significantly more predictive than a random model for 66 % of genes that showed significant regulation when RNA levels were measured using RNA-seq across a time-course of hypoxic growth. Thus, regulons inferred from ChIP-seq and RNA-seq data can be combined to predict transcriptional changes associated with specific environmental or genetic perturbations.

7.12 Mapping Initiating and Elongating RNAP Using Genome-Scale ChIP Methods

An additional level of complexity that can be added to bacterial regulatory networks is the composition of the initiating RNAP associated with transcription of a given RNA. Most bacteria express multiple σ factors, and some RNAP:σ complexes are subject to TF-mediated regulation. Hence, a regulatory network will not be complete without knowing the specific σ factor associated with a given regulatory TF binding site. As with TFs, ChIP methods and transcription profiling can be used to map σ factor regulons (Nonaka et al. 2006; Reppas et al. 2006; Singh et al. 2014; Wade et al. 2006; Zhao et al. 2005).

7.13 Conclusions

Technological advances in genome-scale approaches have revolutionized our ability to map TF regulons. ChIP-seq and RNA-seq are currently the methods of choice for mapping TF regulons and they can be easily applied to a wide range of bacterial species. However, significant challenges remain. First, intragenic TF binding sites are poorly understood, as is TF regulation of distally encoded genes. Second, although many TFs can be identified accurately as TFs based on amino acid sequence, there are likely to be many unidentified TFs. Third, binding of many TFs to DNA may be affected by environmental conditions that have not been assessed, e.g. temperature, osmolarity. Fourth, DNA structure and chemical composition is modulated by other factors, such as supercoiling and methylation, that are not currently considered when constructing regulatory networks. Fifth, effects on other aspects of gene expression such as RNA stability, translation, post-translational modification, are not currently considered. Nevertheless, despite these challenges, ChIP-seq and RNA-seq data are already able to provide far better predictions than other modeling approaches. Thus, these methods greatly enhance our ability to model complex regulatory networks, including those for whole organisms.

References

Aparicio O, Geisberg JV, Sekinger E, Yang A, Moqtaderi Z, Struhl K (2005) Chromatin immunoprecipitation for determining the association of proteins with specific genomic sequences *in vivo*. In: Ausubel FM, Brent R, Kingston RE, Moore DD, Seidman JG, Smith JA, Struhl K (eds) Current protocols in molecular biology. Wiley, Hoboken, pp 21.23.21–21.23.33

Auerbach RK, Euskirchen G, Rozowsky J, Lamarre-Vincent N, Moqtaderi Z, Lefrançois P, Struhl K, Gerstein M, Snyder M (2009) Mapping accessible chromatin regions using Sono-Seq. Proc Natl Acad Sci U S A 106:14926–14931

Bailey TL, Elkan C (1994) Fitting a mixture model by expectation maximization to discover motifs in biopolymers. Proc Int Conf Intell Syst Mol Biol 2:28–36

Bailey TL, Machanick P (2012) Inferring direct DNA binding from ChIP-seq. Nucleic Acids Res 40, e128

Belitsky BR, Sonenshein AL (2013) Genome-wide identification of *Bacillus subtilis* CodY-binding sites at single-nucleotide resolution. Proc Natl Acad Sci U S A 110:7026–7031

Blat Y, Kleckner N (1999) Cohesins bind to preferential sites along yeast chromosome III, with differential regulation along arms versus the centric region. Cell 98:249–259

Bonocora RP, Fitzgerald DM, Stringer AM, Wade JT (2013) Non-canonical protein–DNA interactions identified by ChIP are not artifacts. BMC Genomics 14:254

Cameron AD, Dorman CJ (2012) A fundamental regulatory mechanism operating through OmpR and DNA topology controls expression of *Salmonella* pathogenicity islands SPI-1 and SPI-2. PLoS Genet 8, e1002615

Cawley S, Bekiranov S, Ng HH, Kapranov P, Sekinger EA, Kampa D, Piccolboni A, Smentchenko V, Cheng J, Williams AJ et al (2004) Unbiased mapping of transcription factor binding sites along human chromosomes 21 and 22 points to widespread regulation of non-coding RNAs. Cell 116:499–509

Cho BK, Kim D, Knight EM, Zengler K, Palsson BO (2014) Genome-scale reconstruction of the sigma factor network in *Escherichia coli*: topology and functional states. BMC Biol 12:4

Churchman LS, Weissman JS (2011) Nascent transcript sequencing visualizes transcription at nucleotide resolution. Nature 469:368–373

Courcelle J, Khodursky A, Peter B, Brown PO, Hanawalt PC (2001) Comparative gene expression profiles following UV exposure in wild-type and SOS-deficient *Escherichia coli*. Genetics 158:41–64

Courey AJ (2001) Cooperativity in transcriptional control. Curr Biol 11:R250–R252

DeRisi JL, Iyer VR, Brown PO (1997) Exploring the metabolic and genetic control of gene expression on a genomic scale. Science 278:680–686

ENCODE, p.c. (2007) Identification and analysis of functional elements in 1 % of the human genome by the ENCODE pilot project. Nature 447:799–816

Galagan JE, Minch K, Peterson M, Lyubetskaya A, Azizi E, Sweet L, Gomes A, Rustad T, Dolganov G, Glotova I et al (2013) The *Mycobacterium tuberculosis* regulatory network and hypoxia. Nature 499:178–183

Grainger DC, Webster CL, Belyaeva TA, Hyde EI, Busby SJ (2004) Transcription activation at the *Escherichia coli* melAB promoter: interactions of MelR with its DNA target site and with domain 4 of the RNA polymerase sigma subunit. Mol Microbiol 51:1297–1309

Harbison CT, Gordon DB, Lee TI, Rinaldi NJ, Macisaac KD, Danford TW, Hannett NM, Tagne JB, Reynolds DB, Yoo J et al (2004) Transcriptional regulatory code of a eukaryotic genome. Nature 431:99–104

Iyer VR, Horak CE, Scafe CS, Botstein D, Snyder M, Brown PO (2001) Genomic binding sites of the yeast cell-cycle transcription factors SBF and MBF. Nature 409:533–538

Kallipolitis BH, Norregaard-Madsen M, Valentin-Hansen P (1997) Protein–protein communication: structural model of the repression complex formed by CytR and the global regulator CRP. Cell 89:1101–1109

Kim D, Hong JS, Qiu Y, Nagarajan H, Seo JH, Cho BK, Tsai SF, Palsson BØ (2012) Comparative analysis of regulatory elements between *Escherichia coli* and *Klebsiella pneumoniae* by genome-wide transcription start site profiling. PLoS Genet 8, e1002867

Laub MT, Chen SL, Shapiro L, McAdams HH (2002) Genes directly controlled by CtrA, a master regulator of the Caulobacter cell cycle. Proc Natl Acad Sci U S A 99:4632–4637

Lee TI, Rinaldi NJ, Robert F, Odom DT, Bar-Joseph Z, Gerber GK, Hannett NM, Harbison CT, Thompson CM, Simon I et al (2002) Transcriptional regulatory networks in *Saccharomyces cerevisiae*. Science 298:799–804

Lieb JD, Liu X, Botstein D, Brown PO (2001) Promoter-specific binding of Rap1 revealed by genome-wide maps of protein–DNA association. Nat Genet 28:327–334

Lun DS, Sherrid A, Weiner B, Sherman DR, Galagan JE (2009) A blind deconvolution approach to high-resolution mapping of transcription factor binding sites from ChIP-seq data. Genome Biol 10:R142

McClure R, Balasubramanian D, Sun Y, Bobrovskyy M, Sumby P, Genco CA, Vanderpool CK, Tjaden B (2013) Computational analysis of bacterial RNA-Seq data. Nucleic Acids Res 41, e140

Myers KS, Yan H, Ong IM, Chung D, Liang K, Tran F, Keles S, Landick R, Kiley PJ (2013) Genome-scale analysis of *Escherichia coli* FNR reveals complex features of transcription factor binding. PLoS Genet 9, e1003565

Nonaka G, Blankschien M, Herman C, Gross CA, Rhodius VA (2006) Regulon and promoter analysis of the *E. coli* heat shock factor, Sigma 32, reveals a multifaceted cellular response to heat stress. Genes Dev 20:1776–1789

Park PJ (2009) ChIP-seq: advantages and challenges of a maturing technology. Nat Rev Genet 10:669–680

Park D, Lee Y, Bhupindersingh G, Iyer VR (2013a) Widespread misinterpretable ChIP-seq bias in yeast. PLoS One 8, e83506

Park DM, Akhtar MS, Ansari AZ, Landick R, Kiley PJ (2013b) The bacterial response regulator ArcA uses a diverse binding site architecture to regulate carbon oxidation globally. PLoS Genet 9, e1003839

Qi LS, Larson MH, Gilbert LA, Doudna JA, Weissman JS, Arkin AP, Lim WA (2013) Repurposing CRISPR as an RNA-guided platform for sequence-specific control of gene expression. Cell 152:1173–1183

Qian Z, Dimitriadis EK, Edgar R, Eswaramoorthy P, Adhya S (2012) Galactose repressor mediated intersegmental chromosomal connections in *Escherichia coli*. Proc Natl Acad Sci U S A 109:11336–11341

Reid JL, Iyer VR, Brown PO, Struhl K (2000) Coordinate regulation of yeast ribosomal protein genes is associated with targeted recruitment of Esa1 histone acetylase. Mol Cell 6:1297–1307

Ren B, Robert F, Wyrick JJ, Aparicio O, Jennings EG, Simon I, Zeitlinger J, Schreiber J, Hannett N, Kanin E et al (2000) Genome-wide location and function of DNA binding proteins. Science 290:2306–2309

Reppas NB, Wade JT, Church G, Struhl K (2006) The transition between transcriptional initiation and elongation in *E. coli* is highly variable and often rate-limiting. Mol Cell 24:747–757

Rhee HS, Pugh BF (2011) Comprehensive genome-wide protein–DNA interactions detected at single-nucleotide resolution. Cell 147:1408–1419

Robertson G, Hirst M, Bainbridge M, Bilenky M, Zhao Y, Zeng T, Euskirchen G, Bernier B, Varhol R, Delaney A et al (2007) Genome-wide profiles of STAT1 DNA association using chromatin immunoprecipitation and massively parallel sequencing. Nat Methods 4:651–657

Salama RA, Stekel DJ (2010) Inclusion of neighboring base interdependencies substantially improves genome-wide prokaryotic transcription factor binding site prediction. Nucleic Acids Res 38, e135

Schena M, Yamamoto KR (1988) Mammalian glucocorticoid receptor derivatives enhance transcription in yeast. Science 241:965–967

Schweikert C, Brown S, Tang Z, Smith PR, Hsu DF (2012) Combining multiple ChIP-seq peak detection systems using combinatorial fusion. BMC Genomics 13:S12

Sharma CM, Hoffmann S, Darfeuille F, Reignier J, Findeiss S, Sittka A, Chabas S, Reiche K, Hackermüller J, Reinhardt R et al (2010) The primary transcriptome of the major human pathogen *Helicobacter pylori*. Nature 464:250–255

Shen-Orr SS, Milo R, Mangan S, Alon U (2002) Network motifs in the transcriptional regulation network of *Escherichia coli*. Nat Genet 31:64–68

Shimada T, Ishihama A, Busby SJ, Grainger DC (2008) The *Escherichia coli* RutR transcription factor binds at targets within genes as well as intergenic regions. Nucleic Acids Res 36:3950–3955

Singh N, Wade JT (2014) Identification of regulatory RNA in bacterial genomes by genome-scale mapping of transcription start sites. Methods Mol Biol 1103:1–10

Singh S, Singh N, Bonocora RP, Fitzgerald DM, Wade JT, Grainger DC (2014) Widespread suppression of intragenic transcription initiation by H-NS. Genes Dev 28:214–219

Solomon MJ, Larsen PL, Varshavsky A (1988) Mapping protein–DNA interactions *in vivo* with formaldehyde: evidence that histone H4 is retained on a highly transcribed gene. Cell 53:937–947

Stringer AM, Currenti SA, Bonocora RP, Petrone BL, Palumbo MJ, Reilly AE, Zhang Z, Erill I, Wade JT (2014) Genome-scale analyses of *Escherichia coli* and *Salmonella enterica* AraC reveal non-canonical targets and an expanded core regulon. J Bacteriol 196:660–671

Teytelman L, Thurtle DM, Rine J, van Oudenaarden A (2013) Highly expressed loci are vulnerable to misleading ChIP localization of multiple unrelated proteins. Proc Natl Acad Sci U S A 110:18602–18607

Trapnell C, Williams BA, Pertea G, Mortazavi A, Kwan G, van Baren MJ, Salzberg SL, Wold BJ, Pachter L (2010) Transcript assembly and quantification by RNA-Seq reveals unannotated transcripts and isoform switching during cell differentiation. Nat Biotechnol 28:511–515

Valouev A, Johnson DS, Sundquist A, Medina C, Annton E, Batzoglou S, Myers RM, Sidow A (2008) Genome-wide analysis of transcription factor binding sites based on ChIP-Seq data. Nat Methods 5:829–834

Vega VB, Cheung E, Palanisamy N, Sung WK (2009) Inherent signals in sequencing-based Chromatin-ImmunoPrecipitation control libraries. PLoS One 4, e5241

Wade JT, Grainger DC (2014) Pervasive transcription: illuminating the gloomy corners of bacterial transcription. Nat Rev Microbiol 12:647–653

Wade JT, Belyaeva TA, Hyde EI, Busby SJ (2001) A simple mechanism for co-dependence on two activators at an *Escherichia coli* promoter. EMBO J 20:7160–7167

Wade JT, Reppas NB, Church GM, Struhl K (2005) Genomic analysis of LexA binding reveals the permissive nature of the *Escherichia coli* genome and identifies unconventional target sites. Genes Dev 19:2619–2630

Wade JT, Roa DC, Grainger DC, Hurd D, Busby SJW, Struhl K, Nudler E (2006) Extensive functional overlap between s factors in *Escherichia coli*. Nat Struct Mol Biol 13:806–814

Wade JT, Struhl K, Busby SJ, Grainger DC (2007) Genomic analysis of protein–DNA interactions in bacteria: insights into transcription and chromosome organization. Mol Microbiol 65:21–26

Wodicka L, Dong H, Mittman M, Ho MH, Lockhart DJ (1997) Genome-wide expression monitoring in *Saccharomyces cerevisiae*. Nat Biotechnol 15:1359–1367

Zhang A, Altuvia S, Tiwari A, Argaman L, Hengge-Aronis R, Storz G (1998) The OxyS regulatory RNA represses *rpoS* translation and binds the Hfq (HF-I) protein. EMBO J 17:6061–6068

Zhao K, Liu M, Burgess RR (2005) The global transcriptional response of *Escherichia coli* to induced Sigma 32 protein involves Sigma 32 regulon activation followed by inactivation and degradation of Sigma 32 *in vivo*. J Biol Chem 280:17758–17768

Chapter 8
Quantitative and Systems-Based Approaches for Deciphering Bacterial Membrane Interactome and Gene Function

Viktor Deineko, Ashwani Kumar, James Vlasblom, and Mohan Babu

Abstract High-throughput genomic and proteomic methods provide a concise description of the molecular constituents of a cell, whereas systems biology strives to understand the way these components function as a whole. Recent developments, such as genome editing technologies and protein epitope-tagging coupled with high-sensitivity mass-spectrometry, allow systemic studies to be performed at an unprecedented scale. Available methods can be successfully applied to various goals, both expanding fundamental knowledge and solving applied problems. In this review, we discuss the present state and future of bacterial cell envelope interactomics, with a specific focus on host–pathogen interactions and drug target discovery. Both experimental and computational methods will be outlined together with examples of their practical implementation.

Keywords Affinity purification coupled with mass spectrometry • Antibiotic resistance • Bacteria • Computational methods • Epistasis • Membrane proteins • Network biology • Proteomics • Proteogenomics • Protein–protein interactions

8.1 Introduction

The last decade was marked by a rapid expansion of "omics" sciences, aimed at studying living organisms at different levels. Transcriptomics studies the entire pool of cellular mRNAs and its variability either between different pathotypes or under different conditions. Advancement of bioinformatic tools, development of exhaustive databases, and the high sensitivity and resolution of modern mass-spectrometers gave birth to shotgun proteomics, allowing several thousand proteins to be identified during a single mass spectrometry (MS) experiment. The improvements in these

V. Deineko • A. Kumar • J. Vlasblom • M. Babu (✉)
Department of Biochemistry, Research and Innovation Centre, University of Regina, Regina, SK, Canada
e-mail: mohan.babu@uregina.ca

N.J. Krogan, M. Babu (eds.), *Prokaryotic Systems Biology*, Advances in Experimental Medicine and Biology 883, DOI 10.1007/978-3-319-23603-2_8

technologies are reflected in the genome online database (GOLD) (Reddy et al. 2015), which now provides access to >4200 sequenced genomes, the majority of which are prokaryote.

The gram-negative bacterium *Escherichia coli* has been used as a model organism in numerous microbiological, biochemical and genetic studies. Recent estimates claim that 75% of the 4225 *E. coli* genes have been experimentally characterized and ascribed a biological function (Hu et al. 2009; Vlasblom et al. 2015). The remaining genes have either been annotated using bioinformatic methods (for example, taking into account sequence similarity) or remained uncharacterized ("orphan" genes). Another means of assigning function is by classifying many proteins into sub-proteomes, depending on their function and location within the cell. This network organization allows us to functionally characterize "orphan" genes by applying high-throughput genomic, bioinformatic, and proteomic approaches to elucidate the network neighborhood of a gene or protein, and integrate this information with complementary physical or functional evidences. In one study of this kind (Hu et al. 2009), 1241 tagged *E. coli* proteins were affinity purified, and potential interactors identified by MS. The resulting data were methodically analyzed, and a functional network was reconstructed using machine learning algorithms. In this way, a large number of orphan genes was classified into functional modules (for example, protein synthesis, envelope assembly, motility, and biofilm), and additional experiments were performed with selected sub-sets of orphan genes to confirm predictions. This study exemplifies the approach of combining complementary data to serve as a methodological guide for further large scale interactomics studies of bacterial species (Vlasblom et al. 2015).

The cell envelope of gram-negative bacteria consists of an outer membrane and an inner membrane delimited by a periplasmic peptidoglycan layer. The inner membrane is composed of phospholipids, whereas the outer membrane contains mainly lipopolysaccharides. Each of these components has its own set of integral and associated proteins (subproteome) (Díaz-Mejía et al. 2009; Silhavy et al. 2010; Needham and Trent 2013). This envelope, together with associated membrane or cytosolic proteins, mediates bacterial morphogenesis, division, uptake of nutrient, and release of metabolites (Silhavy et al. 2010). Moreover, bacterial membrane proteins (MPs) facilitate host–pathogen interactions and play key roles in host colonization, biofilm formation, evasion of the host immune response, and drug resistance (Flores-Mireles et al. 2015; Ribet and Cossart 2015). Currently, 60% of antimicrobial drug targets are represented by MPs (Fruh et al. 2010), yet only 2100 of ~93,000 protein structures currently deposited at the Protein Data Bank (Berman et al. 2013) belong to MPs (Fig. 8.1a).

Comparative genomic studies suggest that roughly 25–30% of all bacterial genes encode proteins of the inner or outer bacterial membrane (Daley et al. 2005; Díaz-Mejía et al. 2009; Pogozheva et al. 2013). For example, in *E. coli*, ~900 transmembrane proteins are integral and ~90 span the outer membrane. Despite their many essential functions (Silhavy et al. 2010; Flores-Mireles et al. 2015; Ribet and Cossart 2015), the *E. coli* membrane proteome is sparsely characterized, and only 50% of a predicted 1133 integral inner membrane proteins have

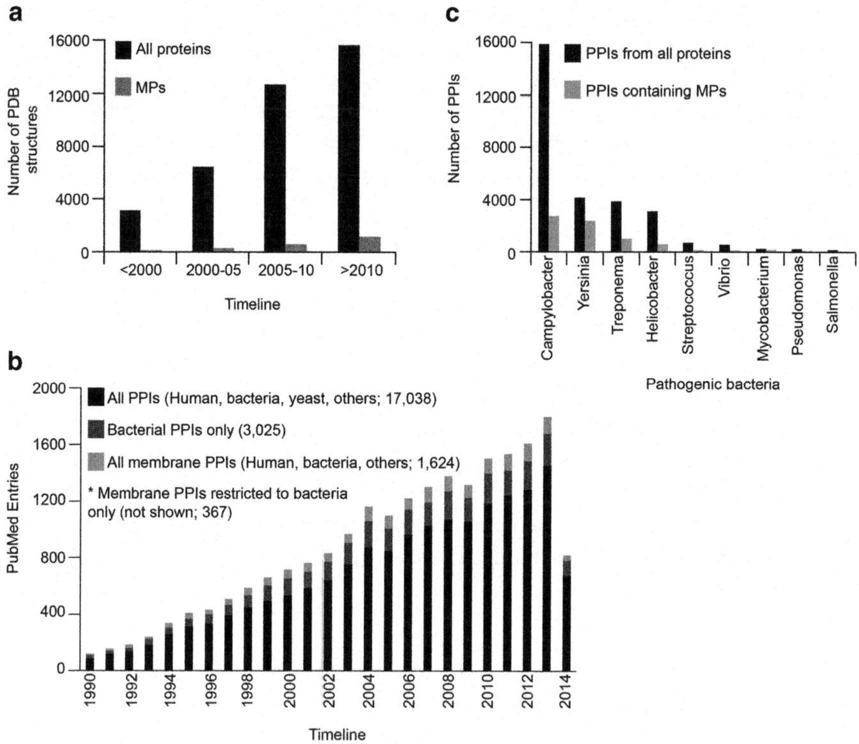

Fig. 8.1 Timeline of membrane proteins (MPs) and their physical interactions in various pathogenic bacteria. (**a**) Number of PDB structures show a steady increase for non-membrane proteins as opposed to MPs over time. (**b**) PubMED index (extracted from public databases as of August, 2014) indicating studies that are focused on PPIs from all (cytosolic and/or membrane) organisms (including humans, bacteria, yeast, and others) or only to bacteria (i.e. cytosolic) over time. PubMED records of bacterial membrane PPIs (367) were not shown in the graph. (**c**) Sparsity of PPIs for MPs is shown across major groups of bacterial pathogens

been experimentally identified by biochemical studies (Bernsel and Daley 2009; Papanastasiou et al. 2013). This is due to a confluence of factors including low abundance, poor solubility, and relatively small portion of charged residues in MPs; any of which can negatively impact experiment outcome.

Further exploration of the membrane proteome and interactome will broaden our understanding of basic bacterial biology, particularly in the areas of transmembrane trafficking, nutrient uptake, dormancy, and quorum sensing. It will also improve the current state of knowledge of stress response and defense mechanisms, and help to reveal the origin of the robustness of microorganisms. Most notably, information about bacterial MP organization and function will assist in identifying targets of novel antimicrobial drugs and vaccine development.

Often antibiotics trigger a cascade of events that finally leads to bactericidal or bacteriostatic action (Davies and Davies 2010; Kohanski et al. 2010). Comparative proteomic studies reveal the sequence of events that leads to bactericidal effects (Lima et al. 2013), and elucidate the mechanisms of action for antibiotics. Moreover, creating proteomic signatures to identify specific responses to different antibiotics will expand our understanding on the clinically relevant antibiotics targeting diverse bacterial envelope pathways and structures (Wenzel and Bandow 2011). This proteomic response based strategy can be tailored towards the search for new antimicrobial compounds which can be used alone or in synergy with existing drugs to cope with emerging pathogens that are resistant to a particular antibiotic.

In this review, we focus on both experimental and computational approaches to dissect the bacterial membrane interactome. These methods are either modified versions of conventional techniques, or were developed for the sole purpose of studying MPs. Reviewed methods include the *E. coli* synthetic genetic array (eSGA) for studying gene–gene interactions, gel-based and gel-free proteomics, affinity purification coupled with liquid chromatography (LC) and mass spectrometry (AP/MS), proteogenomic and chemogenomic approaches as well as computational methods for the discovery of novel protein complexes and networks *in silico*. To demonstrate the applicability of these methods to problems of particular importance, we provide examples from literature wherever appropriate.

8.2 Assessing Membrane Proteome Through Functional Networks and Proteomics

A broad spectrum of experimental techniques is available for studying protein–protein interactions (PPIs) and the complex associations of MPs (Babu et al. 2012; Lam and Stagljar 2012). They include the classical yeast two-hybrid system as well as its modifications: namely, the dual bait yeast two-hybrid (Y2H) system, split-ubiquitin membrane based yeast two-hybrid assay and the bacterial two-hybrid system based on transcription activation, *in vivo* protein cross-linking, fluorescent protein-fragment complementation assays, AP/MS, and co-fractionation of protein complexes (Kerppola 2006; Díaz-Mejía et al. 2009; Babu et al. 2012; Havugimana et al. 2012; Ryan et al. 2013). These methods can be grouped into ones that primarily identify direct 'binary' PPIs, such as yeast two-hybrid screening and protein complementation assays, and those that identify co-complex associations, such as AP/MS (Yu et al. 2008; Rajagopala et al. 2014; Wuchty and Uetz 2014; Caufield et al. 2015). These two classes of methods can be used in combination to effectively complement each other. Unlike the other methods, quantitative MS provides information about the relative or absolute abundance of proteins within a particular sample, and hence it can be used for comparative proteomics, where changes in a MP expression and interactions under variable conditions (for example, under stress conditions or antibiotic treatment) are monitored.

8.2.1 Reconstruction of Interaction Networks Using eSGA

Gene products usually function within functional modules (operons) or pathways (Babu et al. 2009, 2014). By identifying the functional connection of an "orphan" gene to a well-studied metabolic chain or signaling cascade, one can assume that its product performs either the same or similar role. The reconstruction of networks and the connections between sub-networks to enable these inferences can be performed in a high-throughput manner by using eSGA (Butland et al. 2008), which has been applied to study a specific process of interest such as protein synthesis (Gagarinova et al. unpublished), envelope biogenesis (Babu et al. 2011), and genomic integrity (Kumar et al. unpublished). This strategy was originally developed for systems biology studies of *E. coli*, and can be further expanded to other bacteria (Babu et al. 2009). The method is based on epistasis (lack of independence), i.e. an effect of a mutation depends on the presence of other mutations in genome. It is often observed that two single mutant strains display normal phenotype, whereas double mutant strains show significantly impaired viability (i.e. synthetic sickness or lethality), implying that each mutant corresponds to components of redundant pathways, implying that two redundant pathways were impassable by the combination of both mutations (for a more in-depth review on epistatic interactions, see Baryshnikova et al. 2013; Mitra et al. 2013; Ryan et al. 2013).

8.2.2 Gel-Based Methods for Dissecting MP Complexes

Techniques for analyzing membrane interactomes can be performed in vitro, or *ex vivo* using whole cell lysates or partially purified proteins. The latter approaches can be further grouped into gel-based and gel-free methods. Gel-based approaches (e.g. 2D-PAGE, BN-PAGE, BAC/SDS-PAGE, and DIGE) are useful for the separation of complex protein mixtures, and isolation of protein complexes and their subunits. They are used either for quantitative assessment of the proteome, or they can be integrated into an MS protein identification or quantitation pipeline. Here, we briefly describe the method and its most widely used modifications; for a comprehensive review of two-dimensional protein electrophoresis, the reader can consult the expanded review by Curreem et al. (2012).

Two-dimensional polyacrylamide gel electrophoresis (2D-PAGE) utilizes isoelectric focusing and SDS-PAGE in the first and second dimensions, respectively. Immobilized pH gradients separate proteins according to their isoelectric point in the pH range 2.5–12. In the second dimension, denatured proteins denatured proteins are separated by molecular weight and can be identified by MS or quantified (see DIGE). This method allows one to obtain global views of an organism's proteome or sub-proteomes, and to estimate variations in protein expression levels. For example, in the studies of human pathogen *Staphylococcus aureus*, ~700 proteins were identified and traced under different physiological conditions by 2D electrophoresis (Becher et al. 2009).

The hydrophobicity of MPs and their tendency to aggregate significantly limit the applicability of 2D-PAGE. However, two-dimensional benzyldimethyl-n-hexadecylammonium chloride (BAC) or SDS-PAGE can often substitute for the conventional procedure, where the detergent BAC and SDS are used in the first and second dimensions, respectively (Zahedi et al. 2007). 2D-BAC/SDS-PAGE is capable of resolving such difficult proteins as cytochrome-c oxidase subunit I, containing 12 transmembrane domains, and Sec61α, with 10 transmembrane domains. Another alternative approach for 2D-PAGE utilizes combinations of surfactants and chaotropes such as sulfobetaines, non-ionic detergents, dodecylmaltoside, or short chain lecithins for solubilizing MPs (Rabilloud 2009).

Blue native polyacrylamide gel electrophoresis (BN-PAGE) is widely used for the identification of PPIs and isolation of protein complexes from cell lysates and, in particular, from lipid membranes (Schlegel et al. 2010; Dresler et al. 2011; Ngounou Wetie et al. 2013). It is useful for an approximate determination of molecular masses and oligomeric states of native proteins, identification of PPI, isolation of MPs, and determination of multiprotein complex stoichiometry (Dresler et al. 2011). Purified biomembranes are solubilized using mild non-ionic detergents, such as digitonin, dodecyl-β-D-maltopyranoside (DDM), and Triton X-100. Coomassie® Blue G-250 is added to solubilized membranes in the next stage. The binding of this anionic dye to MPs allows them to migrate towards the anode at pH 7.5 during electrophoresis. Coomassie prevents MP aggregation and stabilizes them in aqueous solution due to its negative surface charge. Proteins and protein complexes are resolved in non-denaturing gradient polyacrylamide gels (~4–16%) as distinct blue bands.

Native protein assemblies can be further dissociated into subunits and separated accordingly to their molecular weight in the second dimension by classical SDS-PAGE. One-dimensional BN-PAGE can also be combined with orthogonal BN-PAGE, doubled SDS-PAGE or IEF/SDS 3D PAGE (Salzano et al. 2013). For example, BN-PAGE was successfully used for dissecting porin complexes in the membrane of *Neisseria meningitidis* (often referred to as meningococcus). Such studies are important for vaccine development (Dresler et al. 2011). MP complexes of another pathogenic bacterium, *Helicobacter pylori*, have been studied by 2D BN electrophoresis (Pyndiah et al. 2007), leading to a deeper understanding of *H. pylori* physiology, and the identification of potential vaccine candidates. Complexomic studies of *Mycobacterium bovis*, and *Bacillus Calmette-Guérin* (BCG) MPs were performed by the combination of BN-PAGE and LC–MS (Zheng et al. 2011), where 40 different proteins, organized into 9 complexes, were identified. Lastly, 2D BN/SDS-PAGE has been used for interactomic studies of the classical model organism, *E. coli* (Pan et al. 2010). For example, Stenberg et al. (2005) identified 43 protein complexes within the *E. coli* cell envelope. Applications of BN-PAGE for studying the organization of MP's in complexes are further reviewed detail in Dresler et al. (2011).

Two-dimensional difference gel electrophoresis (2D-DIGE) is a modified form of 2D-PAGE (Rabilloud and Lelong 2011), which allows quantitative assessment of proteome changes with approximately 15% expression difference. Specimens (i.e. tissue homogenates or cell lysates) are covalently labeled with fluorescent cyanine dyes (for example, Cy3-NHS and Cy5-NHS). These spectrally resolvable, positively charged, amine reactive dyes generally do not interfere with electrophoretic mobility of proteins. Hence, identical proteins from samples precisely superimpose with each other on 2D gel images. Due to the high sensitivity of fluorescent labeling and detection, as little as 0.2 fmol of protein can be detected. However, conventional DIGE has limited ability to resolve proteins with hydrophobic transmembrane domains. In this case, experimental conditions should be modified using suitable detergents. Furthermore, gel-based approaches can be coupled with MS. In this case, MP complexes or their sub-units are resolved by PAGE. Proteins of interest are identified as bands or spots on the gel and excised. Tryptic in-gel sample digestion is performed, and the resulting peptide mixture is submitted for identification by MS.

8.2.3 Non-gel Based Methods for High-Throughput Interactomics

Gel-free methods begin with a different workflow to the above. Here the target protein and its interacting partners are purified from crude lysate or from isolated cell membranes. This can be done by immunoaffinity chromatography, when resin-conjugated antibodies are used. Antibodies can be raised against endogenous protein and co-immunoprecipitation is performed using lysed and homogenized bacterial cells. Although large numbers of antibodies are commercially available, the selection of antibodies for bacterial proteins is rather limited. When an antibody is not available for a protein of interest, a chromosomal epitope-tag can be added to the protein. Then the protein can be purified using readily available antibodies to the tag of choice (i.e. anti-HA, anti-FLAG etc). This method is called AP/MS (reviewed in detail in Díaz-Mejía et al. 2009; Ahrens et al. 2010; Kuzmanov and Emili 2013; Ryan et al. 2013). For example, a large-scale PPI network study was done in *E. coli* by pull-down assays using the sequential peptide affinity (SPA)-tagged bait proteins (Hu et al. 2009). The tag consists of triple FLAG, followed by a tobacco etch virus protease cleavage site and a calmodulin binding peptide. This cassette containing the SPA-tag coupled with an antibiotic resistance marker is integrated into bacterial chromosome. The advantages of this method (i.e. the near-endogenous expression level of tagged proteins) have made it standard for bacterial interactomic studies (Díaz-Mejía et al. 2009).

An alternative to chromosomal tagging is cloning of MPs into plasmid vectors, which allows inducible expression of tagged proteins. A large-scale study of *E. coli* MP topology was performed by Daley et al. (2005). In this work, a total of 714 MPs were cloned into alkaline phosphatase (PhoA) and green fluorescent protein (GFP) fusion vectors, and high-confidence topology models were generated for $\sim$600 proteins. Moreover, a PPI system was developed for cases when chromosomal tagging of bacterial protein was difficult or impossible (Pelletier et al. 2008). This system utilized a broad-host-range plasmid vector pBBR1MCS5 for fusion of proteins with fluorescent or affinity tags, and was successfully tested in the gram-negative bacteria *Rhodopseudomonas palustris* CGA010 and *Shewanella oneidensis* MR-1, as well as in *E. coli*, though overall performance was poorer than chromosomal tagging.

Co-fractionation is a powerful tag-free method that can be used for complexomic studies of bacterial MPs, and, particularly, for host–pathogen interactomes. Recently, Havugimana et al. (2012) developed an integrative global proteomic profiling approach based on LC–MS, where cell extracts were separated into fractions that were subsequently analyzed by quantitative tandem MS. Using this technique, a network of 13,993 high-confidence physical interactions involving 3006 stably associated soluble human proteins was re-constructed. Though aimed at detecting physical interactions within human cells, this approach could be potentially applied to detect interactions in various pathogens or in multiple isolates of clinical pathogenic (enteric or diarrheagenic) strains as well. To do so, bacterial cell lysate could be pre-fractionated into membrane and cytosol, and the former fractions are subjected to extensive complementary biochemical fractionation procedures. Stably-interacting proteins that co-fractionate together could then be identified by LC–MS/MS. As with any type of MP's purification or fractionation procedure, care should be taken of detergent selection for effectively solubilizing MPs, while preserving protein complexes in their native state.

8.2.4 Quantitative MS to Track Changes of MPs Expression and Interaction

Current methods enable not only qualitative investigation, but also tracking quantitative changes in proteome or sub-proteomes under changing conditions or environmental stimuli, such as antibiotic treatment, immune response or depletion of nutrients (Hui et al. 2015). First quantitative proteomic experiments were based on 2D-DIGE. A newer and more powerful alternative is quantitative MS (reader is referred to some excellent reviews on this topic; Gstaiger and Aebersold 2009; Domon and Aebersold 2010; Kall and Vitek 2011; Hughes and Krijgsveld 2012; Liebler and Zimmerman 2013), which allows: (1) comparing proteome compositions between different samples (relative quantification); (2) determining sample concentration of certain proteins (absolute quantification); and (3) monitoring changes in protein expression. While this approach has been successful for many different protein types in other model organisms (Gouw et al. 2010; Taniguchi et al. 2010; Miteva et al. 2013), MS-based quantitative proteomics of bacterial MPs is still under development. The main challenges are the low abundance of MPs and difficulty of isolation of cell envelope and membrane associated proteins, as well as incomplete digestion of peptides leading to large peptide sizes that make MS analysis intractable. In this section, we describe the most common approaches that are used for comparative proteomic studies in bacteria.

Metabolic protein labeling with stable isotopes (^{13}C, ^{15}N) is a robust and versatile approach that can be done by growing bacteria on minimal media which includes isotopes such as $^{15}NH_4Cl$ or $(^{15}NH)_2SO_4$, or by isotope-labeled amino acids added to cultivation media (Stable Isotope Labeling by Amino acids in Cell culture—SILAC). $^{16}O/^{18}O$ isotope labeling is of special interest for membrane MS-based proteomics or interactomics (Ye et al. 2009). The main advantages of this method are: simplicity, no restriction due to peptide composition, and the minimal amount of proteins required. Typically, samples are subjected to tryptic digestion in $H_2{}^{18}O$. Two ^{16}O atoms are exchanged for ^{18}O at the C-terminus of peptides during trypsinization, leading to a 4 Da difference in peptide masses across the sample. Complex samples such as cell lysates or cell membrane preparations are fractionated by ion-exchange chromatography after digestion and analyzed further by LC–MS.

Alternative labeling approaches that do not involve isotopes are possible with one of several methods involving chemical modifications of peptides, and include iTRAQ (modifies N-terminal and lysine residues), ICAT (targets thiol groups of cysteine) and ICPL (targets primary amines, similarly to iTRAQ). While a detailed description of these methods is beyond the scope of this work (see reviews for details on this topic; Shui et al. 2009; Petriz and Franco 2014), these were successfully applied for membrane proteomic studies of pathogenic species such as *Mycobacterium avium* (Radosevich et al. 2007) and *Pseudomonas aeruginosa* (Tan et al. 2013). Likewise, quantitative MS was performed on *Acinetobacter baumannii* (Sauer and Kliem 2010; Cabral et al. 2011), a gram-negative opportunistic nosocomial pathogen affecting people with immune deficiency.

As evolution of multidrug resistant strains of *A. baumanii* is a growing problem for hospitals, significant efforts have been directed towards the discovery of new chemical compounds that will help to fight this emerging pathogen (Kanj and Kanafani 2011; Wong et al. 2012). The study of Yun et al. (2011) serves as an illustrative example of quantitative MS applied to pathogenic species. In this work, both isobaric tags and label-free approaches were used for absolute and relative quantitation of the cell wall proteome of *A. baumanii* under antibiotic stress. Out of 484 identified proteins, 302 were confirmed to be inner membrane, outer membrane, or periplasmic proteins. Different pools of proteins were found to be activated upon tetracycline and imipenem treatment. Such studies help to decipher mechanisms of antibiotic resistance and identify future drug targets to combat pathogenic bacteria.

8.2.5 Proteogenomics: A Novel Approach for Studying Emerging Pathogens

Comprehensive proteogenomic studies are important for identification of novel drug targets in multidrug resistant pathogens (Pawar et al. 2012; Bragazzi et al. 2014; Lee et al. 2014). While proteomics (especially quantitative) provides insights into the mechanisms of the bacterial stress response and multiple antibiotic resistance, the usability of publicly available databases for analysis of emerged pathogens is limited, as genomes of multidrug resistant bacteria often differ significantly from well-characterized parent strains. Hence, when reference proteomes are used for interpretation of MS results, a number of proteins within a sample can be overlooked.

Recently, several de novo sequenced genomes of multidrug resistant pathogens were published (Vignaroli et al. 2012; Albersmeier et al. 2014; Ozer et al. 2014; Riedel et al. 2015). These data can further facilitate the study of drug resistance mechanisms and the role of MPs. Lee et al. (2014) performed a showcase proteogenomic study of the multidrug resistance *A. baumannii* DU202, where 144 genes involved in antibiotic resistance were identified in a de novo sequenced genome of this strain. Notably, 14 genes were absent in the genome of *A. baumannii* ATCC 17978, which had been used as a reference in proteomics studies of *A. baumannii* DU202. While *A. baumannii* DU202 is closely related to *A. baumannii* 1656-2, the structure, size, and copy number of *strA* and *strB* streptomycin resistance genes in these strains were different, indicating that the induction patterns of antibiotic resistance genes can be altered in response to antibiotic culture conditions despite their homology in genome sequence. Collectively, the integration of proteomics with genomic resources are invaluable for the discovery of new ORFs and annotation of newly sequenced genomes of the emerging pathogens.

8.3 *In Silico* Prediction of Physical Membrane Protein Interactions

Over the past two decades, the number of PPI studies in bacterial species have increased exponentially (Fig. 8.1b), yet the PPI interactome for MPs in many bacteria is still far from being complete (Fig. 8.1c). MPs are particularly underrepresented, and the number of studies on bacterial membrane PPI are as low as 367 compared to 2658 pubMed entries recorded for PPIs derived from cytosolic proteins (Fig. 8.1b). One of the major reasons for the knowledge gap is the limited applicability of experimental high-throughput methods to MPs. Despite recent developments in experimental high-throughput methods, poorly expressed or highly hydrophobic proteins are still challenging to work with, and powerful computational methods are necessary to complement high-throughput experimental methods such as Y2H and AP/MS. The *in silico* methods (for example, FpClass and others; Lee et al. 2007; Kotlyar et al. 2015) are generally less time consuming, and in some cases can allow interaction predictions even when no or little experimental data is available.

8.3.1 Structure and Sequence Based Methods for Computational Membrane PPI Prediction

Comprehensive knowledge of existing binding motifs can be used to predict putative interaction sites. For example, Bracken et al. (2004) developed a method that included recognition of unstructured or disordered regions of the protein, which are frequently found to be involved in PPIs (Mosca et al. 2012; van der Lee et al. 2014; Wright and Dyson 2015). Although, the integration of multiple sequence alignments into the algorithm improved the confidence of prediction, sequence based methods are mainly useful for the identification of protein binding sites instead of actual PPIs (Ghersi and Sanchez 2011; Mills et al. 2015). Another example is the SLiM (Short Linear Motifs) software including SLiMSearch, SLiMScape, and SLiMFinder, which all find shared motifs in proteins sharing a common attribute such as sub-cellular localizations or an interaction partner (Edwards et al. 2007; Davey et al. 2011; O'Brien et al. 2013). Specifically, SLiMFinder is aimed at identifying small peptide microdomains ~3–10 residues long, which usually occur in intrinsically disordered regions of proteins. These SLiMs are known for mediating many vital PPIs in a wide range of scenarios including post-translational modifications and subcellular localization (Palopoli et al. 2015).

Next, several methods which utilize information about 3D structure have been developed to identify possible interaction sites of proteins and their interactions with other proteins. For example, PRISM (Protein Interactions by Structural Matching) is used for the large scale prediction of PPIs and assembly of protein complex structures (Tuncbag et al. 2011; Baspinar et al. 2014). The rationale behind the PRISM is that if two complementary sides of a template interface are similar to the surfaces of two target proteins, then these two proteins can interact with each other using this template interface architecture (Tuncbag et al. 2011; Baspinar et al. 2014). Another imperative algorithm termed PrePPI (PPI prediction method) combines 3D structural information and other functional evidences, such as co-expression and functional and evolutionary similarity using a Bayesian classifier, in order to predict protein interactions (Zhang et al. 2013) with claimed performance comparable to high-throughput experimental strategies. While its coverage for MP interactions is limited because 3D structures of MPs are harder to solve, the method itself may still prove effective for MPs, as the number of membrane structures available is continually increasing.

MPs can be classified on the basis of their transmembrane domain types, namely, α-helices or β-barrels. While virtually all β-barrel proteins are located in the outer membrane, majority of these outer membrane proteins are functional in oligomeric form, as the oligomerization "hot spot" serves as the key residues for facilitating PPIs (Perica et al. 2012). To date, more than 500 indices for MPs have been compiled and made available at AAINDEX (Amino Acid Index) database (Kawashima et al. 2008) for the scaling of amino acids on the basis of their physiochemical properties such as size, polarity, and secondary structure propensity. The β-barrel TransMembrane eXposure (BTMX) method uses these indices for the prediction of exposure status of transmembrane residues, and identifies the physiochemical properties characteristic to oligomeric interfaces, thus enabling the prediction of outer membrane PPIs (Hayat et al. 2011). Although methods such as these attempt to predict the membrane PPIs, efforts are still in the initial stages and demand further advances.

8.3.2 Genome-Based Computational Methods to Predict Membrane PPI

Genomic features have been used for the prediction of interacting protein pairs, including phylogenetic profiling, gene distance, gene fusions, similarity of phylogenetic trees of proteins and gene order; discussed below. In the simplest model, phylogenetic profiles can be represented as a binary vector where 1 or 0 correspond to presence or absence of gene orthologs in a reference genome, or as a vector of transformed e-value scores of sequence alignments between a gene and its putative orthologs. Here, the rationale is that if a pair of genes are needed to perform a particular, common function, then the orthologs of these two query proteins should be either present or absent simultaneously across species. Consequently, genes of an interacting pair tend to have similar phylogenetic profiles (Hu et al. 2009; Schneider et al. 2013; Babu et al. 2014), and various similarity metrics, such as the Pearson correlation (Beltrao et al. 2010; Ryan et al. 2013; Babu et al. 2014), can therefore be used as quantitative evidence for an interaction.

Several methods have used the conservation of the proximity of genes along the genome (gene distance) between distantly related species to predict interactions (Harrington et al. 2008; Koch et al. 2012). Bacterial genomes are especially suitable for the use of this approach, since functionally related genes often cluster within operons to allow co-transcription (Babu et al. 2014). Gene fusions can also be predictive of interactions, as several functions performed by two different genes in one organism are often performed by a single gene in another organism. Thus, in the reverse case, one may infer that two genes are functionally similar if they are fused in another organism, and hence more likely to interact. While this method can be an accurate predictor, fusion events, in spite of being very informative, are not very frequent.

Phylogenetic tree construction methods assume that interacting proteins tend to co-evolve due to selection imposed by the surrounding environment. The advantage of this method as a predictor of PPIs, as compared to phylogenetic profiles, is that phylogenetic tree approaches are able to more specifically predict physical interactions, while phylogenetic profiles look for both physical as well as functional interactions and cannot differentiate between them (Tillier and Charlebois 2009; Khan et al. 2014). By comparing several genomes, gene order can be used as a fingerprint of PPIs (Enright et al. 1999; Qin et al. 2014). While gene order shows relatively stronger correlation among less evolutionary distant species due to a lack of time for genome reshuffling after divergence of the two organisms from their most recent common ancestor, when it does occur between distant organisms it is unlikely to be due to chance, and can therefore be a high confidence predictor in such cases.

8.4 MP Networks in Unearthing Potential Drug Targets

One of the most important applications of the interaction network is to discover potential membrane drug targets. While imperative nodes of biological network can be potential drug targets, robust pathogen interaction networks withstand the removal of a single protein which leads to ineffectiveness for anticipated drugs. A more direct approach is to study the protein–drug interactome, which is analogous to the protein interactome, except that interactions are detected between a protein and drug (Hopkins 2008; Pujol et al. 2010). Protein–drug interactomes revealed that a drug can interact with multiple membrane or non-membrane proteins, and a protein can be targeted by multiple drugs (Yildirim et al. 2007; Gokhale et al. 2012). Keeping up with this notion, Nichols et al. combined large-scale chemogenomics with quantitative fitness measurements and provided insights into the genes encoding for certain envelope bioprocesses that are resistant or sensitive to multiple drugs (Fig. 8.2a) and synergism between the drugs in *E. coli* (Nichols et al. 2011). The study additionally provided quantitative drug-gene ontology (GO) phenotypes where membrane enriched GO processes were significantly targeted by drugs. Consistent with this, we also found many proteins that localized to the membranes were preferred drug targets, and more frequently ($p \leq 0.005$) showed conditional essentiality (i.e. the gene becomes essential for cell growth under a particular drug condition) when treated with one or more drugs in comparison to non-membrane proteins (Fig. 8.2b). Analogous to PPIs in the interactome, protein structures also provide a better understanding of physical interactions between drugs and its target in the protein–drug interactome. Such mechanistic information helps in designing the new drugs to disrupt specific PPI by synthesizing molecules which can mimic the interaction between a membrane target protein and its interacting partner.

8.5 Concluding Remarks

High-throughput proteomic methods enable simultaneous identification of hundreds of proteins. Using these approaches, the composition of bacterial proteomes can be rapidly determined. Moreover, changes in protein expression levels can be accessed both qualitatively and quantitatively by using isotope-labelled samples for high-resolution mass-spectrometry. While gel-based techniques are still in use, AP/MS has become the method of choice for membrane proteomics as it success-

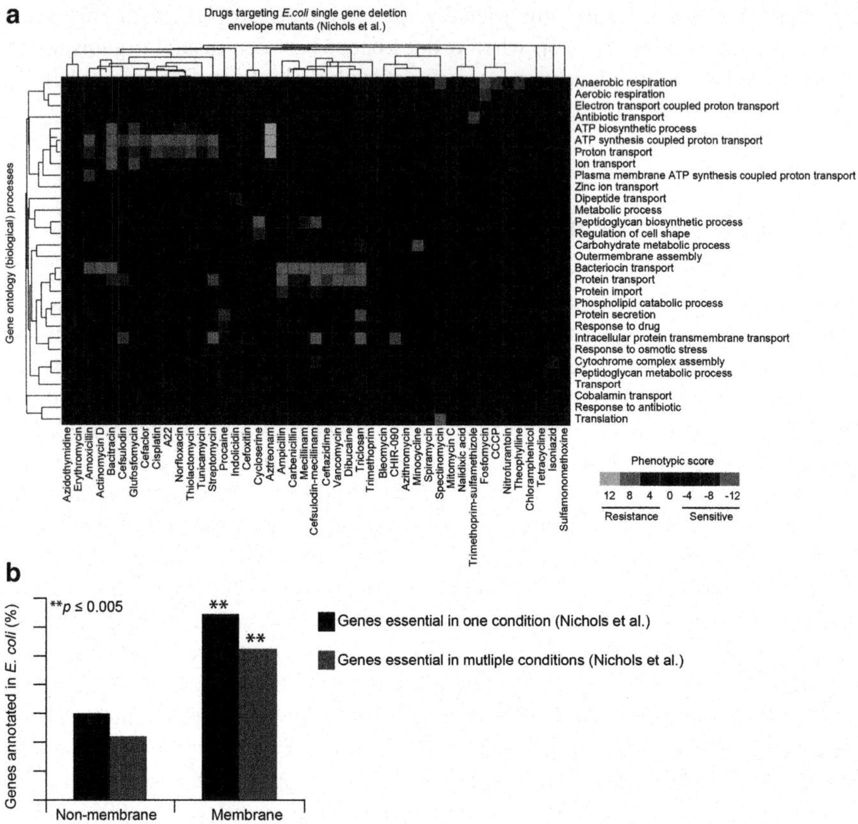

Fig. 8.2 *E. coli* membrane proteins and its association to drugs. (**a**) Heat map showing significantly enriched ($p \leq 0.05$) envelope bioprocesses (from Gene Ontology) of *E. coli* targeted by the indicated drugs (data extracted from Nichols et al. 2011). (**b**) Cell envelope genes of *E. coli* showing strong ($p \leq 0.005$) conditional essentiality when treated with one or more drugs

fully overcomes a number of traditional limitations such as low expression, high hydrophobicity, and poor solubility that make MPs a difficult target for previous proteomics methods. Inner and outer membrane preparations can be analysed as a whole (untargeted proteomics) or particular proteins can be selected.

Gene–gene interaction studies identify genes involved in bacterial cell wall biosynthesis and functioning, including those that are unannotated. Affinity purification or LC fractionation in conjunction with MS can elucidate not only the exact composition of the membrane sub-proteome, but also to reconstruct the entire interaction network. Computational approaches which integrate existing proteomics and genomics knowledge bases help to validate experimental data and extrapolate them to other organisms. For instance, many fundamental findings regarding *E. coli* envelope assembly, composition, and function can be translated to pathogenic bacteria.

Described methods in this review have provided much needed insights into the mechanisms of antibiotic resistance, quorum sensing, adhesion, and host colonization. Numerous studies of membrane proteomes are being performed in order to decipher resistance mechanisms of emerging and multidrug resistant pathogens. New drug targets can be identified by closely monitoring changes in MP expression under antimicrobial treatment. Moreover, so called proteomic signatures of differentially expressed proteins can be created for all major classes of antibiotics. Such information helps to understand the mechanism of action of existing antibiotics as well as rapidly characterize newly synthesized compounds. Membrane and surface proteomics is also employed in the development of vaccines against pathogenic bacteria. Overall, in our view, a combination of proteomics, genomics, and computational methods are essential to generate a plethora of complementary information that together allows reconstructing a systemic picture of structural and functional organization of the bacterial membrane and its interaction with the environment.

Acknowledgements This work was supported in part by funds from the Natural Sciences and Engineering Research Council of Canada (DG-20234) to MB. JV was supported by a Saskatchewan Postdoctoral Research Fellowship. MB holds a Canadian Institute of Health Research New Investigator award.

References

Ahrens CH, Brunner E, Qeli E, Basler K, Aebersold R (2010) Generating and navigating proteome maps using mass spectrometry. Nat Rev Mol Cell Biol 11:789–801

Albersmeier A, Bomholt C, Glaub A, Ruckert C, Soriano F, Fernandez-Natal I, Tauch A (2014) Draft genome sequence of the multidrug-resistant clinical isolate Dermabacter hominis 1368. Genome Announc 2

Babu M, Musso G, Diaz-Mejia JJ, Butland G, Greenblatt JF, Emili A (2009) Systems-level approaches for identifying and analyzing genetic interaction networks in Escherichia coli and extensions to other prokaryotes. Mol Biosyst 12:1439–1455

Babu M, Díaz-Mejía JJ, Vlasblom J, Gagarinova A, Phanse S, Graham C, Arnold R, Yousif F, Ding H, Xiong X, Nazarians-Armavil A, Alamgir M, Ali M, Pogoutse O, Pe'er A, Parkinson J, Golshani A, Whitfield C, Wodak SJ, Moreno-Hagelsieb G, Greenblatt JF, Emili A (2011) Genetic interaction maps in Escherichia coli reveal functional crosstalk among cell envelope biogenesis pathways. PLoS Genet 7, e1002377

Babu M, Vlasblom J, Pu S, Guo X, Graham C, Bean BDM, Burston HE, Vizeacoumar FJ, Snider J, Phanse S, Fong V, Tam YYC, Davey M, Hnatshak O, Bajaj N, Chandran S, Punna T, Christopolous C, Wong V, Yu A, Zhong G, Li J, Stagljar I, Conibear E, Wodak SJ, Emili A, Greenblatt JF (2012) Interaction landscape of membrane-protein complexes in Saccharomyces cerevisiae. Nature 489:585–589

Babu M, Arnold R, Bundalovic-Torma C, Gagarinova A, Wong KS, Kumar A, Stewart G, Samanfar B, Aoki H, Wagih O, Vlasblom J, Phanse S, Lad K, Yeou Hsiung Yu A, Graham C, Jin K, Brown E, Golshani A, Kim P, Moreno-Hagelsieb G, Greenblatt J, Houry WA, Parkinson J, Emili A (2014) Quantitative genome-wide genetic interaction screens reveal global epistatic relationships of protein complexes in Escherichia coli. PLoS Genet 10, e1004120

Baryshnikova A, Costanzo M, Myers CL, Andrews B, Boone C (2013) Genetic interaction networks: toward an understanding of heritability. Annu Rev Genomics Hum Genet 14:111–133
Baspinar A, Cukuroglu E, Nussinov R, Keskin O, Gursoy A (2014) PRISM: a web server and repository for prediction of protein–protein interactions and modeling their 3D complexes. Nucleic Acids Res 42:W285–W289
Becher D, Hempel K, Sievers S, Zuhlke D, Pane-Farre J, Otto A, Fuchs S, Albrecht D, Bernhardt J, Engelmann S, Volker U, van Dijl JM, Hecker M (2009) A proteomic view of an important human pathogen—towards the quantification of the entire Staphylococcus aureus proteome. PLoS One 4, e8176
Beltrao P, Cagney G, Krogan NJ (2010) Quantitative genetic interactions reveal biological modularity. Cell 141:739–745
Berman HM, Coimbatore Narayanan B, Di Costanzo L, Dutta S, Ghosh S, Hudson BP, Lawson CL, Peisach E, Prlic A, Rose PW, Shao C, Yang H, Young J, Zardecki C (2013) Trendspotting in the Protein Data Bank. FEBS Lett 587:1036–1045
Bernsel A, Daley DO (2009) Exploring the inner membrane proteome of Escherichia coli: which proteins are eluding detection and why? Trends Microbiol 17:444–449
Bracken C, Iakoucheva LM, Romero PR, Dunker AK (2004) Combining prediction, computation and experiment for the characterization of protein disorder. Curr Opin Struct Biol 14:570–576
Bragazzi NL, Pechkova E, Nicolini C (2014) Proteomics and proteogenomics approaches for oral diseases. Adv Protein Chem Struct Biol 95:125–162
Butland G, Babu M, Díaz-Mejía JJ, Bohdana F, Phanse S, Gold B, Yang W, Li J, Gagarinova AG, Pogoutse O, Mori H, Wanner BL, Lo H, Wasniewski J, Christopolous C, Ali M, Venn P, Safavi-Naini A, Sourour N, Caron S, Choi JY, Laigle L, Nazarians-Armavil A, Deshpande A, Joe S, Datsenko KA, Yamamoto N, Andrews BJ, Boone C, Ding H, Sheikh B, Moreno-Hagelseib G, Greenblatt JF, Emili A (2008) eSGA: E. coli synthetic genetic array analysis. Nat Methods 5:789–795
Cabral MP, Soares NC, Aranda J, Parreira JR, Rumbo C, Poza M, Valle J, Calamia V, Lasa I, Bou G (2011) Proteomic and functional analyses reveal a unique lifestyle for Acinetobacter baumannii biofilms and a key role for histidine metabolism. J Proteome Res 10:3399–3417
Caufield JH, Abreu M, Wimble C, Uetz P (2015) Protein complexes in bacteria. PLoS Comput Biol 11, e1004107
Curreem SO, Watt RM, Lau SK, Woo PC (2012) Two-dimensional gel electrophoresis in bacterial proteomics. Protein Cell 3:346–363
Daley DO, Rapp M, Granseth E, Melen K, Drew D, von Heijne G (2005) Global topology analysis of the Escherichia coli inner membrane proteome. Science 308:1321–1323
Davey NE, Haslam NJ, Shields DC, Edwards RJ (2011) SLiMSearch 2.0: biological context for short linear motifs in proteins. Nucleic Acids Res 39:W56–W60
Davies J, Davies D (2010) Origins and evolution of antibiotic resistance. Microbiol Mol Biol Rev 74:417–433
Díaz-Mejía JJ, Babu M, Emili A (2009) Computational and experimental approaches to chart the Escherichia coli cell-envelope-associated proteome and interactome. FEMS Microbiol Rev 33:66–97
Domon B, Aebersold R (2010) Options and considerations when selecting a quantitative proteomics strategy. Nat Biotechnol 28:710–721
Dresler J, Klimentova J, Stulik J (2011) Bacterial protein complexes investigation using blue native PAGE. Microbiol Res 166:47–62
Edwards RJ, Davey NE, Shields DC (2007) SLiMFinder: a probabilistic method for identifying over-represented, convergently evolved, short linear motifs in proteins. PLoS One 2, e967
Enright AJ, Iliopoulos I, Kyrpides NC, Ouzounis CA (1999) Protein interaction maps for complete genomes based on gene fusion events. Nature 402:86–90
Flores-Mireles AL, Walker JN, Caparon M, Hultgren SJ (2015) Urinary tract infections: epidemiology, mechanisms of infection and treatment options. Nat Rev Microbiol 13:269–284

Fruh V, Zhou Y, Chen D, Loch C, Ab E, Grinkova YN, Verheij H, Sligar SG, Bushweller JH, Siegal G (2010) Application of fragment-based drug discovery to membrane proteins: identification of ligands of the integral membrane enzyme DsbB. Chem Biol 17:881–891

Ghersi D, Sanchez R (2011) Beyond structural genomics: computational approaches for the identification of ligand binding sites in protein structures. J Struct Funct Genomics 12:109–117

Gokhale A, Perez-Cornejo P, Duran C, Hartzell HC, Faundez V (2012) A comprehensive strategy to identify stoichiometric membrane protein interactomes. Cell Logist 2:189–196

Gouw JW, Krijgsveld J, Heck AJ (2010) Quantitative proteomics by metabolic labeling of model organisms. Mol Cell Proteomics 9:11–24

Gstaiger M, Aebersold R (2009) Applying mass spectrometry-based proteomics to genetics, genomics and network biology. Nat Rev Genet 10:617–627

Harrington ED, Jensen LJ, Bork P (2008) Predicting biological networks from genomic data. FEBS Lett 582:1251–1258

Havugimana PC, Hart GT, Nepusz T, Yang H, Turinsky AL, Li Z, Wang PI, Boutz DR, Fong V, Phanse S, Babu M, Craig SA, Hu P, Wan C, Vlasblom J, Dar V-N, Bezginov A, Clark GW, Wu GC, Wodak SJ, Tillier ERM, Paccanaro A, Marcotte EM, Emili A (2012) A census of human soluble protein complexes. Cell 150:1068–1081

Hayat S, Walter P, Park Y, Helms V (2011) Prediction of the exposure status of transmembrane beta barrel residues from protein sequence. J Bioinform Comput Biol 9:43–65

Hopkins AL (2008) Network pharmacology: the next paradigm in drug discovery. Nat Chem Biol 4:682–690

Hu P, Janga SC, Babu M, Diaz-Mejia JJ, Butland G, Yang W, Pogoutse O, Guo X, Phanse S, Wong P, Chandran S, Christopoulos C, Nazarians-Armavil A, Nasseri NK, Musso G, Ali M, Nazemof N, Eroukova V, Golshani A, Paccanaro A, Greenblatt JF, Moreno-Hagelsieb G, Emili A (2009) Global functional atlas of Escherichia coli encompassing previously uncharacterized proteins. PLoS Biol 7, e96

Hughes C, Krijgsveld J (2012) Developments in quantitative mass spectrometry for the analysis of proteome dynamics. Trends Biotechnol 30:668–676

Hui S, Silverman JM, Chen SS, Erickson DW, Basan M, Wang J, Hwa T, Williamson JR (2015) Quantitative proteomic analysis reveals a simple strategy of global resource allocation in bacteria. Mol Syst Biol 11:784

Kall L, Vitek O (2011) Computational mass spectrometry-based proteomics. PLoS Comput Biol 7, e1002277

Kanj SS, Kanafani ZA (2011) Current concepts in antimicrobial therapy against resistant gram-negative organisms: extended-spectrum beta-lactamase-producing Enterobacteriaceae, carbapenem-resistant Enterobacteriaceae, and multidrug-resistant Pseudomonas aeruginosa. Mayo Clin Proc 86:250–259

Kawashima S, Pokarowski P, Pokarowska M, Kolinski A, Katayama T, Kanehisa M (2008) AAindex: amino acid index database, progress report 2008. Nucleic Acids Res 36:D202–D205

Kerppola TK (2006) Visualization of molecular interactions by fluorescence complementation. Nat Rev Mol Cell Biol 7:449–456

Khan I, Chen Y, Dong T, Hong X, Takeuchi R, Mori H, Kihara D (2014) Genome-scale identification and characterization of moonlighting proteins. Biol Direct 9:30

Koch EN, Costanzo M, Bellay J, Deshpande R, Chatfield-Reed K, Chua G, D'Urso G, Andrews BJ, Boone C, Myers CL (2012) Conserved rules govern genetic interaction degree across species. Genome Biol 13:R57

Kohanski MA, Dwyer DJ, Collins JJ (2010) How antibiotics kill bacteria: from targets to networks. Nat Rev Microbiol 8:423–435

Kotlyar M, Pastrello C, Pivetta F, Lo Sardo A, Cumbaa C, Li H, Naranian T, Niu Y, Ding Z, Vafaee F, Broackes-Carter F, Petschnigg J, Mills GB, Jurisicova A, Stagljar I, Maestro R, Jurisica I (2015) *In silico* prediction of physical protein interactions and characterization of interactome orphans. Nat Methods 12:79–84

Kuzmanov U, Emili A (2013) Protein–protein interaction networks: probing disease mechanisms using model systems. Genome Med 5:37

Lam MH, Stagljar I (2012) Strategies for membrane interaction proteomics: no mass spectrometry required. Proteomics 12:1519–1526

Lee D, Redfern O, Orengo C (2007) Predicting protein function from sequence and structure. Nat Rev Mol Cell Biol 8:995–1005

Lee SY, Yun SH, Lee YG, Choi CW, Leem SH, Park EC, Kim GH, Lee JC, Kim SI (2014) Proteogenomic characterization of antimicrobial resistance in extensively drug-resistant Acinetobacter baumannii DU202. J Antimicrob Chemother 69:1483–1491

Liebler DC, Zimmerman LJ (2013) Targeted quantitation of proteins by mass spectrometry. Biochemistry 52:3797–3806

Lima TB, Pinto MF, Ribeiro SM, de Lima LA, Viana JC, Gomes Junior N, Candido Ede S, Dias SC, Franco OL (2013) Bacterial resistance mechanism: what proteomics can elucidate. FASEB J 27:1291–1303

Mills CL, Beuning PJ, Ondrechen MJ (2015) Biochemical functional predictions for protein structures of unknown or uncertain function. Comput Struct Biotechnol J 13:182–191

Miteva YV, Budayeva HG, Cristea IM (2013) Proteomics-based methods for discovery, quantification, and validation of protein–protein interactions. Anal Chem 85:749–768

Mitra K, Carvunis AR, Ramesh SK, Ideker T (2013) Integrative approaches for finding modular structure in biological networks. Nat Rev Genet 14:719–732

Mosca R, Pache RA, Aloy P (2012) The role of structural disorder in the rewiring of protein interactions through evolution. Mol Cell Proteomics 11:M111.014969

Needham BD, Trent MS (2013) Fortifying the barrier: the impact of lipid A remodelling on bacterial pathogenesis. Nat Rev Microbiol 11:467–481

Ngounou Wetie AG, Sokolowska I, Woods AG, Roy U, Loo JA, Darie CC (2013) Investigation of stable and transient protein–protein interactions: past, present, and future. Proteomics 13:538–557

Nichols RJ, Sen S, Choo YJ, Beltrao P, Zietek M, Chaba R, Lee S, Kazmierczak KM, Lee KJ, Wong A, Shales M, Lovett S, Winkler ME, Krogan NJ, Typas A, Gross CA (2011) Phenotypic landscape of a bacterial cell. Cell 144:143–156

O'Brien KT, Haslam NJ, Shields DC (2013) SLiMScape: a protein short linear motif analysis plugin for Cytoscape. BMC Bioinformatics 14:224

Ozer EA, Fitzpatrick MA, Hauser AR (2014) Draft genome sequence of Acinetobacter baumannii strain ABBL099, a multidrug-resistant clinical outbreak isolate with a novel multilocus sequence type. Genome Announc 2

Palopoli N, Lythgow KT, Edwards RJ (2015) QSLiMFinder: improved short linear motif prediction using specific query protein data. Bioinformatics 31:2284–2293

Pan JY, Li H, Ma Y, Chen P, Zhao P, Wang SY, Peng XX (2010) Complexome of Escherichia coli envelope proteins under normal physiological conditions. J Proteome Res 9:3730–3740

Papanastasiou M, Orfanoudaki G, Koukaki M, Kountourakis N, Sardis MF, Aivaliotis M, Karamanou S, Economou A (2013) The Escherichia coli peripheral inner membrane proteome. Mol Cell Proteomics 12:599–610

Pawar H, Sahasrabuddhe NA, Renuse S, Keerthikumar S, Sharma J, Kumar GS, Venugopal A, Sekhar NR, Kelkar DS, Nemade H, Khobragade SN, Muthusamy B, Kandasamy K, Harsha HC, Chaerkady R, Patole MS, Pandey A (2012) A proteogenomic approach to map the proteome of an unsequenced pathogen – Leishmania donovani. Proteomics 12:832–844

Pelletier DA, Hurst GB, Foote LJ, Lankford PK, McKeown CK, Lu TY, Schmoyer DD, Shah MB, Hervey WJt, McDonald WH, Hooker BS, Cannon WR, Daly DS, Gilmore JM, Wiley HS, Auberry DL, Wang Y, Larimer FW, Kennel SJ, Doktycz MJ, Morrell-Falvey JL, Owens ET, Buchanan MV (2008) A general system for studying protein–protein interactions in Gram-negative bacteria. J Proteome Res 7:3319–3328

Perica T, Chothia C, Teichmann SA (2012) Evolution of oligomeric state through geometric coupling of protein interfaces. Proc Natl Acad Sci U S A 109:8127–8132

Petriz BA, Franco OL (2014) Application of cutting-edge proteomics technologies for elucidating host-bacteria interactions. Adv Protein Chem Struct Biol 95:1–24

Pogozheva ID, Tristram-Nagle S, Mosberg HI, Lomize AL (2013) Structural adaptations of proteins to different biological membranes. Biochim Biophys Acta 1828:2592–2608

Pujol A, Mosca R, Farres J, Aloy P (2010) Unveiling the role of network and systems biology in drug discovery. Trends Pharmacol Sci 31:115–123

Pyndiah S, Lasserre JP, Menard A, Claverol S, Prouzet-Mauleon V, Megraud F, Zerbib F, Bonneu M (2007) Two-dimensional blue native/SDS gel electrophoresis of multiprotein complexes from Helicobacter pylori. Mol Cell Proteomics 6:193–206

Qin T, Matmati N, Tsoi LC, Mohanty BK, Gao N, Tang J, Lawson AB, Hannun YA, Zheng WJ (2014) Finding pathway-modulating genes from a novel Ontology Fingerprint-derived gene network. Nucleic Acids Res 42, e138

Rabilloud T (2009) Membrane proteins and proteomics: love is possible, but so difficult. Electrophoresis 30(Suppl 1):S174–S180

Rabilloud T, Lelong C (2011) Two-dimensional gel electrophoresis in proteomics: a tutorial. J Proteomics 74:1829–1841

Radosevich TJ, Reinhardt TA, Lippolis JD, Bannantine JP, Stabel JR (2007) Proteome and differential expression analysis of membrane and cytosolic proteins from Mycobacterium avium subsp. paratuberculosis strains K-10 and 187. J Bacteriol 189:1109–1117

Rajagopala SV, Sikorski P, Kumar A, Mosca R, Vlasblom J, Arnold R, Franca-Koh J, Pakala SB, Phanse S, Ceol A, Hauser R, Siszler G, Wuchty S, Emili A, Babu M, Aloy P, Pieper R, Uetz P (2014) The binary protein–protein interaction landscape of Escherichia coli. Nat Biotechnol 32:285–290

Reddy TB, Thomas AD, Stamatis D, Bertsch J, Isbandi M, Jansson J, Mallajosyula J, Pagani I, Lobos EA, Kyrpides NC (2015) The Genomes OnLine Database (GOLD) v.5: a metadata management system based on a four level (meta)genome project classification. Nucleic Acids Res 43:D1099–D1106

Ribet D, Cossart P (2015) How bacterial pathogens colonize their hosts and invade deeper tissues. Microbes Infect/Institut Pasteur 17:173–183

Riedel T, Bunk B, Thurmer A, Sproer C, Brzuszkiewicz E, Abt B, Gronow S, Liesegang H, Daniel R, Overmann J (2015) Genome resequencing of the virulent and multidrug-resistant reference strain Clostridium difficile 630. Genome Announc 3

Ryan CJ, Cimermancic P, Szpiech ZA, Sali A, Hernandez RD, Krogan NJ (2013) High-resolution network biology: connecting sequence with function. Nat Rev Genet 14:865–879

Salzano AM, Novi G, Arioli S, Corona S, Mora D, Scaloni A (2013) Mono-dimensional blue native-PAGE and bi-dimensional blue native/urea-PAGE or/SDS-PAGE combined with nLC-ESI-LIT-MS/MS unveil membrane protein heteromeric and homomeric complexes in Streptococcus thermophilus. J Proteomics 94:240–261

Sauer S, Kliem M (2010) Mass spectrometry tools for the classification and identification of bacteria. Nat Rev Microbiol 8:74–82

Schlegel S, Klepsch M, Wickstrom D, Wagner S, de Gier JW (2010) Comparative analysis of cytoplasmic membrane proteomes of Escherichia coli using 2D blue native/SDS-PAGE. Methods Mol Biol 619:257–269

Schneider A, Seidl MF, Snel B (2013) Shared protein complex subunits contribute to explaining disrupted co-occurrence. PLoS Comput Biol 9, e1003124

Shui W, Gilmore SA, Sheu L, Liu J, Keasling JD, Bertozzi CR (2009) Quantitative proteomic profiling of host–pathogen interactions: the macrophage response to Mycobacterium tuberculosis lipids. J Proteome Res 8:282–289

Silhavy TJ, Kahne D, Walker S (2010) The bacterial cell envelope. Cold Spring Harb Perspect Biol 2:a000414

Stenberg F, Chovanec P, Maslen SL, Robinson CV, Ilag LL, von Heijne G, Daley DO (2005) Protein complexes of the Escherichia coli cell envelope. J Biol Chem 280:34409–34419

Tan SY, Chua SL, Chen Y, Rice SA, Kjelleberg S, Nielsen TE, Yang L, Givskov M (2013) Identification of five structurally unrelated quorum-sensing inhibitors of Pseudomonas aeruginosa from a natural-derivative database. Antimicrob Agents Chemother 57:5629–5641

Taniguchi Y, Choi PJ, Li GW, Chen H, Babu M, Hearn J, Emili A, Xie XS (2010) Quantifying E. coli proteome and transcriptome with single-molecule sensitivity in single cells. Science 329:533–538
Tillier ER, Charlebois RL (2009) The human protein coevolution network. Genome Res 19:1861–1871
Tuncbag N, Gursoy A, Nussinov R, Keskin O (2011) Predicting protein–protein interactions on a proteome scale by matching evolutionary and structural similarities at interfaces using PRISM. Nat Protoc 6:1341–1354
van der Lee R, Buljan M, Lang B, Weatheritt RJ, Daughdrill GW, Dunker AK, Fuxreiter M, Gough J, Gsponer J, Jones DT, Kim PM, Kriwacki RW, Oldfield CJ, Pappu RV, Tompa P, Uversky VN, Wright PE, Babu MM (2014) Classification of intrinsically disordered regions and proteins. Chem Rev 114:6589–6631
Vignaroli C, Luna GM, Rinaldi C, Di Cesare A, Danovaro R, Biavasco F (2012) New sequence types and multidrug resistance among pathogenic Escherichia coli isolates from coastal marine sediments. Appl Environ Microbiol 78:3916–3922
Vlasblom J, Zuberi K, Rodriguez H, Arnold R, Gagarinova A, Deineko V, Kumar A, Leung E, Rizzolo K, Samanfar B, Chang L, Phanse S, Golshani A, Greenblatt JF, Houry WA, Emili A, Morris Q, Bader G, Babu M (2015) Novel function discovery with GeneMANIA: a new integrated resource for gene function prediction in Escherichia coli. Bioinformatics 31:306–310
Wenzel M, Bandow JE (2011) Proteomic signatures in antibiotic research. Proteomics 11:3256–3268
Wong WR, Oliver AG, Linington RG (2012) Development of antibiotic activity profile screening for the classification and discovery of natural product antibiotics. Chem Biol 19:1483–1495
Wright PE, Dyson HJ (2015) Intrinsically disordered proteins in cellular signalling and regulation. Nat Rev Mol Cell Biol 16:18–29
Wuchty S, Uetz P (2014) Protein–protein interaction networks of E. coli and S. cerevisiae are similar. Sci Rep 4:7187
Ye X, Luke B, Andresson T, Blonder J (2009) 18O stable isotope labeling in MS-based proteomics. Brief Funct Genomic Proteomic 8:136–144
Yildirim MA, Goh KI, Cusick ME, Barabasi AL, Vidal M (2007) Drug-target network. Nat Biotechnol 25:1119–1126
Yu H, Braun P, Yildirim MA, Lemmens I, Venkatesan K, Sahalie J, Hirozane-Kishikawa T, Gebreab F, Li N, Simonis N, Hao T, Rual JF, Dricot A, Vazquez A, Murray RR, Simon C, Tardivo L, Tam S, Svrzikapa N, Fan C, de Smet AS, Motyl A, Hudson ME, Park J, Xin X, Cusick ME, Moore T, Boone C, Snyder M, Roth FP, Barabasi AL, Tavernier J, Hill DE, Vidal M (2008) High-quality binary protein interaction map of the yeast interactome network. Science 322:104–110
Yun SH, Choi CW, Kwon SO, Park GW, Cho K, Kwon KH, Kim JY, Yoo JS, Lee JC, Choi JS, Kim S, Kim SI (2011) Quantitative proteomic analysis of cell wall and plasma membrane fractions from multidrug-resistant Acinetobacter baumannii. J Proteome Res 10:459–469
Zahedi RP, Moebius J, Sickmann A (2007) Two-dimensional BAC/SDS-PAGE for membrane proteomics. Subcell Biochem 43:13–20
Zhang QC, Petrey D, Garzon JI, Deng L, Honig B (2013) PrePPI: a structure-informed database of protein–protein interactions. Nucleic Acids Res 41:D828–D833
Zheng J, Wei C, Zhao L, Liu L, Leng W, Li W, Jin Q (2011) Combining blue native polyacrylamide gel electrophoresis with liquid chromatography tandem mass spectrometry as an effective strategy for analyzing potential membrane protein complexes of Mycobacterium bovis bacillus Calmette-Guerin. BMC Genomics 12:40

Chapter 9
Toward Network Biology in *E. coli* Cell

Hirotada Mori, Rikiya Takeuchi, Yuta Otsuka, Steven Bowden, Katsushi Yokoyama, Ai Muto, Igor Libourel, and Barry L. Wanner

Abstract *E. coli* has been a critically important model research organism for more than 50 years, particularly in molecular biology. In 1997, the *E. coli* draft genome sequence was published. Post-genomic techniques and resources were then developed that allowed *E. coli* to become a model organism for systems biology. Progress made since publication of the *E. coli* genome sequence will be summarized.

Keywords *E. coli* • Network biology • Plasmid clone libraries

9.1 Introduction

The discovery of conjugation in *Escherichia coli* by Lederberg and Tatum in 1946 (Lederberg and Tatum 1946) led to its rapid adoption as a model organism. During the latter half of the twentieth century, studies using *E. coli* were critical in the development of gene theory, and *E. coli* became one of the most comprehensively studied organisms, especially in the field of molecular biology.

The last decade of the twentieth century and the first decade of the twenty-first century saw rapid advances in genome sequencing technologies, revolutionizing biological research. Exploitation of rapid sequencing technologies allowed *E. coli* to become one of the leading research organisms for systems biology.

H. Mori (✉) • R. Takeuchi • Y. Otsuka • S. Bowden • K. Yokoyama • A. Muto
Graduate School of Biological Sciences, Nara Institute of Science and Technology, 8916-5 Takayama, Ikoma, Nara 630-0101, Japan
e-mail: hmori@gtc.naist.jp; takeriki0502@gmail.com; y-otsuka@bs.naist.jp; sbowden@umn.edu; k.yokoyama@hamayaku.ac.jp; muto@bs.naist.jp

I. Libourel
The Biotechnology Institute, University of Minnesota, 1479 Gortner Avenue, St. Paul, MN 55108-6106, USA
e-mail: libourel@umn.eud

B.L. Wanner
Department of Biological Sciences, Purdue University, West Lafayette, IN 47907, USA
e-mail: blwanner@genetics.med.harvard.edu

N.J. Krogan, M. Babu (eds.), *Prokaryotic Systems Biology*, Advances in Experimental Medicine and Biology 883, DOI 10.1007/978-3-319-23603-2_9

The first *E. coli* genome was determined in 1997 by researchers in the USA and Japan. Around the same time, systematic approaches in so-called omics, such as transcriptome analysis, were developed by Schena et al. (1995). In 1999, *E. coli* was used for the first microarray analysis of a bacterium, which was performed using PCR fragments amplified from genomic DNA with primer sets encompassing all predicted open reading frames (ORFs) (Richmond et al. 1999; Tao et al. 1999).

During the late 1990s, a plasmid clone library (Hudson et al. 1997) and systematic deletion strain library (Winzeler et al. 1999) were established as the first comprehensive genomic resources for *Saccharomyces cerevisiae.* These systematically constructed biological resources were useful for individual molecular biology research projects and were invaluable for omics approaches in the field of systems biology. The new wave of systems biology analysis expanded dramatically in the first decade of the twenty-first century. *E. coli* research also exploited the technological and conceptual developments in biology, and comprehensive experimental resources were constructed that has a big advantage to use *E. coli* as a prominent model research organism for systems biology.

9.2 Resource Construction

After the publication of the first draft of the *E. coli* genome in 1997, we continued clarification of ambiguous sequences (Hayashi et al. 2006) and improvement of genome annotation (Riley et al. 2006). We then used the annotated genome sequence to construct comprehensive plasmid clone libraries (Kitagawa et al. 2005), random Tn insertion mutant libraries (Miki et al. 2008), and targeted knockout libraries (Baba et al. 2006). These resources support systematic omics research.

9.2.1 Plasmid Clone Libraries

9.2.1.1 ASKA Plasmid ORF GFP Fusion Clone Library

All of the annotated *E. coli* ORFs were amplified and cloned into multi-copy plasmid vector pCA24N. Plasmids were designed to express a 6× His-tag and an eGFP fusion protein at the N- and C-termini of the expressed protein, respectively. *Sfi*I cloning sites with different cohesive ends were generated at the ends of the inserted ORF sequences, which allowed subsequent unidirectional cloning into other vectors such as the Gateway entry vector described below (Kitagawa et al. 2005). This cloning strategy is summarized in Fig. 9.1a.

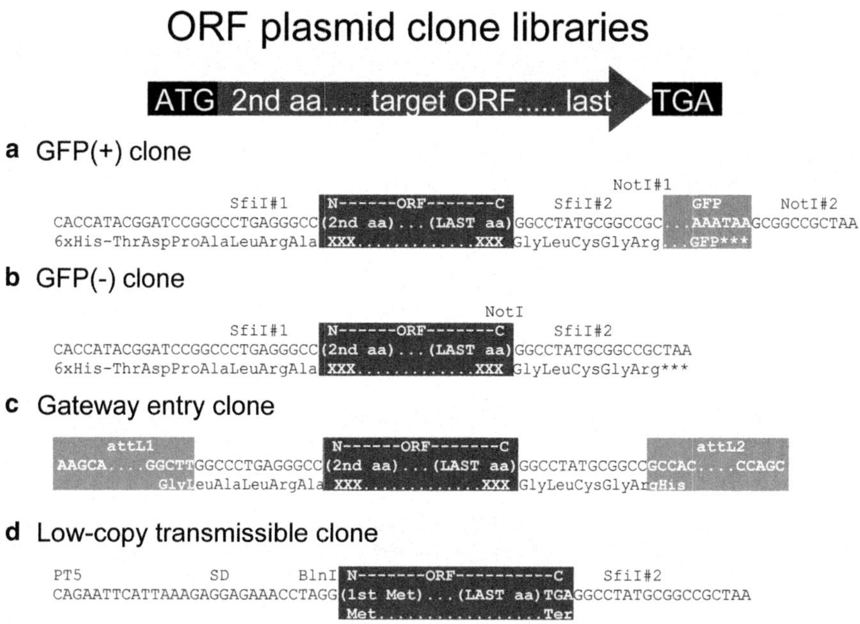

Fig. 9.1 Plasmid clone library containing predicted open reading frames (ORFs). (**a**) ASKA plasmid clones. Whole predicted ORFs were amplified with the exception of the initiation codon. Amplified fragments were purified by EtOH precipitation and cloned into the *Stu*I site of plasmid pCA24N to generate flanking *Sfi*I sites. After the GFP fusion library was established, a non-GFP fusion library was created by *Not*I digestion and religation to remove the GFP sequence. Further details of library construction and validation are described in Kitagawa et al. (2005). (**b**) Gateway entry clones. ORFs were recovered from the ASKA library by *Sfi*I digestion and then cloned into the pAZ677 vector to construct an entry clone library (Rajagopala et al. 2010)

9.2.1.2 Gateway Entry Clone Library

To increase the flexibility of the clone library, *Sfi*I fragments from ASKA-based plasmids were transferred into our modified Gateway entry vector. Gateway cloning technology uses the site-specific recombination capacity of phage lambda to exchange DNA fragments between vectors available as the Gateway Cloning Technology. The fragment of interest can be inserted into an entry vector, which allows the gene of interest to be subsequently shuttled into different destination vectors with different expression parameters. This technology was used on a genomic scale with a yeast-two-hybrid system destination vector for the comprehensive analysis of protein–protein interactions in *E. coli* (Rajagopala et al. 2014).

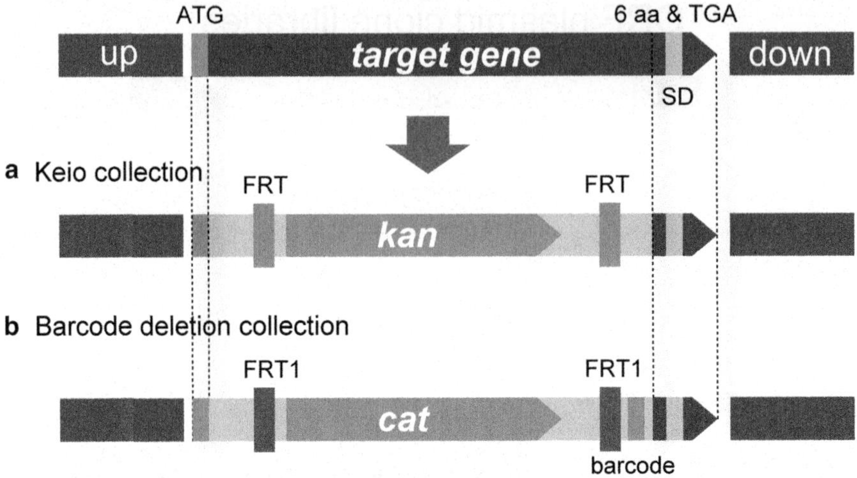

Fig. 9.2 Construction of single deletion *E. coli* strains. (**a**) Chromosomal structure of the initial single gene knockout library (Keio collection, Baba et al. 2006). Lambda RED recombinase was used to replace the region encoding amino acids from the 2nd through the 7th from the termination codon with a kanamycin resistance cassette. (**b**) The same method was used to construct the second library, with the exception that a chloramphenicol resistance cassette was used instead of the kanamycin resistance gene

9.2.2 Keio Collection

Until 2000, *E. coli* was thought to be one of the more difficult organisms to genetically modify by homologous recombination due to its high exonuclease activity on transformed linear DNA. Datsenko and Wanner (2000) used lambda RED recombinase to combat this limitation, and *E. coli* can now be genetically manipulated with levels of ease similar to those of *Saccharomyces cerevisiae* and *Bacillus subtilis*. We used this system to construct *E. coli* deletion strains in which the regions from 2nd to the 7th amino acid coding region from the C-terminus. This maintained translational signals for downstream ORFs. Successful deletion strains were constructed for all annotated ORFs, except essential ones, by replacement of the target region with the kanamycin resistance gene (Baba et al. 2006). An overview of the deletion strategy is shown in Fig. 9.2.

9.3 Application of Comprehensive Resources in Omics Analyses

The availability of comprehensive experimental resources for *E. coli* has accelerated the systematic omics approaches in transcriptomics, proteomics, metabolomics, physiomics, etc. Initially, we constructed full-length cDNA-type microarrays using

PCR fragments amplified by common primer set of vector region and plasmid clones as a template and performed transcriptome analyses (Oshima et al. 2002a, b).

Systematic protein–protein interaction analyses were also applications of resource libraries and were performed using pull-down assay (Arifuzzaman et al. 2006) and chromosomal TAP-tagged strains (Butland et al. 2005). To make our plasmid clone libraries more flexible, we developed an *E. coli* Gateway entry clone library (Rajagopala et al. 2010) and this was used for yeast-two-hybrid analysis of protein–protein interactions (Rajagopala et al. 2014). Chen et al. (2008) used our plasmid ORF clone library for a different approach. In this 2008 study, proteins expressed from the plasmid library were affinity-purified and immobilized onto glass slides, and the glass slides were then used to assess protein–protein interactions (Chen et al. 2008), and the protein chip was expanding both basic science and applied clinical direction (Chen et al. 2009; Sutandy et al. 2013).

His-tagged ORF clone library also provided us tools to easily analyze the *E. coli* enzymes, whose physiological function were not cleared (Gonzalez et al. 2006; Kuznetsova et al. 2006; Proudfoot et al. 2004) and to hunt orphan gene (Melnick et al. 2004).

For metabolomics analyses, the development of the systematic identification and quantification method using capillary electrophoresis and mass-spectrometer has accelerated the metabolomics fields (Soga et al. 2002a, b) and quantitative measurements of mRNAs, proteins and metabolites levels of central metabolic enzyme genes deletion strains had been one of the leading achievements in the field of metabolomics in early stage (Ishii et al. 2007).

Technology innovation in the last two decades since the initiation of the genome project has been so quick and the modern biology has now been expanding to multidisciplinary fields. Sequencing technology is one of the typical examples. Direct sequencing is now replacing microarray technologies for genomic and transcriptomic analyses and rapidly decreasing costs. Sequencing also allows precise binding sites of DNA-binding proteins, determination of DNA modifications and now chromosomal conformation can be analyzed by sequencing (Umbarger et al. 2011).

As mentioned above, still the comprehensively constructed resources are clearly valuable tools not only for the systematic approaches but also individual targeted researches. We are keeping our efforts of quality control of resources already constructed and of development of improved or new resources. Such information will be opened through our database, http://ecoli.naist.jp/.

9.4 Network Biology

Network biology, which examines relationships between cellular components and their activities, is becoming increasingly important. Gene regulatory network analysis using high-density membranes or microarrays was one of the first types of network biology to be developed after the onset of genome sequencing (Schena et al. 1995; Richmond et al. 1999; Tao et al. 1999). As mentioned previously, we constructed microarrays from our plasmid clone library using full-length PCR-amplified coding DNAs. We used transcription factor knockout strains with our arrays to examine *E. coli* regulatory networks (Oshima et al. 2002a, b; Eguchi et al. 2003; Biville et al. 2003; Yamagishi et al. 2002; Nakahigashi et al. 2002). Currently, vast amounts of raw microarray data are stored in publicly available databases such as the Gene Expression Omnibus (GenBank), and these data can be downloaded and used for bioinformatic network analysis. As with most omics technologies, transcriptional regulatory networks and analytical techniques are developing at a rapid pace.

Epistasis, which means genetic interaction in a broad sense and how genes affect one another, has been studied for at least a century. The first definition of "epistasis" was proposed by Bateson (1909) and was based on phenotypic observations of dihybrid crosses in which some phenotypes appeared to affect other mutations (Phillips 2008). In 1919, Fisher defined epistasis quantitatively as any statistical deviation from the additive combination of two genetic loci on their phenotypic effects. The definition of epistasis remains controversial, in part because of the expanding interest in epistatic effects in a number of fields, including classical genetics, population genetics, evolutionary genetics, molecular biology, and systems biology, among others (Mani et al. 2008).

Although widely studied, *E. coli* was not used as a study organism for classical genetics until the discovery of conjugation by Lederberg in 1946 (Lederberg and Tatum 1946). To our knowledge, the first report describing epistasis in *E. coli* was conducted by Murinus and Morris in 1974 and concerned synthetic lethality of double mutations, demonstrated the functional connection between Dam methylase and DNA recombination (RecBC) or DNA polymerase (PolA) (Marinus and Morris 1974). Many subsequent synthetic lethal and suppression studies have clarified the cellular networks in *E. coli*.

In general, epistasis analysis provides functionally rich data and can sometimes reveal surprising interactions that would not otherwise be identified through prediction. The availability of genomic data raises the possibility of systematic reverse genetic analysis. In 2001, Tong et al. developed a method for the systematic construction of double mutants in *Saccharomyces cerevisiae* (Tong et al. 2001). In this method, conjugation between single gene deletion strains of opposite mating type was used to generate a double-knockout strain. This approach was extended to other unicellular organisms such as *Schizosaccharomyces pombe* (Roguev et al. 2007) and *E. coli* (Butland et al. 2008; Typas et al. 2008).

9.5 Development of Tools and Resources

For the high-throughput construction of double-knockout strains for genetic interaction analysis, two types of deletion collections carrying different antibiotic resistance and conversion tool for host cell sexuality are required. Mating type determinants in *E. coli* involve plasmid-based conjugation machinery evolutionarily derived from the secretion system (Alvarez-Martinez and Christie 2009). And we used *oriT* and *tra* genes operon of *incF* group conjugative F plasmid for this purpose.

9.5.1 Development of the Tool Converting Host Cells to Hfr

For comprehensive genetic interaction analysis, the pseudo-Hfr construction system (Francois et al. 1990), and Hfr Cavalli (Bachmann 1972), were initially used for genetic interaction studies (Butland et al. 2008; Typas et al. 2008). To make more flexible Hfr strains, however, we developed the new tool for conversion of the host cell sexuality using F plasmid (kind gift from Dr. Sampei, G) and named CIP. Our plasmid system involves a chimeric plasmid consisting of the *tra* operon, transfer origin *oriT*, an artificial replication origin fragment containing an antibiotic resistance gene, and a 300 bp chromosomal fragment as a target site for recombination. Ten fragments from different chromosomal sites in both *E. coli* chromosomal orientations were established as target sites for the creation of Hfrs. All are validated for activities of conjugation and transmission, and for efficiency of double-deletion strain construction after mating. Further details of this plasmid and library are forthcoming (Takeuchi et al. and Otsuka et al., in preparation). Schematic view how to construct Hfr using CIPs are illustrated in Fig. 9.3.

9.5.2 Construction of the Second Deletion Library

Two distinct deletion libraries with different selective markers are required for construction of double knockouts. In 2006, we established a single gene knockout strain library (Keio collection; Baba et al. 2006). As shown in Fig. 9.2, the deletion of Keio collection was designed deletions from the 2nd through the 7th codon from the C-terminus, leaving the ORF start codon and translational signal for the downstream gene intact. For multiple deletions, and to eliminate a polar effect on downstream gene expression, two site-specific recombination FRT sites were introduced for the generation of in-frame deletions after FLP-FRT recombination (Baba et al. 2006).

Construction and validation of a second deletion collection have been completed and the manuscript is now in preparation (Otsuka et al.). The new strain library and the host cell conversion tools will be available shortly. The new strain library carries chloramphenicol resistance gene to allow double selection to select double genes knockout strain after conjugation with Keio collection carrying kanamycin resistance. The method how to generate a double genes deletion strain is illustrated in Fig. 9.4.

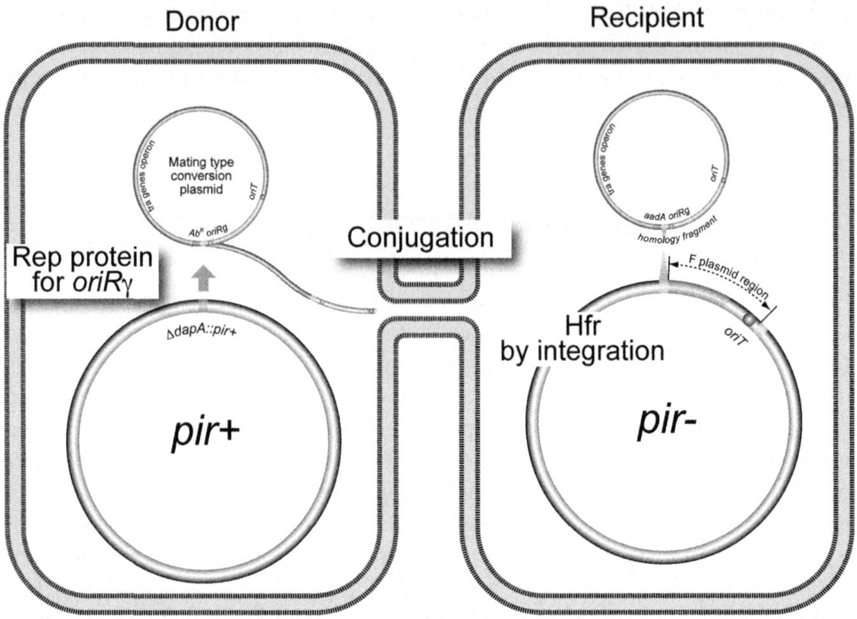

Fig. 9.3 Tool for switching *E. coli* cell type. Sequences responsible for conjugation and DNA transfer were joined to a fragment carrying *oriR*γ, an antibiotic resistance gene, and a chromosomal fragment responsible for recombination targeting. The REP protein is supplied by the chromosomal *pir* gene. The host strain carrying this plasmid works as a donor cell. After mating with an F-derived plasmid-free cell, the plasmid can be integrated into the homologous region on the recipient chromosome because the plasmid cannot replicate without the *pir* gene. In this case, the integrant cell may function as a high frequency recombination (Hfr) strain

9.5.3 High-Throughput Growth Monitoring

Systematic analysis requires high-throughput assessment methods, however, accuracy often suffers and assay design can be challenging. A variety of systematic biases must be considered when developing high-throughput methods, including position effects, competition with neighboring colonies, plate-specific effects, and experimental batch effects (Levin-Reisman et al. 2010). To address these issues, we applied normalization to ccd-captured fixed-time point images of colonies on agar plates. We also developed a system, named Colony-live, for quantitative measurement of colony growth over time using commercially available transmitting light scanners (Takeuchi et al. 2014). This system allowed three separate quantitative parameters to be captured for further analysis, namely, lag time, growth rate, and growth saturation. The Scanning system, real captured plate image, and time-series colony images from the plate image are shown in Fig. 9.5.

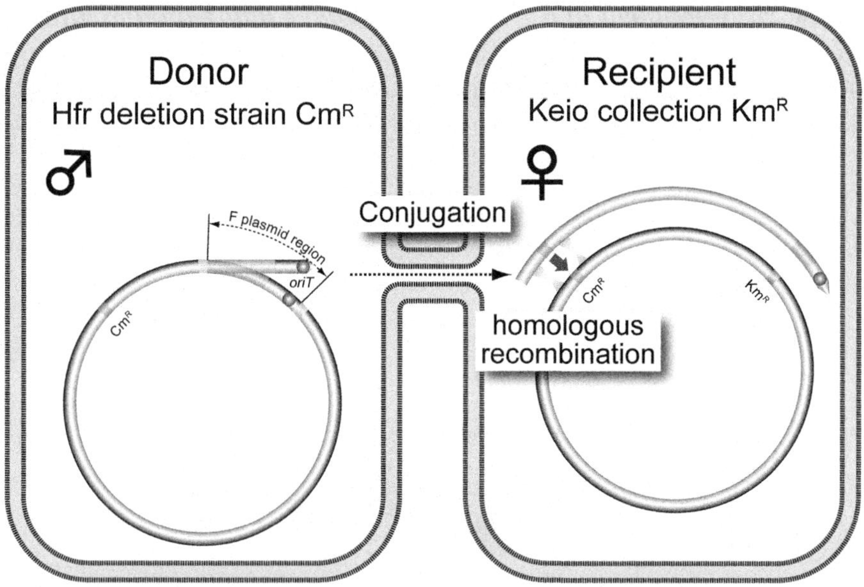

Fig. 9.4 Construction of double-knockout strains by conjugation. Conversion of the deletion strains in the second library (Fig. 9.2) to high frequency recombination (Hfr) using the tool shown in Fig. 9.3 will allow the resultant deletion strains to conjugate with the single gene deletion strains from the Keio collection. Double-knockout strains can be isolated using dual antibiotic selection

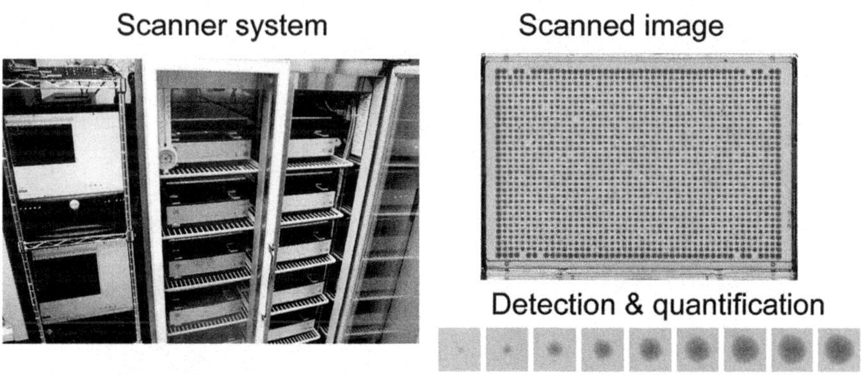

Fig. 9.5 High-throughput system. A high-throughput time-series growth monitoring system was constructed using commercially available scanners and computers. A 1536 colony high-density image and colony quantification are shown

9.5.4 Development of a High-Throughput Conjugation Method

E. coli has approximately 4000 ORFs, which means that a comprehensive double-knockout collection in both directions would comprise 16 million strains. With the exception of a study that used phage transduction (Nakahigashi et al. 2009), all systematic genetic interaction analyses in unicellular organisms to date have used the highly efficient conjugation method. We used a commercially available robotic stamping system (Singer Rotor, UK) to facilitate library development. A schematic of the experimental procedure is shown in Fig. 9.6.

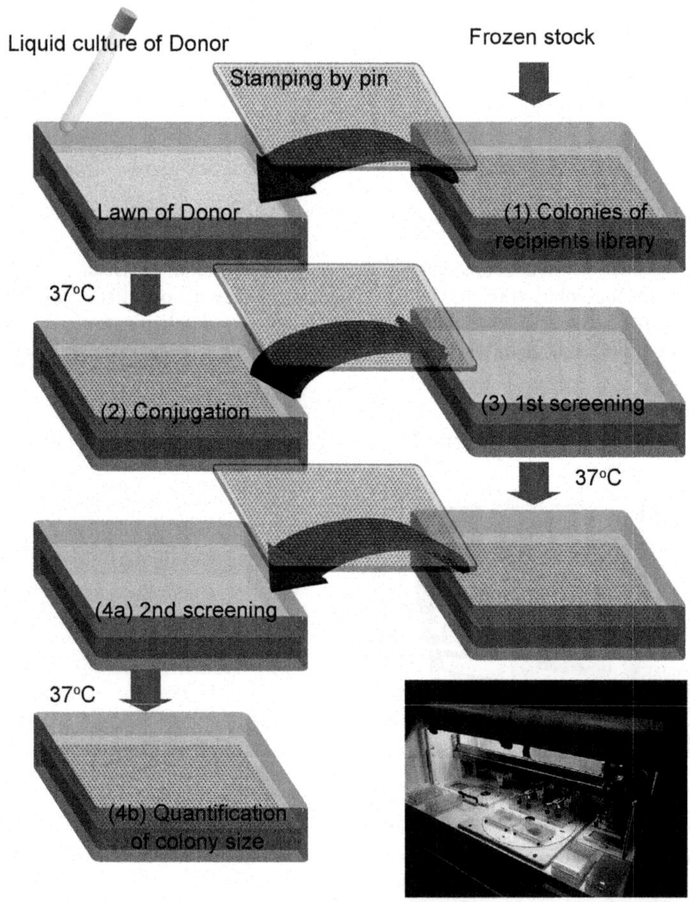

Fig. 9.6 High-throughput stamping system for conjugation. A stamping robotic system developed by Singer Inc. RoTor system was used for automation. Four stamping steps were required: (*1*) seed colony plate from frozen glycerol stock, (*2*) conjugation, (*3*) first screen to kill donor, and (*4a*) and (*4b*) second double antibiotic screen and measurement by scanning

9.6 Future Perspectives

Whilst the high-throughput study of genetic interactions and synthetic lethality phenotypes remains laborious, several emerging technologies may facilitate this research in the near future. For example, advances in microfluidics and emulsion PCR have already enabled single-cell and single molecule PCR reactions (Zhu et al. 2012). Indeed, methods such as Linking Emulsion PCR and Emulsion Haplotype Fusion PCR (Wetmur et al. 2005) have demonstrated that it is possible to directly sequence multiple mutations present within an individual chromosome. Some of these technologies may be utilized to enable multiplex analysis of complex libraries of cells containing many mutations. Microfluidics is an extremely high-throughput technology and holds great promise for advancing synthetic lethality and genetic interaction research.

In addition to facilitating higher sample throughput through miniaturization and automation, disruptive progress may come from advances in readout technology. In microfluidics, for instance, fitness measurements may transition from visualization of single colony growth-rate observations to OD measurements. More profoundly, a transition to parallel sequencing technologies to establish mutant fitness enables multiplexing of fitness measurements. Bar-seq (Smith et al. 2009) was the first such technology. More than a decade ago, genome-wide single gene mutant yeast libraries were created in which full-length genes were replaced with a barcode. Individual mutant fitness estimates were performed by observing changes in barcode prevalence by deep sequencing after brief growth of the pooled mutants. Related transposon-based methods such as Tn-seq have made this approach more accessible to non-model species (van Opijnen and Camilli 2013; van Opijnen et al. 2009). The implementation of sequencing-based fitness measurements for double mutants, which is needed for gene-interaction studies, is hampered by difficulties in tracking two mutations at the same time, however; hence, a sequence associated with a given deleted gene could come from any double mutant in which that gene had been removed. Future technology will need to address this problem by providing a single readout that encodes both mutations, such as a specific barcode for each double mutant.

Acknowledgments I thank to Dr. Gen-ichi Sampei for kindly giving us the F plasmid. This work was supported by the Create Revolutionary Technological Seeds for Science and Technology (CREST) from JST (Japan Science and Technology Agency), Grant-in-Aid for Scientific Research (A), Innovation Areas "Biosynthetic machinery" from the Ministry of Education, Culture, Sports, Science and Technology of Japan, and the Exploratory Research for Advanced Technology (ERATO).

References

Alvarez-Martinez CE, Christie PJ (2009) Biological diversity of prokaryotic type IV secretion systems. Microbiol Mol Biol Rev 73(4):775–808

Arifuzzaman M et al (2006) Large-scale identification of protein–protein interaction of Escherichia coli K-12. Genome Res 16(5):686–691

Baba T et al (2006) Construction of Escherichia coli K-12 in-frame, single-gene knockout mutants: the Keio collection. Mol Syst Biol 2:2006.0008

Bachmann BJ (1972) Pedigrees of some mutant strains of Escherichia coli K-12. Bacteriol Rev 36(4):525–557

Bateson W (1909) Mendel's principles of heredity. Cambridge Univ. Press, Cambridge

Biville F et al (2003) Escherichia coli response to exogenous pyrophosphate and analogs. J Mol Microbiol Biotechnol 5(1):37–45

Butland G et al (2005) Interaction network containing conserved and essential protein complexes in Escherichia coli. Nature 433(7025):531–537

Butland G et al (2008) eSGA: E. coli synthetic genetic array analysis. Nat Methods 5(9):789–795

Chen CS et al (2008) A proteome chip approach reveals new DNA damage recognition activities in Escherichia coli. Nat Methods 5(1):69–74

Chen CS et al (2009) Identification of novel serological biomarkers for inflammatory bowel disease using Escherichia coli proteome chip. Mol Cell Proteomics 8(8):1765–1776

Datsenko KA, Wanner BL (2000) One-step inactivation of chromosomal genes in Escherichia coli K-12 using PCR products. Proc Natl Acad Sci U S A 97(12):6640–6645

Eguchi Y et al (2003) Transcriptional regulation of drug efflux genes by EvgAS, a two-component system in Escherichia coli. Microbiology 149(Pt 10):2819–2828

Francois V, Conter A, Louarn JM (1990) Properties of new Escherichia coli Hfr strains constructed by integration of pSC101-derived conjugative plasmids. J Bacteriol 172(3):1436–1440

Gonzalez CF et al (2006) Molecular basis of formaldehyde detoxification. Characterization of two S-formylglutathione hydrolases from Escherichia coli, FrmB and YeiG. J Biol Chem 281(20):14514–14522

Hayashi K et al (2006) Highly accurate genome sequences of Escherichia coli K-12 strains MG1655 and W3110. Mol Syst Biol 2:2006.0007

Hudson JR Jr et al (1997) The complete set of predicted genes from Saccharomyces cerevisiae in a readily usable form. Genome Res 7(12):1169–1173

Ishii N et al (2007) Multiple high-throughput analyses monitor the response of E. coli to perturbations. Science (New York, N.Y.) 316(5824):593–597

Kitagawa M et al (2005) Complete set of ORF clones of Escherichia coli ASKA library (a complete set of E. coli K-12 ORF archive): unique resources for biological research. DNA Res 12(5):291–299

Kuznetsova E et al (2006) Genome-wide analysis of substrate specificities of the Escherichia coli haloacid dehalogenase-like phosphatase family. J Biol Chem 281(47):36149–36161

Lederberg J, Tatum EL (1946) Gene recombination in Escherichia coli. Nature 158(4016):558

Levin-Reisman I et al (2010) Automated imaging with ScanLag reveals previously undetectable bacterial growth phenotypes. Nat Methods 7(9):737–739

Mani R, St Onge RP, Hartman JL, Giaever G, Roth FP (2008) Defining genetic interaction. Proc Natl Acad Sci U S A 105(9):3461–3466

Marinus MG, Morris NR (1974) Biological function for 6-methyladenine residues in the DNA of Escherichia coli K12. J Mol Biol 85(2):309–322

Melnick J et al (2004) Identification of the two missing bacterial genes involved in thiamine salvage: thiamine pyrophosphokinase and thiamine kinase. J Bacteriol 186(11):3660–3662

Miki T, Yamamoto Y, Matsuda H (2008) A novel, simple, high-throughput method for isolation of genome-wide transposon insertion mutants of Escherichia coli K-12. Methods Mol Biol 416:195–204

Nakahigashi K et al (2002) HemK, a class of protein methyl transferase with similarity to DNA methyl transferases, methylates polypeptide chain release factors, and hemK knockout induces defects in translational termination. Proc Natl Acad Sci U S A 99(3):1473–1478
Nakahigashi K et al (2009) Systematic phenome analysis of Escherichia coli multiple-knockout mutants reveals hidden reactions in central carbon metabolism. Mol Syst Biol 5:306
Oshima T et al (2002a) Transcriptome analysis of all two-component regulatory system mutants of Escherichia coli K-12. Mol Microbiol 46(1):281–291
Oshima T et al (2002b) Genome-wide analysis of deoxyadenosine methyltransferase-mediated control of gene expression in Escherichia coli. Mol Microbiol 45(3):673–695
Phillips PC (2008) Epistasis—the essential role of gene interactions in the structure and evolution of genetic systems. Nat Rev Genet 9(11):855–867
Proudfoot M et al (2004) General enzymatic screens identify three new nucleotidases in Escherichia coli. Biochemical characterization of SurE, YfbR, and YjjG. J Biol Chem 279(52):54687–54694
Rajagopala SV et al (2010) The Escherichia coli K-12 ORFeome: a resource for comparative molecular microbiology. BMC Genomics 11(1):470
Rajagopala SV et al (2014) The binary protein–protein interaction landscape of Escherichia coli. Nat Biotechnol 32(3):285–290
Richmond CS, Glasner JD, Mau R, Jin H, Blattner FR (1999) Genome-wide expression profiling in Escherichia coli K-12. Nucleic Acids Res 27(19):3821–3835
Riley M et al (2006) Escherichia coli K-12: a cooperatively developed annotation snapshot—2005. Nucleic Acids Res 34(1):1–9
Roguev A, Wiren M, Weissman JS, Krogan NJ (2007) High-throughput genetic interaction mapping in the fission yeast Schizosaccharomyces pombe. Nat Methods 4(10):861–866
Schena M, Shalon D, Davis RW, Brown PO (1995) Quantitative monitoring of gene expression patterns with a complementary DNA microarray. Science 270(5235):467–470
Smith AM et al (2009) Quantitative phenotyping via deep barcode sequencing. Genome Res 19(10):1836–1842
Soga T et al (2002a) Simultaneous determination of anionic intermediates for Bacillus subtilis metabolic pathways by capillary electrophoresis electrospray ionization mass spectrometry. Anal Chem 74(10):2233–2239
Soga T et al (2002b) Pressure-assisted capillary electrophoresis electrospray ionization mass spectrometry for analysis of multivalent anions. Anal Chem 74(24):6224–6229
Sutandy FX, Qian J, Chen CS, Zhu H (2013) Overview of protein microarrays. Curr Protoc Protein Sci/editorial board, John E. Coligan ... [et al.] Chapter 27:Unit 27 21
Takeuchi R et al (2014) Colony-live – a high-throughput method for measuring microbial colony growth kinetics – reveals diverse growth effects of gene knockouts in Escherichia coli. BMC Microbiol 14(1):171
Tao H, Bausch C, Richmond C, Blattner FR, Conway T (1999) Functional genomics: expression analysis of Escherichia coli growing on minimal and rich media. J Bacteriol 181(20):6425–6440
Tong AH et al (2001) Systematic genetic analysis with ordered arrays of yeast deletion mutants. Science (New York, N.Y.) 294(5550):2364–2368
Typas A et al (2008) High-throughput, quantitative analyses of genetic interactions in E. coli. Nat Methods 5(9):781–787
Umbarger MA et al (2011) The three-dimensional architecture of a bacterial genome and its alteration by genetic perturbation. Mol Cell 44(2):252–264
van Opijnen T, Camilli A (2013) Transposon insertion sequencing: a new tool for systems-level analysis of microorganisms. Nat Rev Microbiol 11(7):435–442
van Opijnen T, Bodi KL, Camilli A (2009) Tn-seq: high-throughput parallel sequencing for fitness and genetic interaction studies in microorganisms. Nat Methods 6(10):767–772
Wetmur JG et al (2005) Molecular haplotyping by linking emulsion PCR: analysis of paraoxonase 1 haplotypes and phenotypes. Nucleic Acids Res 33(8):2615–2619

Winzeler EA et al (1999) Functional characterization of the S. cerevisiae genome by gene deletion and parallel analysis. Science (New York, N.Y.) 285(5429):901–906

Yamagishi K et al (2002) Conservation of translation initiation sites based on dinucleotide frequency and codon usage in Escherichia coli K-12 (W3110): non-random distribution of A/T-rich sequences immediately upstream of the translation initiation codon. DNA Res 9(1):19–24

Zhu Z et al (2012) Single-molecule emulsion PCR in microfluidic droplets. Anal Bioanal Chem 403(8):2127–2143

Chapter 10
Genetic Interaction Scoring Procedure for Bacterial Species

Omar Wagih and Leopold Parts

Abstract A genetic interaction occurs when the phenotype of an organism carrying two mutant genes differs from what should have been observed given their independent influence. Such unexpected outcome indicates a mechanistic connection between the perturbed genes, providing a key source of functional information about the cell. Large-scale screening for genetic interactions involves measuring phenotypes of single and double mutants, which for microorganisms is usually done by automated analysis of images of ordered colonies. Obtaining accurate colony sizes, and using them to identify genetic interactions from such screens remains a challenging and time-consuming task. Here, we outline steps to compute genetic interaction scores in *E. coli* by measuring colony sizes from plate images, performing normalisation, and quantifying the strength of the effect.

Keywords eSGA • Genetic interaction • Normalisation • Image analysis • Colony

10.1 Introduction

The fundamental goal of genetics is to understand how variation in the genome leads to changes in the phenotype (Lehner 2013). A single mutation can have a very different manifestation depending on genotype at other loci (gene-gene interactions), or the environment (gene-environment interactions), and these conditional genetic effects often explain a large proportion of phenotype variability. A clean approach for identifying such gene-gene interactions is to systematically perturb pairs of loci in the genome, and compare the observed phenotype to the expected combined effect of individual perturbations. This idea has been used for large-scale

O. Wagih (✉) • L. Parts
EMBL-EBI (South Building), Wellcome Trust Genome Campus, Saffron Walden CB10 1SD, United Kingdom

N.J. Krogan, M. Babu (eds.), *Prokaryotic Systems Biology*, Advances in Experimental Medicine and Biology 883, DOI 10.1007/978-3-319-23603-2_10

interaction mapping in budding yeast (Tong et al. 2001, 2004; Costanzo et al. 2010; Bandyopadhyay et al. 2010; Bean et al. 2014), *S. pombe* (Roguev et al. 2007), *C. elegans* (Fortunato and Fraser 2005; Byrne et al. 2007), and *E. coli* (Butland et al. 2008; Babu et al. 2014, 2011). The rich data from these global studies have provided insight into biological functions of genes, revealed components of protein complexes and pathways and helped describe the cross-talk between different functional modules in the cell (Babu et al. 2014; Hu et al. 2009; Babu et al. 2011; Baryshnikova et al. 2010).

In yeast, the Synthetic Genetic Array (SGA) technique has been successfully used on a large scale for combining two mutations in one background by mating (Tong et al. 2001). An analogous method, taking advantage of the conjugation process for exchange of genetic material in bacteria, was developed for *E. coli* (eSGA) (Butland et al. 2008). Typically, an eSGA screen is carried out by robotic machinery capable of undertaking automatic pinning steps, to produce a plate filled with an ordered grid of colonies (Fig. 10.1). The growth characteristics of the colonies can then be used as a phenotypic readout for the strains on the plate, usually quantified as strain fitness (relative growth compared to a reference).

Obtaining genetic interaction scores from the raw image data is a challenging process. First, colony sizes need to be accurately measured from plate images,

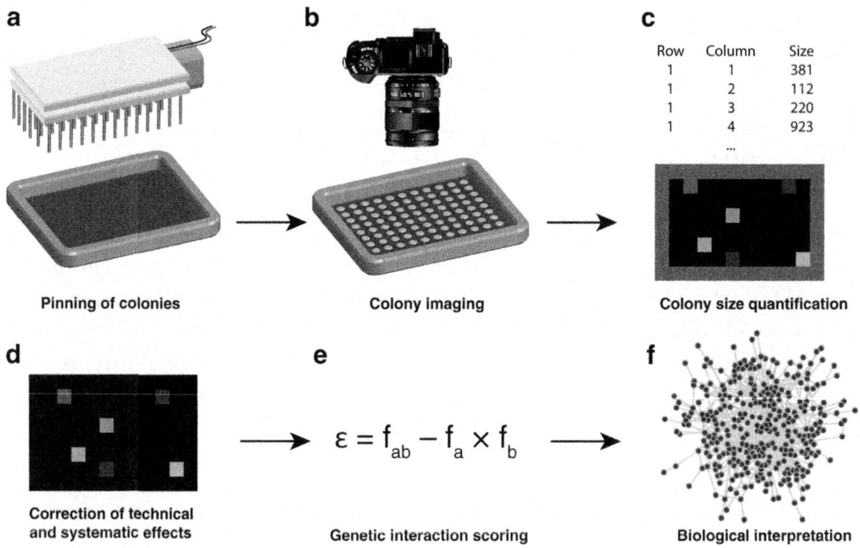

Fig. 10.1 The eSGA analysis workflow. (**a**) Colonies are pinned onto a plate containing media with the aid of robotic machinery and (**b**) photographed using a high-resolution camera. (**c**) Colony size is quantified from plate images and (**d**) systematic and technical effects are eliminated through normalisation and filtering of spurious colonies. (**e**) Quantitative scores are computed to measure the difference between the observed and expected double mutant fitness. This corresponds to the magnitude and sign of the genetic interaction. (**f**) Resulting scores are used to reveal functional information about the cell

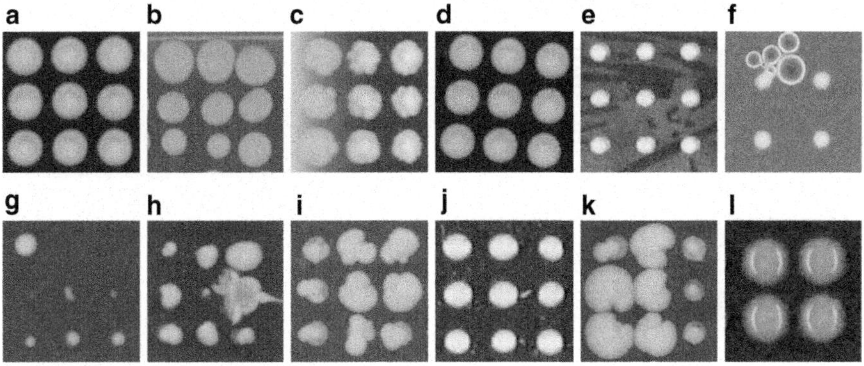

Fig. 10.2 Plate image artefacts. There are several artefacts that can interfere with the image analysis step. (**a**) Best-case scenario of growing colonies. (**b**) Interference of plate edges with the colony. (**c**) Uneven lighting. (**d**) Rotated colonies. (**e**) Artefacts in the background. (**f**) Bubbles surrounding colonies. (**g**) Sparsely growing colonies. (**h**) Cracked plates. (**i**) Irregular colony growth outside the defined boundary. (**j**) Speckles or noise around the colony. (**k**) Spillage of colonies onto one another. (**l**) Glossy effect due to mucoidity of bacterial colonies

which often include artefacts (Fig. 10.2). Second, the measured sizes have to be corrected to take into account experimental or technical biases. For example, sizes of colonies that are closer to the border of the plate, and therefore have fewer colony neighbours, are typically larger due to better nutrient accessibility (Fig. 10.1). Third, technical outliers that cannot be corrected need to be filtered out. Finally, a score to quantify the sign and magnitude of the genetic interaction can be computed from the difference between the observed and expected strain fitnesses. The choice of the scoring scheme reflects mechanistic assumptions of how the genetic interaction works, and can produce substantial differences in the resulting findings (Mani et al. 2008).

In this chapter, we present a process to compute genetic interaction scores for eSGA screens by analysing eSGA images, normalising and filtering the estimated strain fitnesses, and calculating the interaction scores. We use the gitter image analysis tool and the SGAtools suite for this purpose (Wagih and Parts 2014; Wagih et al. 2013). We give the rationale behind each of the steps and address the challenges and limitations faced throughout.

10.2 Quantifying Colony Sizes from Images

Here, we describe how to obtain colony sizes from image files. We first discuss the usual steps in image analysis pipelines, and how to optimise the acquisition setup to avoid problematic images, and then present a workflow based on the gitter image analysis tool (Wagih and Parts 2014).

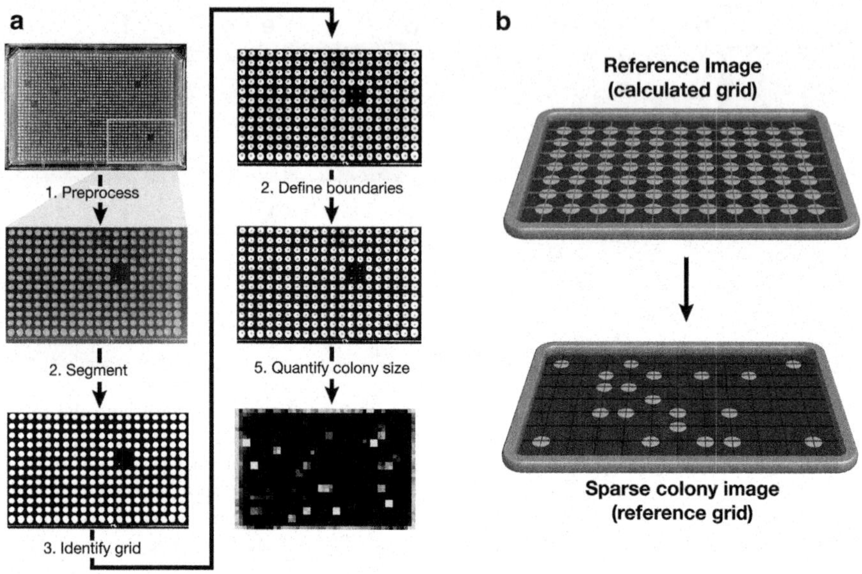

Fig. 10.3 The typical image analysis workflow. (**a**) Overview of the five steps for processing a plate image. First, the raw image is preprocessed and segmented. Next, the grid of colonies is identified and the boundaries of colonies are computed. Pixels classified as belonging to the colony that fall within these boundaries are then summed up for a quantified colony size and can be visualised as a heatmap for additional validation. (**b**) An example of processing with the aid of a reference image. Here, the computed grid from a reference image with well growing colonies (*orange lines*) is used to define a grid for an image with sparse colonies (*magenta lines*)

10.2.1 Image Analysis Steps

The bottleneck of analysing genetic screens is accurately quantifying colony sizes from plate images. This typically involves five main steps (Fig. 10.3a):

1. Prepossessing the raw image: the image undergoes several preprocessing steps. This involves conversion of the image to grayscale, automatic rotation, noise reduction and background correction.
2. Segmenting the image: pixels of the preprocessed image are classified as belonging to the colony or background.
3. Detecting the grid of colonies: given the format of the plate, the grid that best fits the colonies is identified.
4. Identifying colony boundaries: the boundaries defining each colony are computed.
5. Quantifying colony size: the colony size is calculated as the number of colony pixels inside the boundary.

The quantified colony sizes can be visualised as heatmaps, allowing for an additional layer of manual verification.

There are several factors that make the image analysis step challenging (Fig. 10.2). First, colonies may be mis-pinned, slightly rotated, interfered by plate edges, or of low density, making it difficult to correctly identify the underlying grid. Second, variations in lighting across the plate or non-uniform background agar complicates segmentation of the image. Last, noise speckles in the image, glossy colonies, or irregular morphology may cause over- or underestimation of the size. It is therefore important to carry out appropriate controls and replicate experiments to help automatically filter out erroneous results.

10.2.2 Quantifying Colony Sizes Using Gitter

The majority of these issues are addressed in the gitter image analysis package available for the R programming language. gitter is easy to install, fast, accurate and provides several visualisation tools. To begin using gitter, the R programming language (version 3 or higher) is required (http://www.r-project.org/). First, the bioconductor package `EBImage` (Pau et al. 2010) is installed in the R console:

```
source("http://bioconductor.org/biocLite.R")
biocLite("EBImage")
```

The `gitter` R package is then installed and loaded as follows:

```
install.packages("gitter")
library(gitter)
```

`gitter` has two image processing modes. The singular mode allows processing individual images and the batch mode allows the processing of a large number of images from the same batch (*i.e.* typically carried out by the same person using the same robot and camera). We describe each mode in detail below.

10.2.3 Processing a Single Image

To process a single image, the path to the image on the file system is passed to the function `gitter` along with the plate format as follows:

```
image.file = "/path/to/image.jpg"
dat = gitter(image.file, plate.format=384)
```

Here, `gitter` processes an image of a 384-format plate located at `/path/to/image.jpg`. The results are stored in the `dat` variable as an R `data.frame` object and has a five column format (Table 10.1). Additionally, the contents of `dat` and the segmented image showing the boundaries for each colony are saved to the working directory. Resulting data can then be visualised as a heatmap using the plot function:

```
plot(dat, type="heatmap")
```

Table 10.1 A sample of gitter's output data file

Row	Col	Size	Circularity	Flags
1	1	3102	0.5276	S,C
1	2	1615	0.8479	S
1	3	1851	0.8355	–
1	4	1834	0.8244	–
1	5	1815	0.7612	–
1	6	1665	0.8256	–
1	7	1894	0.8257	S
1	8	1829	0.8543	–

There are five tab-delimited columns. 1. row: row of the colony, 2. col: column of the colony, 3. size: quantified colony size computed as the number of pixels within the colony boundary, 4. circularity: the roundness of the colony and 5. flags: comma-separated list of one letter codes which flag colonies considered spurious because they spill over one another or have a low circularity. The S flag indicates colony spillage of colonies onto one another or edge interference, whereas the C flag indicates low colony circularity

Running `help(gitter)` provides columns can be processed. a full list of the other available options and features, which can be added to a `gitter` call. For example, images can be automatically rotated, speckles and noise can be reduced, or a plate with a non-standard number of rows and

10.2.4 Processing Multiple Images

The function `gitter.batch` can be used to process more than one image. This is similar to processing a single image, but instead, the path to the directory containing the images is provided. The batch function is executed as follows:

```
image.directory = "/path/to/directory"
gitter.batch(image.directory)
```

Here, `image.directory` is the directory containing the images or a list of paths to all images, to be processed in a batch. The same options that are passed to the single mode function can also be passed to `gitter.batch`.

Images that have a low density of colonies are difficult to analyse, as the correct grid of colonies is not apparent. The batch function can still process such images with the aid of a reference image that has well growing colonies. If the grid on the reference image can be easily identified, gitter can use this information to calculate the location of colonies on the sparse plate (Fig. 10.3b). The reference image processing feature can be used through the batch function as follows:

```
reference.image = "/path/to/reference"
gitter.batch(image.directory, reference.image)
```

Here, `reference.image` is the path to the reference image, typically taken from the same batch. As for single plate processing, data files and segmented images showing colony boundaries are saved to the working directory.

More examples, tutorials, and documentation can be found at gitter.ccbr.utoronto.ca.

10.3 Normalisation

Global biases and variation between experiments makes it difficult to compare colony sizes from different plates to one another directly. Therefore, normalisation of the sizes within and across plates is required to eliminate systematic effects. Failure to do so may result in over and under-estimated colony sizes leading to false positive and negative genetic interaction calls.

There are two types of confounding influences on colony size - reproducible systematic effects that can be corrected, and unexpected experimental outcomes that need to be filtered out (Fig. 10.4a). The systematic effects due to plate and colony positions in the plate are assumed to act in a multiplicative manner. Thus, correcting for them involves estimating a multiplicative constant α for each effect. In the following, we first describe the common corrections that help accurately estimate the fitness f (relative phenotype) of the strain in row i column j (in a plate with n rows and m columns) from the observed colony size C_{ij} by calculating the correction factors for plate, row, column, and surrounding region:

$$f_{ij} = \alpha_{\text{plate}}\alpha_{\text{row }i}\alpha_{\text{column }j}\alpha_{\text{spot }i,j}C_{ij}$$

The plate correction factor α_{plate} scales colony sizes such that the typical fitness is 1. Under no confounding effects, the other correction factors are 1. We also describe the filters can be applied based on unexpected colony features, such as cases where the colony is too large, differs from replicates or has a non-circular shape. In general, the need for all the correction factors and filters should be established from looking at plate images. For example, large colonies should not be filtered out in suppression screens, or non-circular ones in screens of flocculant strains. The following methods are those implemented in SGAtools (Wagih et al. 2013) and we describe them in more detail.

10.3.1 Corrections

10.3.1.1 Plate Effect

Quantified colony sizes are often not directly comparable between plates. The images can vary in resolution, or come from experiments with slightly different conditions and incubation times (Baryshnikova et al. 2010; Collins et al. 2006). As a result, the colony size distributions differ between plates, and have to be corrected before quantitative comparisons can be made (Fig. 10.4b).

One way to perform the correction is to match the typical colony size in each plate. There are several ways to estimate what the standard colony size is; here, we use the median of colony sizes in the 60 % innermost rows and columns, termed plate middle median (PMM) μ (Baryshnikova et al. 2010). This choice ignores the atypically large colony sizes in the outer rows and columns (see row-column effect). The plate correction factor α for the colony sizes in the plate is then given by:

$$\alpha_{\text{plate}} = \frac{1}{\mu} \tag{10.1}$$

Thus, plate-normalised colony sizes $\alpha_{\text{plate}} C_{ij}$ reflect their relative magnitude compared to the standard plate colony size of 1.

10.3.1.2 Spatial Effect

Size gradients, as well as plate regions with consistently smaller or larger colony sizes are often observed. Such effects are usually due to an uneven distribution of media across the plate (Baryshnikova et al. 2010). Correcting these spatial effects is required for similar size distributions across all plate regions.

To do so, a technique known as spatial smoothing is used. This involves correcting each colony size using other colony sizes in the local neighbourhood to calculate the expected size of the colony. Median smoothing in a 7 × 7 window is first applied at each spot, calculating a robust estimate of its expected size based on nearby colonies. Next, average smoothing with a 9 × 9 window is applied to further reduce the variability between the expectations of neighbouring spots. The choices of window sizes are arbitrary, and picked to reflect the typical scale of local correlations (Baryshnikova et al. 2010). The spatial correction coefficient is the deviation of the calculated neighbourhood expectation $\hat{C}_{i,j}$ of the colony size from the plate average:

$$\alpha_{\text{spot } i,j} = \frac{\bar{C}}{\hat{C}_{i,j}}, \tag{10.2}$$

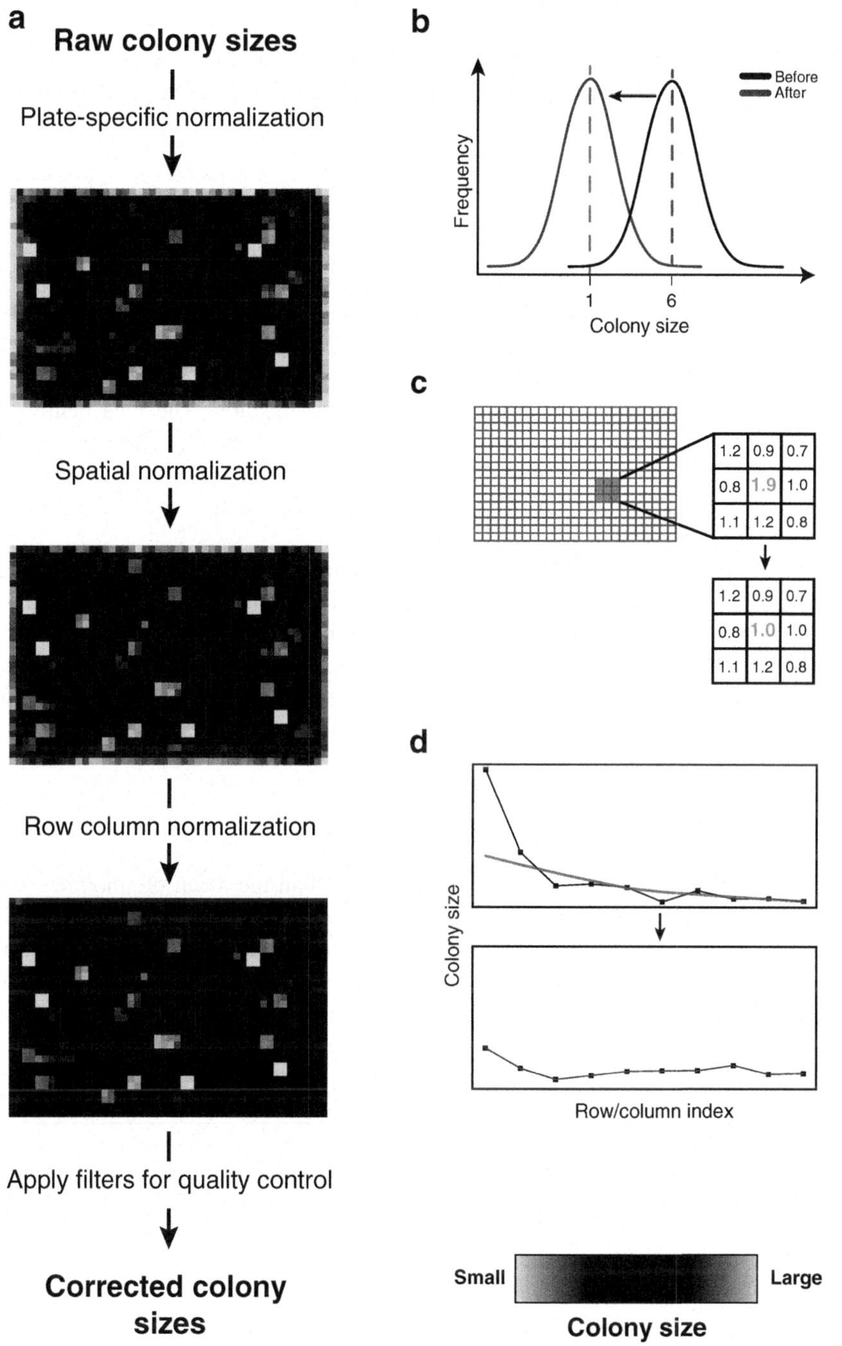

Fig. 10.4 Correction and filtering of colony sizes. Colony sizes undergo several normalisation and filtering steps to eliminate systematic effects and dubious measurements. An overview of the steps

where $\bar{C} = \frac{\sum_{i,j} \hat{C}_{i,j}}{mn}$. This correction makes the colonies that are large, but also have large neighbours, smaller. At the same time, if a colony is large compared to its neighbours, the relative size remains unchanged due to the broad smoothing operations applied.

10.3.1.3 Row-Column Effect

Colonies closer to the border of the plate have fewer neighbours, resulting in better nutrient availability, and usually larger size (Fig. 10.4a) (Baryshnikova et al. 2010; Collins et al. 2006). This effect is more pronounced for outside colonies, varies between rows and columns, and can be accounted for by adjusting the colony size to be relative to the expectation at the particular location. The idea behind the correction is similar to that of the spatial correction, but in this case, the smoothing to calculate the expectation is applied to entire rows and columns, instead of patches of colonies. We describe the correction for rows, $\alpha_{\text{row}i}$. The corrections for the columns, $\alpha_{\text{column}j}$ are computed in a similar manner.

First, the colony sizes are smoothed in each row i using locally weighted scatterplot smoothing (LOWESS, Cleveland 1979) within a window of size 2k+1, corresponding to the nearest 10 % of the colonies in the row, to obtain $\theta_{i,c} = \text{LOWESS}(C_{i,m-k}, C_{i,m-k+1}, \ldots, C_{i,m+k})$. The correction factor α for row i column j is the deviation of the smoothed values in row i from the plate average ones (Fig. 10.4d):

$$\alpha_{\text{row}i_j} = \frac{\mu_{\theta,i}}{\theta_{i,j}}, \tag{10.3}$$

where $\mu_{\theta,i} = \frac{1}{m} \cdot \sum_{l=1}^{m} \theta_{i,l}$. Colony sizes larger than the mean of smoothed sizes μ_θ will have the coefficient $\alpha < 1$ and therefore, will be corrected to be smaller, whereas larger than average colonies will have $\alpha > 1$, and thus a larger corrected size (Fig. 10.4d). In contrast, colonies with sizes closer to the mean of smoothed sizes will have an $\alpha \approx 1$, and would not be substantially impacted by the row-column correction.

Fig. 10.4 (continued) of this process is shown in (**a**). First, colony sizes are scaled to the standard size of a colony (**b**). The distribution of colony sizes is shown before (*black*) and after (*blue*). *Dashed lines* represent the median colony size. Second, spatial effects, such as gradients of colony sizes are corrected for. This is represented in (**c**), where the size of a colony (*green*) is scaled by the median or mean of the surrounding colony sizes. Third, the row-column effect, where colonies in outer rows are larger, is corrected for (**d**). The colony sizes for an exemplary first row is shown (*black line*) along with the averaged smoothed values used to correct for this effect (*red line*). The corrected colony sizes are shown in the panel below (*blue line*). The last step is to filter out spurious colony size measurements (not shown). Heatmaps represent the colony size after the normalisation step

10.3.2 *Filters*

10.3.2.1 Technical Replicate Outliers

Individual colonies whose sizes differ substantially from other technical replicates of the same strain can be contaminated or missed in pinning, and should be excluded from interaction scoring. Leave-one-out analysis is commonly used to detect such outliers by identifying colonies whose size increases the variation within replicates at least tenfold. This can be tested by checking whether removing one measurement reduces the variance within replicates by at least 90 %. If this is the case, the replicate is considered an outlier and discarded from further analyses.

10.3.2.2 Genetic Linkage

Genes in close proximity to one another in the genome are unlikely to be separated by homologous recombination due to genetic linkage. Unexpectedly small colony sizes for double mutants of nearby genes therefore generate false positive genetic interactions, which are considered spurious and discarded from further analyses (Butland et al. 2008; Baryshnikova et al. 2010). A proximity cutoff of 30 kb is typically used for *E. coli* (Babu et al. 2014; Butland et al. 2008), although this value can vary for different organisms (Baryshnikova et al. 2010).

10.3.2.3 Overgrown Colonies

In some cases, it may be required to eliminate overgrown colonies to avoid false positive results. This filter is carried out after plate normalisation and before all others. We consider a simple threshold of two times the standard colony size, which is the typical threshold previously used for yeast screens (Baryshnikova et al. 2010). A given colony size above this threshold is considered spurious and discarded if at least 75 % of all replicates were also larger than this threshold. Note that in suppression screens, many colonies are expected to be substantially larger than average, and this filter should not be applied.

10.3.2.4 Circularity

Often, image quantification programs will report the circularity of each colony (Wagih and Parts 2014; Lawless et al. 2010; Collins et al. 2006). This is a measure of how closely the shape of the colony resembles a perfect circle and the value ranges from 0 to 1. Colonies below a certain circularity threshold, typically 0.6, can be irregular, mis-quantified, or contaminated, and can be discarded. If the strain used does not form circular colonies, the performance of the threshold should be verified by eye.

10.4 Computing Genetic Interaction Scores

By definition, a genetic interaction occurs when the observed double mutant fitness deviates from the expected. The interaction is positive, if the observed phenotype is better than expected, and negative if it is worse. Each of the developed methods to compute the magnitude and sign of the genetic interaction (Costanzo et al. 2010; Collins et al. 2006; Bean and Ideker 2012) provide a different way to quantify the observed-expected deviation. These methods rely primarily on single and double-mutant fitness measurements. We focus on the ϵ score, one of the first and most commonly used scoring metrics, and describe some alternatives.

10.4.1 Epsilon Score

One well-established scoring metric is the ϵ score, also known as the SGA score (Phillips et al. 2000), which was first used in a large scale experiment by Costanzo et al. (2010) This score is based on a multiplicative model, where the expected fitness $\hat{f}_{a,b}$ for a double mutant of alleles a and b is the product of the measured fitnesses for strains with the individual mutations f_a and f_b. The single mutant fitness is measured either from single mutant strains, or estimated from all the double mutant pairs, e.g. assuming that the majority of the double mutants have the same fitness as the single one. As described above, the fitness is usually estimated as the relative growth compared to a reference, and is therefore expected to be centred on 1. The ϵ score is calculated as the difference between the observed and expected double mutant growth:

$$\epsilon = f_{a,b} - \hat{f}_{a,b} = f_{a,b} - f_a \cdot f_b \quad (10.4)$$

A negative ϵ score corresponds to a negative genetic interaction and a positive ϵ score corresponds to a positive genetic interaction.

10.4.2 Alternative Scoring Metrics

While the ϵ score is most commonly used, there exist several other scoring metrics. Here, we briefly describe the S-score and the relative fitness.

10.4.2.1 S-Score

The interaction score, commonly known as the S-score was introduced by Collins et al. (2006) and takes a slightly different approach. Here, the double mutant fitness is compared against control double mutant fitness using a variation of the t-statistic, which measures the deviation of an observed value from its expectation. For a double mutant, let there be n observed fitness values with an estimated mean $\hat{\mu}$, and m observed control fitnesses with a mean $\hat{\mu}_0$ for the same double mutant. The controls can be computed from the corresponding single mutants, of measured in another condition for gene-environment interaction screens. The standard two sample t-statistic for the difference between the means is:

$$t = \frac{\hat{\mu} - \hat{\mu}_0}{\sqrt{\text{var}(\hat{\mu} - \hat{\mu}_0)}} = \frac{\hat{\mu} - \hat{\mu}_0}{\sqrt{\text{var}(\hat{\mu}) + \text{var}(\hat{\mu}_0)}} \tag{10.5}$$

The S-score is constructed based on the t-statistic, with the following modifications:

1. $\hat{\mu}$ and $\hat{\mu}_0$ are computed as the median of all double mutant fitnesses in a set of replicates, instead of the mean.
2. $\text{var}(\hat{\mu}_0)$ is computed from the median of variances for double mutant fitnesses in a set of replicates.
3. $\text{var}(\hat{\mu})$ is computed from the maximum of the variance of normalised colony sizes for the double mutant of interest.

In addition to these modifications, Collins *et al.* impose lower bounds on $\text{var}(\hat{\mu})$ as similar measurements in replicates showed unusually small standard deviations, resulting in false positive genetic interactions.

The S-score is then typically scaled to a range of -20 to 20 to make subtle changes in the score more apparent. Similar to the ϵ score, a positive value indicates a positive genetic interaction and a negative value a negative genetic interaction. While the S-score is a relatively widely-used alternative to the epsilon score, it reflects the confidence that there is a non-zero effect, but not the magnitude of the effect. This is less intuitive for comparing interaction effect sizes.

10.4.2.2 Relative Fitness

The relative fitness score, r is the log fold-change in growth rate compared to the expectation (Travisano and Lenski 1996; Gros et al. 2009; Martin et al. 2007; MacLean 2010):

$$\begin{aligned} r_{a,b} &= \log_2 \frac{f_{a,b}}{\hat{f}_{a,b}} = \log_2 \frac{f_{a,b}}{f_a \cdot f_b} = \\ &= \log_2(f_{a,b}) - \log_2(f_a) - \log_2(f_b) \end{aligned} \tag{10.6}$$

An r score of 0 represents no change in growth, and thus no genetic interaction. A score of -1 would represent a two-fold decrease in fitness in the double mutant compared to the single mutant expectation, and thus a negative genetic interaction. Similarly, a score of 1 would represent a two-fold increase in fitness, and thus a positive genetic interaction.

10.4.3 Quality Control

Several steps need to be taken to improve the quality of the scored data.

Visual confirmation. Visual checks should be applied throughout the scoring process. First, the segmentation and colony size quantification results should be verified for a random sample of images to make sure the starting data are of high quality. Second, the plate heatmaps should be checked after normalisation to ensure all large confounding effects are taken into account. After calculating the interaction scores, a random sample of colonies corresponding to different strata of score magnitudes should be verified by eye to confirm they are not due to unexpected artefacts. A simple computational way to do so is to look at the heat maps of plate colony sizes using gitter's plot function, SGAtools visualisation utility, or other means (Fig. 10.4a).

Concordance of technical and biological replicates. True interactions ought to be reproducible between replicates. There are many approaches for making use of multiple measurements of the same strains to establish the reproducibility, we will focus on one of them. To test whether the observed score could be explained by discordant outliers, we compute a p-value of observing a more extreme effect using a one-sample t-test based on the variability between replicates. If the replicates are variable, large scores are more frequent, and a large p-value (e.g. > 0.05) indicates that they likely do not correspond to real effects. These scores are considered unreliable and are excluded from further analyses. The remaining scores are averaged based on the replicates.

Picking a score cutoff. In addition to the p-value, an ϵ cutoff that minimises false positive genetic interactions is computed. One way of doing this is to analyse previously published genetic interactions as a gold standard and compute the positive predictive value (PPV) or precision (Baryshnikova et al. 2010) as follows:

$$PPV = \frac{TP}{TP + FP} \tag{10.7}$$

For a given ϵ, TP (true positives) is the number of tested genetic interactions that have a score larger than ϵ and exist in the literature, while FP (false positives) is the number of tested genetic interactions with a score larger than ϵ, that does not exist in the literature. The ϵ cutoff is then picked to maximise the PPV. If there is insufficient genetic interaction data available, a similar approach can be applied to GO term biological processes (BP). For a given score cutoff, a hypergeometric enrichment is computed for all BPs and pairs of BPs. The cutoff which results in the most enrichment in the BPs can then be used in analysis.

Such methods cannot be used to generate a universal score cutoff that is constant across studies. The scoring scheme used, variability in the replicate concordance, and the frequency of genetic interactions in the studied processes will result in cutoff values that vary from experiment to experiment. Ultimately, secondary assays should be performed to validate the function of individual interactions.

10.5 Concluding Remarks

Genetic interactions have been proven to be a valuable resource of functional information. Automated methods to carry out large scale screens have made them more accessible, but their analysis to quantify the magnitude of the genetic interaction remains a non-trivial task. We have described how to quantify colony sizes from plate images, normalise systematic confounders or technical effects and score the genetic interactions. Together, the presented methods provide a complete pipeline to analyse eSGA screens. With many exciting recent advances (Babu et al. 2011, 2014; Bean and Ideker 2012; Bean et al. 2014; Guénolé et al. 2013), we expect that the application of these approaches will reveal much about the workings of the cell.

Acknowledgements Canadian Institute for Advanced Research Global Fellowship and Marie Curie International Outgoing Fellowship.

References

Babu M, Díaz-Mejía JJ, Vlasblom J, Gagarinova A, Phanse S, Graham C, Yousif F, Ding H, Xiong X, Nazarians-Armavil A, Alamgir M, Ali M, Pogoutse O, Pe'er A, Arnold R, Michaut M, Parkinson J, Golshani A, Whitfield C, Wodak SJ, Moreno-Hagelsieb G, Greenblatt JF, Emili A (2011) Genetic interaction maps in Escherichia coli reveal functional crosstalk among cell envelope biogenesis pathways. PLoS Genet 7(11):e1002377

Babu M, Arnold R, Bundalovic-Torma C, Gagarinova A, Wong KS, Kumar A, Stewart G, Samanfar B, Aoki H, Wagih O, Vlasblom J, Phanse S, Lad K, Yu AYH, Graham C, Jin K, Brown E, Golshani A, Kim P, Moreno-Hagelsieb G, Greenblatt J, Houry WA, Parkinson J, Emili A (2014) Quantitative genome-wide genetic interaction screens reveal global epistatic relationships of protein complexes in Escherichia coli. PLoS Genet 10(2):e1004120

Bandyopadhyay S, Mehta M, Kuo D, Sung M-K, Chuang R, Jaehnig EJ, Bodenmiller B, Licon K, Copeland W, Shales M, Fiedler D, Dutkowski J, Guénolé A, van Attikum H, Shokat KM, Kolodner RD, Huh W-K, Aebersold R, Keogh M-C, Krogan NJ, Ideker T (2010) Rewiring of genetic networks in response to DNA damage. Science 330(6009):1385–1389

Baryshnikova A, Costanzo M, Kim Y, Ding H, Koh J, Toufighi K, Youn J-Y, Ou J, San Luis B-J, Bandyopadhyay S, Hibbs M, Hess D, Gingras A-C, Bader GD, Troyanskaya OG, Brown GW, Andrews B, Boone C, Myers CL (2010) Quantitative analysis of fitness and genetic interactions in yeast on a genome scale. Nat Methods 7(12):1017–1024

Bean GJ, Ideker T (2012) Differential analysis of high-throughput quantitative genetic interaction data. Genome Biol 13(12):R123

Bean GJ, Jaeger PA, Bahr S, Ideker (2014) Development of ultra-high-density screening tools for microbial omics. PLoS ONE 9(1):e85177

Butland G, Babu M, Díaz-Mejía JJ, Bohdana F, Phanse S, Gold B, Yang W, Li J, Gagarinova AG, Pogoutse O, Mori H, Wanner BL, Lo H, Wasniewski J, Christopolous C, Ali M, Venn P, Safavi-Naini A, Sourour N, Caron S, Choi J-Y, Laigle L, Nazarians-Armavil A, Deshpande A, Joe S, Datsenko KA, Yamamoto N, Andrews BJ, Boone C, Ding H, Sheikh B, Moreno-Hagelseib G, Greenblatt JF, Emili A (2008) eSGA: E. coli synthetic genetic array analysis. Nat. Methods 5(9):789–795

Byrne AB, Weirauch MT, Wong V, Koeva M, Dixon SJ, Stuart JM, Roy PJ (2007) A global analysis of genetic interactions in Caenorhabditis elegans. J Biol 6(3):8

Cleveland W (1979) Robust locally weighted regression and smoothing scatterplots. J Am Stat Assoc 74(368):829–836
Collins SR, Schuldiner M, Krogan NJ, Weissman JS (2006) A strategy for extracting and analyzing large-scale quantitative epistatic interaction data. Genome Biol 7(7):R63
Costanzo M, Baryshnikova A, Bellay J, Kim Y, Spear ED, Sevier CS, Ding H, Koh JLY, Toufighi K, Mostafavi S, Prinz J, Onge RPS, VanderSluis B, Makhnevych T, Vizeacoumar FJ, Alizadeh S, Bahr S, Brost RL, Chen Y, Cokol M, Deshpande R, Li Z, Lin Z-Y, Liang W, Marback M, Paw J, San Luis B-J, Shuteriqi E, Tong AHY, van Dyk N, Wallace IM, Whitney JA, Weirauch MT, Zhong G, Zhu H, Houry WA, Brudno M, Ragibizadeh S, Papp B, Pál C, Roth FP, Giaever G, Nislow C, Troyanskaya OG, Bussey H, Bader GD, Gingras A-C, Morris QD, Kim PM, Kaiser CA, Myers CL, Andrews BJ, Boone C (2010) The genetic landscape of a cell. Science 327(5964):425–431
Fortunato A, Fraser AG (2005) Uncover genetic interactions in Caenorhabditis elegans by RNA interference. Biosci Rep 25(5-6):299–307
Gros P-A, Nagard HL, Tenaillon O (2009) The evolution of epistasis and its links with genetic robustness, complexity and drift in a phenotypic model of adaptation. Genetics 182(1): 277–293
Guénolé A, Srivas R, Vreeken K, Wang ZZ, Wang S, Krogan NJ, Ideker T, van Attikum H (2013) Dissection of DNA damage responses using multiconditional genetic interaction maps. Mol Cell 49(2):346–358
Hu P, Janga SC, Babu M, Díaz-Mejía JJ, Butland G, Yang W, Pogoutse O, Guo X, Phanse S, Wong P, Chandran S, Christopoulos C, Nazarians-Armavil A, Nasseri NK, Musso G, Ali M, Nazemof N, Eroukova V, Golshani A, Paccanaro A, Greenblatt JF, Moreno-Hagelsieb G, Emili A (2009) Global functional atlas of Escherichia coli encompassing previously uncharacterized proteins. PLoS Biol 7(4):e96
Lawless C, Wilkinson DJ, Young A, Addinall SG, Lydall DA (2010) Colonyzer: automated quantification of micro-organism growth characteristics on solid agar. BMC Bioinf 11:287
Lehner B (2013) Genotype to phenotype: lessons from model organisms for human genetics. Nat Rev Genet 14(3):168–178
MacLean RC (2010) Predicting epistasis: an experimental test of metabolic control theory with bacterial transcription and translation. J Evol Biol 23(3):488–493
Mani R, Onge RPS, Hartman JL, Giaever G, Roth FP (2008) Defining genetic interaction. Proc Natl Acad Sci USA 105(9):3461–3466
Martin G, Elena SF, Lenormand T (2007) Distributions of epistasis in microbes fit predictions from a fitness landscape model. Nat Genet 39(4):555–560
Pau G, Fuchs F, Sklyar O, Boutros M, Huber W (2010) EBImage–an R package for image processing with applications to cellular phenotypes. Bioinformatics 26(7):979–981
Phillips P, Otto S, Whitlock M (2000) The evolutionary importance of gene interactions and variability of epistasis effects. In: Wolf JB, Brodie ED, Wade MJ (eds) Epistasis and the evolutionary process. Oxford University Press, Oxford, pp 20–38
Roguev A, Wiren M, Weissman JS, Krogan NJ (2007) High-throughput genetic interaction mapping in the fission yeast Schizosaccharomyces pombe. Nat Methods 4(10):861–866
Tong AH, Evangelista M, Parsons AB, Xu H, Bader GD, Pagé N, Robinson M, Raghibizadeh S, Hogue CW, Bussey H, Andrews B, Tyers M, Boone C (2001) Systematic genetic analysis with ordered arrays of yeast deletion mutants. Science 294(5550):2364–2368
Tong AHY, Lesage G, Bader GD, Ding H, Xu H, Xin X, Young J, Berriz GF, Brost RL, Chang M, Chen Y, Cheng X, Chua G, Friesen H, Goldberg DS, Haynes J, Humphries C, He G, Hussein S, Ke L, Krogan N, Li Z, Levinson JN, Lu H, Ménard P, Munyana C, Parsons AB, Ryan O, Tonikian R, Roberts T, Sdicu A-M, Shapiro J, Sheikh B, Suter B, Wong SL, Zhang LV, Zhu H, Burd CG, Munro S, Sander C, Rine J, Greenblatt J, Peter M, Bretscher A, Bell G, Roth FP, Brown GW, Andrews B, Bussey H, Boone C (2004) Global mapping of the yeast genetic interaction network. Science 303(5659):808–813

Travisano M, Lenski RE (1996) Long-term experimental evolution in Escherichia coli. IV. Targets of selection and the specificity of adaptation. Genetics 143(1):15–26

Wagih O, Parts L (2014) Gitter: a robust and accurate method for quantification of colony sizes from plate images. G3 (Bethesda) 4(3):547–552

Wagih O, Usaj M, Baryshnikova A, VanderSluis B, Kuzmin E, Costanzo M, Myers CL, Andrews BJ, Boone CM, Parts L (2013) SGAtools: one-stop analysis and visualization of array-based genetic interaction screens. Nucleic Acids Res 41(Web Server issue):W591–W596

Chapter 11
Mapping the Protein–Protein Interactome Networks Using Yeast Two-Hybrid Screens

Seesandra Venkatappa Rajagopala

Abstract The yeast two-hybrid system (Y2H) is a powerful method to identify binary protein–protein interactions *in vivo*. Here we describe Y2H screening strategies that use defined libraries of open reading frames (ORFs) and cDNA libraries. The array-based Y2H system is well suited for interactome studies of small genomes with an existing ORFeome clones preferentially in a recombination based cloning system. For large genomes, pooled library screening followed by Y2H pairwise retests may be more efficient in terms of time and resources, but multiple sampling is necessary to ensure comprehensive screening. While the Y2H false positives can be efficiently reduced by using built-in controls, retesting, and evaluation of background activation; implementing the multiple variants of the Y2H vector systems is essential to reduce the false negatives and ensure comprehensive coverage of an interactome.

Keywords Yeast two-hybrid system • Protein–protein interactions • Pooled library screening • Two-hybrid array

11.1 Introduction

Specific interactions between proteins form the basis of most biological processes, thus the knowledge of an organism's protein interaction network provides insights into the function(s) of individual proteins, the structure of functional complexes, and eventually, the organization of the entire cell. The Protein-protein interactions (PPIs) can be identified by a multitude of experimental methods. However, a vast majority of the PPIs available today are generated by yeast two-hybrid (Y2H) method and affinity purification or co-fractionation coupled to mass spectrometry (AP-MS) (Kerrien et al. 2012). It is important to note that these methods yield different types of information; Y2H analyses reveal binary interactions, including transient interactions, whereas the AP-MS approaches report multiple interactions connecting

S.V. Rajagopala (✉)
J. Craig Venter Institute, 9704 Medical Center Drive, Rockville, MD 20850, USA
e-mail: rajgsv@gmail.com

N.J. Krogan, M. Babu (eds.), *Prokaryotic Systems Biology*, Advances in Experimental Medicine and Biology 883, DOI 10.1007/978-3-319-23603-2_11

all of the proteins in fairly stable complexes. Protein interactome analysis on a genome scale was first achieved by using yeast two-hybrid (Y2H) screens (Ito et al. 2001; Uetz et al. 2000) and next by large-scale mass spectrometric analysis of affinity-purified protein complexes (Gavin et al. 2002; Ho et al. 2002). Here we describe the high-throughput Y2H screening strategies, applied to map high-quality proteome-scale interactome networks of model organisms and pathogenic infectious agents.

11.2 Yeast Two-Hybrid System

The Y2H system is a genetic screening extensively used to identify binary protein–protein interactions *in vivo* (in yeast cells). The system was originally developed by Stanley Fields in 1989 (Fields and Song 1989). The principle of the assay relied on major discoveries on transcription initiation (Brent and Ptashne 1985). In general the protein domains can be separated and recombined and can retain their properties. In particular, transcription factors can frequently be split into the DNA-binding domain (DBD) and activation domains (ADs). In the two-hybrid system, a DNA-binding domain (in this case, from the yeast Gal4 protein) is fused to a protein generally called *bait ("B")* for which one wants to find interacting partners. A transcriptional activation domain (from the yeast Gal4 protein) is then fused to one or more ORFs (*preys*) (Fig. 11.1). The bait and prey fusion proteins are then co-expressed in the same yeast cell. If, the two proteins *bait* and *prey* interact, a transcription factor is reconstituted which in turn activates the reporter gene(s) (Fig. 11.1). The expression of the reporter gene allows the yeast cell to grow under certain conditions. For example, the HIS3 reporter encodes imidazoleglycerolphosphate (IGP) dehydratase, a critical enzyme in histidine biosynthesis. In a screening yeast strain lacking an

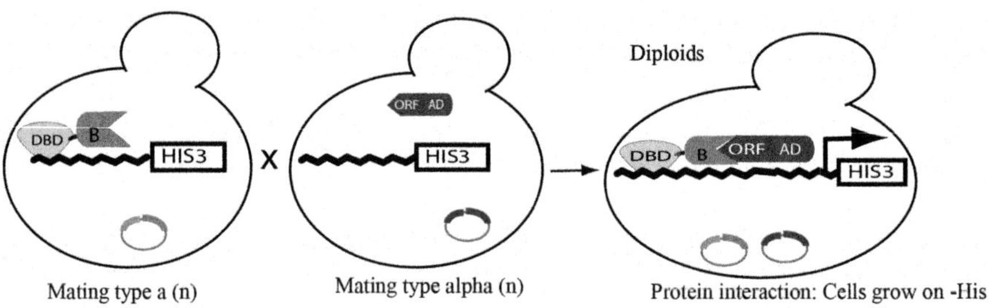

Fig. 11.1 Yeast two-hybrid principle: A protein of interest 'B' is expressed in yeast as a fusion to a Gal4p DNA-binding domain (DBD, "bait"; *circles* denote expression plasmids). Another protein or library of proteins of interest 'ORF' is fused to Gal4p transcriptional activation domain (AD, "prey"). The two yeast strains are mated to combine the two plasmids expressing bait and prey fusion proteins in the same cell (diploid). If, proteins 'B' and 'ORF' interact in the resulting diploids cells, they reconstitute a transcription factor which activates a reporter gene (HIS3) and therefore allows the cell to grow on selective synthetic media (media lacking histidine)

endogenous copy of HIS3, expression of a HIS3 reporter gene is driven by a promoter that contains a Gal4p-binding site, so the bait protein fusion can bind to it. However, the bait fusion does not contain a transcriptional activation domain it remains inactive. If, a *prey* protein with an attached activation domain binds to the *bait* protein, this activation domain can recruit the basal transcription machinery, and expression of the reporter gene ensues. Thus, these cells can now grow in the absence of histidine in the media because they can synthesize their own.

11.3 Y2H Applications

Initially, the two-hybrid system was invented to demonstrate the association of two proteins (Fields and Song 1989). Later, it was demonstrated that completely new protein interactions can be identified with this system. Over time, it has become clear that the ability to perform unbiased large-scale library screens is the most powerful application of the system. In recent years, Y2H method has been extensively applied to map high-quality proteome-scale binary interactome networks of server model organism, including human proteome and may pathogenic infectious agents. In a recent study, Rolland et al., published a human interactome map, which is based on a systematic Y2H screening of 13,000 human proteins that uncovered 14,000 PPIs (Rolland et al. 2014). Similarly, several large-scale Y2H projects have been successful in systematically mapping binary interactome landscape of *Escherichia coli* (Rajagopala et al. 2014), *Saccharomyces cerevisiae* (Uetz et al. 2000; Yu et al. 2008), *Caenorhabditis elegans* (Gong et al. 2004), *Drosophila melanogaster* (Giot et al. 2003); these studies have shown that most of the proteins in a cell are actually connected to each other. The Y2H screening has also been implemented on many pathogenic infectious agents, to name a few, the Kaposi sarcoma-associated herpesvirus (Uetz et al. 2006), varicella-zoster (Uetz et al. 2006; Stellberger et al. 2010), Epstein–Barr (Calderwood et al. 2007), SARS (von Brunn et al. 2007), influenza (Shapira et al. 2009) viruses, the *Campylobacter jejuni* (Parrish et al. 2007), *Helicobacter pylori* (Hauser et al. 2014; Rain et al. 2001) and *Treponema pallidum* (Titz et al. 2008) bacteria, and *Trypanosoma brucei* (Lacomble et al. 2009) parasite. These interactome maps enhance our knowledge on these infectious agents and suggest potential therapeutic targets. Another potential of the Y2H method is to map host–pathogen protein interactions, which has been achieved for Epstein–Barr (Calderwood et al. 2007), hepatitis C (de Chassey et al. 2008), influenza (Shapira et al. 2009) and dengue (Khadka et al. 2011) viruses as well as mapping the interactions of bacterial effectors proteins with the it's host (Memisevic et al. 2013). These studies have the potential to both fundamentally change our understanding of how pathogens (virus/bacteria) modulate the host proteome and aid the development of countermeasures to control infections/diseases. Likewise, the two-hybrid screens, can also be adapted to a variety of related questions, such as the identification of mutants that avert or advance interactions (Schwartz et al. 1998; Wang et al. 2012), the screening for drugs that affect protein interactions (Vidal and Endoh 1999;

Vidal and Legrain 1999), the identification of RNA-binding proteins (SenGupta et al. 1996), or the semiquantitative determination of binding affinities (Estojak et al. 1995). The system can also be exploited to map binding domains (Rain et al. 2001; Ester and Uetz 2008), to study protein folding (Raquet et al. 2001), or to map interactions within a protein complex, for example, spliceosome (Hegele et al. 2012), proteasome (Cagney et al. 2001), flagellum (Rajagopala et al. 2007).

11.4 High-Throughput Yeast Two-Hybrid Screens

11.4.1 Array-Based Screening

In an array screening, a number of pre-defined prey proteins are tested for interactions with a bait protein. Typically the bait protein is expressed in one haploid yeast strain and the prey is expressed in another haploid yeast strain of different mating type (Fig. 11.1). The two strains are then mated so that the two proteins are expressed in the resulting diploid cell (Fig. 11.2). The screenings are done side-by-side under identical conditions with several prey proteins, and negative controls, so they can be well controlled, i.e., compared with control assays. In an array, usually each element (prey) is sequence validated and therefore it is immediately clear which two proteins are interacting when positives are selected. Most importantly, all the assays are done in an ordered array, so that background signals can be easily distinguished from true signals (Fig. 11.2, step 3). However, to perform the array screens the prey library need to be made upfront. This can be done for a subset of genes or for a whole genome (i.e., all ORFs of a genome). The array-based Y2H screenings are ideal for small genomes, for example mapping the interactions of phages (Rajagopala et al. 2011), virus (Uetz et al. 2006; Shapira et al. 2009; Khadka et al. 2011) and mapping the interactions within a protein complex (Rajagopala et al. 2007, 2012). Hands-on time and the amount of used resources grow exponentially with the number of tested proteins; this is a disadvantage for large genome sizes. However, cloned ORFeome collections of whole genomes become increasingly available for several organisms and modern cloning systems also allow direct transfer of entry clones into many specialized vectors (Walhout et al. 2000). One of the first applications of such clone collections is often a high-throughput protein interaction screening.

11.4.2 Pooled Array Screening

The pooling strategy has the potential to accelerate screening but require sequencing capacity and/or extensive pairwise Y2H screening. In the pooled array screening, preys of known identity (systematically cloned or sequenced cDNA library clones)

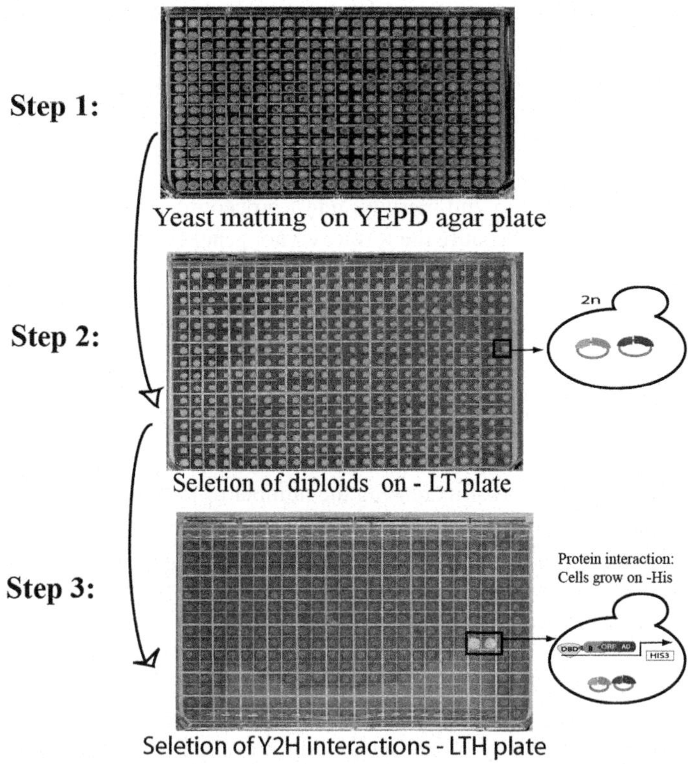

Fig. 11.2 Array-based yeast two-hybrid screens. **Step 1**: Yeast mating combines the bait and prey plasmids. The bait (DNA-binding domain (DBD) fusion) liquid culture is pinned onto YEPDA agar plates using a 384-pin pinning tool, and then the prey array (activation-domain (AD) fusion) is pinned on top of the baits using the sterile pinning tool. The mating plates are incubated at 30 °C for 16 h. **Step 2**: The cells from the yeast mating plates are transferred onto –Leu –Trp medium plates using a sterile 384-pin pinning tool. Only diploid cells will grow on the media lacking leucine and tryptophan agar plates and ensures that both the prey and bait plasmids are combined in the diploid yeast cells. **Step 3**: The diploid cells are transferred onto –Leu –Trp –His medium plates for protein interaction detection. If the bait and prey proteins interact, and an active transcription factor is reconstituted and transcription of a reporter gene is activated. Thus, the cells can grow on selective media plates

are combined and tested as pools against bait strains. The identification of the interacting protein pair commonly requires either sequencing preys in the positive yeast colonies or retesting of all members of the respective pool clones. Prof. Vidal lab at the Dana-Farber Cancer Institute, Boston MA, employed a pooling strategy for several large-scale interactome mapping projects (Rolland et al. 2014; Yu et al. 2008; Rual et al. 2005). Often, they tested each bait against pools of ~188 preys (mini-libraries) and the identity of the interacting prey in the mini-libraries was identified by sequencing the prey PCR amplicons by end-read sequencing (Rual et al. 2005), or 'stitched' the interacting bait and preys together into a single amplicon and sequenced using next-generation sequencing technology (Rolland et al. 2014; Yu et al. 2008, 2011). Stelzl et al. tested pools of 8 baits against a systematic library of individual preys and identified interactions by a 2nd interaction mating (Stelzl et al. 2005). Likewise, Jin et al. proposed an "smart-pool-array" system, which allows the deconvolution of the interacting pairs through the definition of overlapping bait pools (Jin et al. 2007), and thus usually does not depend on sequencing or a 2nd pairwise retest procedure. Preferably, the preys are pooled rather than baits, because the former do not generally result in self-activation of transcription.

11.4.3 Pooled Library Screening

The pooled library screening strategy significantly accelerates screening but might also have the disadvantage of increasing the number of false negatives and multiple sampling is essential to achieve a reasonable sampling sensitivity rate (Rajagopala et al. 2014; Yu et al. 2008) and they require significant sequencing capacity. Similar, to pooled array screening the prey library is constructed by systematically cloning each of the sequenced ORFeome or cDNA library clones into Y2H prey vector(s). In a recent study we implemented this approach to map the *E. coli* interactome network (Rajagopala et al. 2014). In this study all the prey yeast strains ($\sim$4000) were combined into a single pool and tested against each bait strains. After Y2H screening the identity of the interacting preys are identified by sequencing (Fig. 11.3).

11.4.4 Random Library Screening

Random library screens do not require systematic cloning of all prey constructs, however, the prey library must be created. Therefore, the complete DNA sequence of the genome of interest is no prerequisite. Random prey libraries can be made using genomic DNA or cDNA based libraries. For genomic libraries, the genomic DNA of interest is randomly cut, size-selected, and the resulting fragments ligated into one or more two-hybrid prey vector(s). Previous yeast two-hybrid and bacterial two-hybrid screening projects used random genomic DNA libraries (Rain et al. 2001; Joung et al. 2000). A cDNA library is made through reverse transcription of mRNA collected from specific cell types or whole organisms. To simplify the task even more, many cDNA libraries are commercially available. For example, Clontech has a collection of human and tissue-specific cDNA libraries. However the bait clones that need to be screened with a random library need to be made independently.

Similar to pooled library screens, in a random library screen a library of prey proteins is tested for interactions with a bait protein. Similar to pooled library

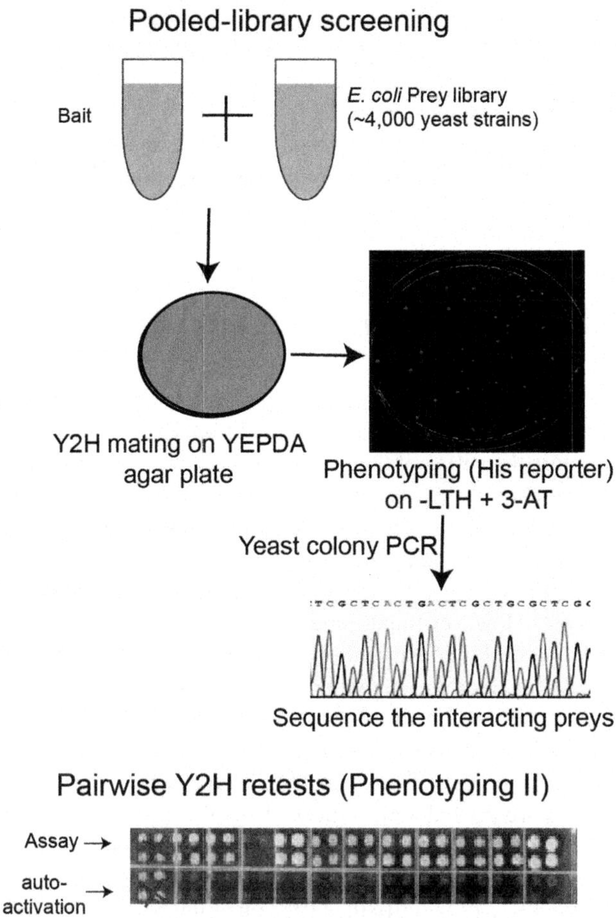

Fig. 11.3 Pooled-library yeast two-hybrid screens: A haploid yeast strain expressing a single protein as a DBD fusion is mixed with the yeast haploid strains expressing a prey library (systematical cloned). The bait and prey (1:1 ratio) culture is plated on YEPDA agar plate and the plates are incubated at 30 °C for 6 h or overnight at room temperature. During this process (yeast mating) both the prey and bait plasmids are combined in the diploid yeast cells. The cells from the mating plates are collected and transferred onto –Leu –Trp –His medium plates (supplemented with different concentration of 3-AT) for protein interaction detection, and plates are incubated at 30 °C for 4–6 days. The identity of interacting prey is identified by yeast colony PCR of positive yeast colonies, followed by DNA sequencing of the PCR product. The Y2H interactions obtained from the pool screening are subjected to pairwise retest (Phenotyping II) using fresh archival yeast stock, the screening was performed as quadruplet. Interactions which are not reproduced or showed signal in the auto-activation test (marked in *red*) should be removed from the interaction list

screens the bait protein is expressed in one yeast strain and the prey is expressed in another yeast strain of different mating type. The two strains are then mated so that the two proteins are expressed in the resulting diploid cell. The diploids are plated on interaction selective medium where only yeast cells having bait and its interacting prey will grow. The prey is identified by isolating the prey plasmids, PCR amplification of the insert, and sequencing (Sect. 11.6.11). The major limitation of the random library screening is unavailability of the indusial prey clones to perform pairwise Y2H retest or other validation assays, for example, validate a subset of interactions using orthogonal assays. Thus, evaluating the quality of PPIs relies on computation methods.

11.4.5 Adapting Next-generation Sequencing

The major shortfall of the high-throughput protein-protein interactome datasets is low coverage (Rajagopala et al. 2014; Yu et al. 2008). Even for the well-studied bacterium *E. coli*, more than 50 % of the interactome remains to be mapped (Rajagopala et al. 2014). An impotent step for high-throughput interactome-mapping methods using Y2H is determining the identities of the interacting proteins. Adapting the next-generation DNA sequencing technologies (Bennett et al. 2005; Margulies et al. 2005) as opposed to Sanger technology, would substantially increase throughput and decrease cost. However, next-generation DNA sequencing technologies are not readily applicable for identification of interacting pairs. Yu et al. describe a massively parallel interactome-mapping strategy that incorporates next-generation DNA sequencing and test the strategy in a high-throughput Y2H system (Yu et al. 2011). The methodology, termed Stitch-seq, which used PCR approach to amplify and stitch the bait and prey ORF or cDNA inserts in to a single amplican. Then the PCR products are pooled and sequenced by next-generation DNA sequencing to produce stitched interacting sequence tags. The sequencing produced an average read length of 207 bases, which are 125 bases longer than the 82-bp linker sequence between bait and preys. To identify the ORFs encoding pairs of interacting proteins, they selected reads that contained the linker sequence and also covered at least 15 bases of ORF-specific sequences on both ends of the linker. These reads could unambiguously identify pairs of unique bait and prey ORFs, after matching these sequences to human ORFeome collection used for the study. This general scheme can be readily extended to increase throughput and decrease cost for other interactome-mapping methods, particularly for binary protein-protein interaction assays

11.4.6 Analysis of Y2H Data

Analysis of raw results significantly improves the data quality of the protein interaction set. It is important to consider at least the following three parameters to obtain a high-quality Y2H data. **Auto-activation**: Is the background self-activation strength of the tested bait. The protein interaction strength of interacting pairs must be significantly higher than with all other (background) pairs. Ideally, no activation (i.e. no colony growth) should be observed in non-interacting pairs or vector control. **Reproducibility**: The protein interactions that are not reproduced in a pairwise retest experiment should be discarded. **Sticky preys**: For each prey the number of different interacting baits (prey count) is counted; preys interacting with a large number of baits are non-specific ("sticky" preys) and thus may have no biological relevance. The cut-off number depends also on the nature of baits and the number of baits screened: if a large family of related proteins are screened it is expected that many of them find the same prey. As a general guideline the number of baits interacting with a certain prey should not be larger than 5–10 % of the bait number, in an unbiased set of baits or genome-wide screenings. Furthermore, more sophisticated statistical evaluations of the raw can be adapted, i.e., using logistic regression approach that uses statistical and topological descriptors to predict the biological relevance of protein-protein interactions obtained from high-throughput screens as well as integrating known and predicted interactions from a variety of sources (Bader et al. 2004; von Mering et al. 2007).

11.4.7 False Negatives and Multiple-Variants of Y2H System

Although Y2H screens have been among the most powerful methods to detect binary protein-protein interactions, a limitation of the technology is the high incidence of false negative interactions (true interactions that are not detected) which is on the order of 70–90 % (Rajagopala et al. 2014; Yu et al. 2008). The interactome studies that have implemented proteome-scale Y2H screening in *E. coli* and yeast are shown to have identified 20–25 % of the PPIs (Rajagopala et al. 2014; Yu et al. 2008). In a previous studies, Rajagopala et al, have investigated underlying causes for this high degree of false negatives and uncovered that the structural constrains and expression levels of recombinant proteins play a major role (Rajagopala et al. 2009). Traditionally, Y2H screens have been performed using N-terminal fusion proteins of DNA-binding and activation domains. To mitigate the structural constrains Stellberger et al. constructed two new vectors that allows to make both C-terminal fusion proteins of DNA-binding and activation domains and showed that permutations of C- and N-terminal Y2H vectors detect different subsets of interactions (Stellberger et al. 2010). A study by Chen et al. benchmarked several variants of two-hybrid vectors (i.e., pGBGT7g-pGADCg, pGBGT7g-pGADT7g, pDEST32-pDEST22, pGBKCg-pGADT7g and pGBKCg-pGADCg) using a human positive reference set and a random reference set of protein-protein interaction pairs (Braun et al. 2009; Chen et al. 2010). In addition to each vector pair, they tested each protein both as activation (prey) and DNA-binding domain fusion (bait), including C-terminal fusions in pGBKCg and pGADCg. This way, they tested each protein pair in ten different configurations (Chen et al. 2010). This study clearly demonstrates that different Y2H variants (multiple-variants) detect markedly different subsets of interactions in the same interactome.

All ten different configurations of bait-prey fusions were required to detect 73 of 92 interactions (79.3 %), whereas individual vector pairs detected only 23.3 out of 92 interactions (25.3 %) on average (Chen et al. 2010). Furthermore, recent studies demonstrate the general effectiveness of the multiple-variants of Y2H system in detecting true direct binary interactions and topology among the protein complex subunits (Rajagopala et al. 2012). Having multiple-variants of Y2H vectors that detect different subsets of interactions will be of great value to generate more comprehensive protein interactions data set, thus future interactome projects must adopt multiple Y2H vector systems with proper controls and adequate stringency.

11.4.8 Quality of Y2H Interaction Data

Like any other assay system, the two-hybrid system has the potential to produce false positives. The false positives may be of technical or biological nature. A "**technical**" **false positive** is an apparent two-hybrid interaction that is not based on the assembly of two hybrid proteins (that is, the reporter gene(s) gets activated without a protein–protein interaction between bait and prey). Frequently, such false positives are associated with bait proteins that act as transcriptional activators. Some bait or prey proteins may affect general colony viability and hence enhance the ability of a cell to grow under selective conditions and activate the reporter gene. Mutations or other random events of unknown nature may be invoked as potential explanations as well. A number of procedures have been developed to identify or avoid false positives, including the utilization of multiple reporters, independent methods of specificity testing, or simply retesting the interactions in a pairwise Y2H assays to make sure that the interaction is reproducible (Rajagopala et al. 2014; Yu et al. 2008; Koegl and Uetz 2007).

A **biological false positive** involves a true two-hybrid interaction with no physiological relevance. Those include the partners that can physically interact but that are never in close proximity to one another in the cell because of distinct subcellular localization or expression at different times during the life cycle. Examples may include paralogs that are expressed in different tissues or at different developmental stages. The problem is that the "false positive" nature can rarely be proven, as there may be unknown conditions under which these proteins do interact with a biological purpose. Overall, few technical false positives can be explained mechanistically, although many may simply do interact non-physiologically. While it often remains difficult to prove the biological significance of an interaction, many studies have attempted to validate them by independent methods. Validating an interaction by other methods certainly increases the probability that it is biologically significant. In a recent study, Rajagopala et al. assess the quality of Y2H interaction by evaluating 114 randomly selected Y2H interactions in two different methods, i.e., coimmunoprecipitation and luminescence-based mammalian interactome mapping (LUMIER) assays and confirmed ~86 % of the Y2H interactions by at least one of these biochemical methods (Rajagopala et al. 2014). Similarly, when subsets of the large-scale human Y2H interactomes were evaluated about 65 % of them could be verified by independent orthogonal methods (Rual et al. 2005; Stelzl et al. 2005).

11.4.9 Integration of AP-MS and Y2H Data

It is important to note that affinity-purification followed by mass spectrometry (AP-MS) derived information about protein complexes does not provide information about the internal topology of multiprotein assemblies. Protein complexes are often interpreted as if the proteins that co-purify are interacting in a particular manner, consistent with either a spoke or matrix model (Goll and Uetz 2006). The yeast two-hybrid and other orthogonal assays detect direct binary interactions. Combination of both methods will give a better picture of protein complex topology and an experimentally derived confidence score for each interaction. In a recent study Rajagopala et al. compiled a list of 227 *E. coli* protein complexes that have three or more components as identified by AP-MS studies (Rajagopala et al. 2014). They identified the binary interactions between subunits of these complexes using proteome-scale Y2H data set and literature-curated binary interactions. Integrating these two data sets were able to map 745 binary interactions in 203 complexes, which deduce a putative complete internal topology for 15 multiprotein complexes. For another 45 complexes they determined the putative internal topology of a sub-complex with at least three subunits. However, even the combination of both methods is usually not sufficient to establish accurate topology as some interactions may be too weak to be detected individually.

11.4.10 Proteome-Scale Y2H Screening

Making an entire proteome library of an organism that can be screened *in vivo* under uniform conditions is a challenge. When proteins are screened on a genome scale, automated robotic procedures are necessary. The Y2H screening protocols described here have been extensively tested with human, yeast, bacterial, and viral proteins, but they can be applied to any other genome. Different high-throughput methods used to generate Y2H clones, i.e., proteins with AD fusions (preys) and the DBD fusions (baits), proteome-scale Y2H screening are included below. Usually, the processes starts with construction of the prey and bait libraries (Protocol 11.6.1–11.6.6); bait auto-activation tests (Protocol 11.6.7) followed by high-throughput array-based Y2H screening (Protocol 11.6.8) or pooled library screening (Protocol 11.6.9) including the selection of positives and identifying the interaction proteins by sequence (Protocol 11.6.11). Finlay, conducting the pairwise Y2H retests (Protocol 11.6.10) to make sure that the interactions are reproducible.

11.5 Materials

11.5.1 Yeast Media

1. **YEPD liquid medium:** 10 g yeast extract, 20 g peptone, 20 g glucose, dissolve in 1 L sterile water, and autoclave.
2. **YEPDA liquid medium:** 10 g yeast extract, 20 g peptone, 20 g glucose, dissolve in 1 L sterile water, and autoclave. After autoclaving, cool the medium to 60–70 °C, and then add 4 ml of 1 % adenine solution (see below).
3. **YEPDA agar medium:** 10 g yeast extract, 20 g peptone, 20 g glucose, 16 g agar, dissolve in 1 L sterile water and autoclave. After autoclaving, cool the medium to 60–70 °C, and then add 4 ml of 1 % adenine solution. Pour 40 ml into each sterile Omnitray plate (Nunc) under sterile hood, and let them solidify.
4. **Medium concentrate**: 8.5 g yeast nitrogen base, 25 g ammonium sulfate, 100 g glucose, 7 g dropout mix (see below). Make up to 1 L with sterile water, and filter-sterilize (Millipore).

11.5.2 Yeast Minimal Media

1. For 1 L of selective medium, autoclave 16 g agar in 800 ml water, cool the medium to 60–70 ° C, and then add 200 ml medium concentrate. Depending on the minimal media plates, the missing amino acids and/or 3AT (3-amino-1,2,4-triazole) solution should be added.
2. For media lacking tryptophan plates (-Trp): Add 8.3 ml leucine and 8.3 ml histidine from the amino acid stock solution (see below).
3. For media lacking leucine plates (-Leu): Add 8.3 ml tryptophan and 8.3 ml histidine solution from the amino acid stock solution (see below).
4. For media lacking tryptophan and leucine plates (-Leu –Trp): Add 8.3 ml histidine from the stock solution (see below).
5. For media lacking tryptophan, leucine, and histidine plates (-Leu –Trp –His): Nothing needs to be added.
6. For –Leu –Trp -His + 3 mM 3AT plates: Add 6 ml of 3AT (3-amino-1,2, 4-triazole, 0.5 M) to a final concentration of 3 mM.
7. **Dropout mix** (-His, -Leu, -Trp): Mix 1 g methionine, 1 g arginine, 2.5 g phenylalanine, 3 g lysine, 3 g tyrosine, 4 g isoleucine, 5 g glutamic acid, 5 g aspartic acid, 7.5 g valine, 10 g threonine, 20 g serine, 1 g adenine, and 1 g uracil and store under dry, sterile conditions.
8. **Amino acid stock solutions**

   Adenine solution (1 %): Dissolve 10 g of adenine in 1 L, 0.1 M NaOH solution and sterile filter.

Histidine solution (His): Dissolve 4 g of histidine in 1 L sterile water and sterile filter.

Leucine solution (Leu): Dissolve 7.2 g of leucine in 1 L sterile water and sterile filter.

Tryptophan solution (Trp): Dissolve 4.8 g of tryptophan in 1 L sterile water and sterile filter

11.5.3 Reagents for Yeast Transformation

1. Salmon sperm DNA (Carrier DNA): Dissolve 7.75 mg/ml salmon sperm DNA (Sigma) in sterile water, autoclave for 15 min at 121 °C, and store at −20 °C.
2. Dimethylsulfoxide (DMSO, Sigma).
3. Competent yeast strains, e.g., AH109 (for baits), and Y187 (for preys).
4. 0.1 M Lithium acetate (LiOAc).
5. Yeast minimal media plates (depending on the selective markers on the yeast expression plasmid).
6. 96PEG solution: Mix 45.6 g PEG (Sigma), 6.1 ml of 2 M LiOAc, 1.14 ml of 1 M Tris, pH 7.5, and 232 μl 0.5 M EDTA. Make up to 100 ml with sterile water and autoclave.
7. Plasmid clones (i.e., bait and prey clones)

11.5.4 Reagents for Bait Auto-Activation Test

1. YEPDA liquid medium.
2. –Trp –Leu ("–LT") selective media agar plates (see Sect. 11.5.2).
3. Selective media agar plates without Trp, Leu, and His ("–LTH"), but with different concentrations of 3-AT, e.g., 0 mM, 1 mM, 3 mM, 10 mM, and 50 mM (–LTH/3-AT plates).
4. Bait strains that need to be tested and prey strains carrying the prey vector (empty vector), e.g., Y187 strain with pGADT7g plasmid.

11.5.5 Array-Based Y2H Screening (Work Station)

1. 20 % (v/v) bleach (1 % sodium hypochlorite).
2. 95 % (v/v) ethanol.
3. Single-well microtiter plate (e.g., OmniTray; Nalge Nunc) containing solid YEPD + adenine medium (see Sect. 11.5.1), –Leu –Trp, –His –Leu –Trp, and –His –Leu –Trp + different concentrations of 3AT.

4. 384-Pin replicator for Beckman Biomek FX work station.
5. Bait liquid culture (DBD fusion-expression yeast strain).
6. Yeast prey array on solid YEPDA plates.

11.5.6 Reagents for Pairwise Y2H Retests

1. 96-well microtiter plates (U- or V-shaped).
2. YEPDA medium and YEPDA agar in Omnitrays (Nunc).
3. Selective agar media plates (–LT, –LTH with 3-AT).
4. Prey yeast strain carrying empty prey plasmid, e.g., pGADT7g in Y187 strain.
5. Bait and prey strains to be retested.

11.6 Protocols

11.6.1 Construction of Y2H Libraries for Proteome-Scale Screening

After the set of proteins or entire ORFeome to be included in the systematic array or library is defined, the coding genes need to be cloned into several Y2H bait and prey expression vectors. In order to facilitate the cloning of a large number to proteins, site-specific recombination-based systems are commonly used (e.g., Gateway cloning, Fig. 11.4) (Walhout et al. 2000).

11.6.2 Gateway Cloning

Adapting Gateway (Life Technologies) technology provides a fast and efficient way of cloning the ORFs (Walhout et al. 2000). It is based on the site-specific recombination properties of bacteriophage lambda (Landy 1989); recombination is mediated between so-called attachment sites (att) of DNA molecules: between attB and attP sites or between attL and attR sites. In the first step of cloning the gene of interest is inserted into a specific Gateway entry vector by recombining a PCR product of the ORF flanked by attB sites with the attP sites of a pDONR vector (Life Technologies). The resulting entry clone plasmid contains the gene of interest flanked by attL recombination sites. These attL sites can be recombined with attR sites on a destination vector, resulting in a plasmid for functional protein expression in a specific host. For example, a Gateway entry clone (pDONR vector) can be subsequently cloned into multiple Y2H vectors (Table 11.1) and other Gateway compatible expression vectors as required.

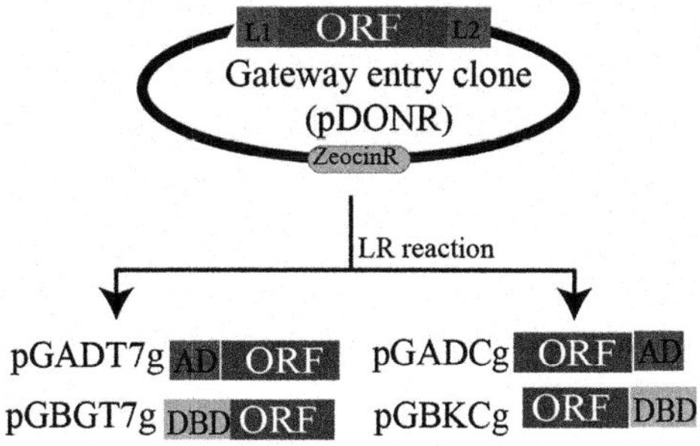

Fig. 11.4 Generating Y2H baits and preys using Gateway cloning. The Gateway-based Y2H expression clones are made by sub-cloning the ORFs of interest from a Gateway entry vector (pDONR/zeo or pDONR201) into the Y2H expression vectors (Table 11.1) by Gateway LR reaction (Walhout et al. 2000). It is possible to simultaneously transfer a single ORF from an entry vector to four target vectors using a single LR reaction, as long as the resulting expression plasmids can be separated using prototrophic markers specific to each vector (Stanyon et al. 2003). We have used up to three Y2H expression vectors (i.e., pGADT7g, pGBGT7g, and pGBKCg, they carry Ampicillin, Gentamycin and Kanamycin resistance respectively) in a LR reaction

Table 11.1 Reagents for a yeast two-hybrid screen

Bait and prey vectors							
		Gal4-fusion		Selection			
Vector	Promoter	DBD	AD	Yeast	Bacterial	Ori	Source
pDEST22	fl-ADH1	–	N-term	Trp1	Amp.	CEN	Life Technologies
pDEST32	fl-ADH1	N-term	–	Leu2	Gent.	CEN	Life Technologies
pGBKT7g	t-ADH1	N-term	–	Trp1	Kan.	2 μ	Uetz et al. (2006)
pGBGT7g	t-ADH1	N-term	–	Trp1	Gent.	2 μ	Rajagopala et al. (2014)
pGBKCatg	t-ADH1	C-term		Trp1	Kan.	2 μ	Rajagopala et al. (2014)
pGADT7g	fl-ADH1	–	N-term	Leu2	Amp.	2 μ	Uetz et al. (2006)
pGBKCg	t-ADH1	C-term	–	Trp1	Kan.	2 μ	Stellberger et al. (2010)
pGADCg	fl-ADH1	–	C-term	Leu2	Amp.	2 μ	Stellberger et al. (2010)
Yeast strains							
Bait yeast strain		AH109 (Clontech)					
Prey yeast strain		Y187 (Clontech)					
Media and instruments							
Yeast media		YEPDA, selective liquid media and agar plates					
Pin tool		Optional but necessary when large number are tested					

Fl full-length, *N/C-term.* N/C-terminal (fusion), *Amp.* Ampicillin, *Kan.* Kanamycin, *Gen.* Gentamicin

11.6.3 ORFeome Collections

The starting point of a systematic proteome-scale Y2H screening is the construction of an ORFeome. An ORFeome represents all ORFs of a genome; in some cases selected gene set is individually cloned into Gateway entry vector. More and more ORFeomes are available and can be directly used for generating the Y2H bait and

Table 11.2 Yeast strains and their genotypes

Yeast strains	Genotypes
Y187	MATα, ura3- 52, his3- 200, ade2- 101, trp1- 901, leu2- 3, 112, gal4Δ, met–, gal80Δ,URA3::GAL1UAS -GAL1TATA -lacZ (after Harper et al. 1993)
AH109	MATa, trp1-901, leu2-3, 112, ura3-52, his3-200, gal4Δ, gal80Δ, LYS2::GAL1UAS-GAL1TATA-HIS3, GAL2UAS-GAL2TATA-ADE2, URA3::MEL1UAS-MEL1 TATA-lacZ (after James et al. 1996)

prey constructs. These ORFeome range from small viral genomes, e.g., KSHV and VZV (Uetz et al. 2006), Phages (Rajagopala et al. 2011), to several bacterial genomes such as *E. coli, Helicobacter pylori, Bacillus anthracis* or *Yersinia pestis* (Rajagopala et al. 2010). Clone sets of multicellular eukaryotes, e.g. *C. elegans* (Lamesch et al. 2004), human (Rual et al. 2004), or plant (Gong et al. 2004), have also been described. However, not all genes of interest are already available in entry vectors.

11.6.4 The Prey Array

The Y2H array is set up from an ordered set of AD-containing strains (preys). The prey constructs are assembled by transfer of the ORFs from entry vectors into specific prey vectors by recombination. Several prey vectors for the Gateway system are available. In our lab we primarily use the Gateway-compatible pGADT7g, pGADCg vectors, a derivative of pGADT7 (Clontech), and pDEST22 (Life Technologies) (Table 11.1). These prey constructs are transformed into haploid yeast cells using yeast transformation protocol (described in Protocol 11.6.6), e.g. the Y187 strain (mating type alpha) (Table 11.2). Finally, individual yeast colonies, each carrying one specific prey construct, are arrayed on agar plates in a 96- or 384-format usually as duplicates or quadruplicates.

11.6.5 The Bait Strains

Similar to prey construction, the bait clones are also constructed by recombination-based transfer of the ORFs into specific bait vectors. Bait vectors used in our lab are the Gateway technology compatible pGBGT7g, pGBKCg vectors, a derivative of pGADT7 vector (Clontech) and pDEST32 (Life Technologies) (Table 11.1). The bait constructs are also transformed into haploid yeast cells (Protocol 11.6.6), e.g. the AH109 strain (mating type a) (Table 11.2). After auto-activation testing, the baits can be tested for interactions against the Y2H prey array or pooled prey library. It is important to note that bait and prey must be transformed into yeast strains of opposite mating types to combine bait and prey plasmids by yeast mating and to co-express the recombinant proteins in diploids. Bait and prey plasmids can go into either mating type. However, this decision also depends on existing bait or prey libraries to which the new library may be screened later. Moreover, at least one of the haploid strains must contain a two-hybrid reporter gene (here, HIS3 gene under GAL4 promoter).

11.6.6 Yeast Transformation

This method is recommended for the high-throughput transformation of the bait or prey plasmid clones into corresponding yeast strains. This protocol is suitable for ~1000 transformations; it can be scaled up and down as required. Selection of the transformed yeast cells requires synthetic media plates (leucine- or tryptophan-free agar media depending on the selective marker on the Y2H plasmid).

11.6.6.1 Prepare Competent Yeast Cells

1. Inoculate 250 ml YEPD liquid medium with freshly grown yeast strains on YEPD agar medium in a 1 L flask and grow in a shaker (shaking at 200 rpm) at 30 °C. Remove the yeast culture from the shaker when the cell density reaches OD 0.8–1. This usually takes 12–16 h.
2. Spin the cells at 2000 × *g* for 5 min at room temperature; pour off the supernatant.
3. Dissolve the cell pellet in 30 ml of LiOAc (0.1 M); make sure pellet is completely dissolved and there are no cell clumps.
4. Transfer the cells into a 50-ml Falcon tube and spin the cells at 2000 × *g* for 5 min at room temperature
5. Pour off the supernatant, and dissolve the cell pellet in 10 ml LiOAc (0.1 M). **Prepare the yeast transformation mix**: Mixing the following components in a 200-ml sterile bottle:

Component	For 1000 reactions
96PEG	100 ml
Salmon sperm DNA	3.2 ml
DMSO	3.4 ml

6. Add the competent yeast cells prepared above (**steps 1–5**) to the yeast transformation mix; shake the bottle vigorously by hand, or vortex for 1 min.

7. Pipette 100 μl of the yeast transformation mix into a 96-well plate (we generally use Costar 3596 plates) by using a robotic liquid handler (e.g., Biomek FX) or a multistep pipette.
8. Now add 25–50 ng of plasmid; keep one negative control (i.e., only yeast transformation mix).
9. Seal the 96-well plates with plastic or aluminum tape and vortex for 2–3 min. Care should be taken to seal the plates properly; vigorous vortexing might cause cross-contamination.
10. Incubate the plates at 42 °C for 30 min.
11. Spin the 96-well plate for 5 min at 2000 × *g*; discard the supernatant and aspirate by tapping on a cotton napkin a couple of times.
12. Wash the cell pellet with 150 μl sterile H_2O
13. Spin the 96-well plate for 5 min at 2000 × *g*; discard the supernatant
14. Add 25 μl sterile H_2O
15. Transfer 10 μl the cells to selective agar plate to select yeast with transformed plasmid (single-well Omnitrays from Nunc are well suited for robotic automation). As an alternative to the robotic automation, one can use a multichannel pipette to transfer the cells. Allow the yeast spots to dry on the plates.
16. Incubate at 30 °C for 2–3 days. Colonies start appearing after 24 h.

11.6.7 Bait Auto-activation Tests

Prior to the Y2H screening, the bait yeast strains should be examined for auto-activation (self-activation). Auto-activation is defined as detectable bait-dependent reporter gene activation in the absence of any prey interaction partner. Weak to intermediate strength auto-activator baits can be used in two-hybrid array screens because the corresponding bait–prey interactions confer stronger signals than the auto-activation background. In case of the *HIS3* reporter gene, the self-activation background can be suppressed by titrating with 3-Amino-1,2,4-triazole(3-AT), a competitive inhibitor of *HIS3*. Auto-activation of all the baits is examined on selective plates containing different concentrations of 3-AT. The lowest concentration of 3-AT that suppresses growth in this test is used for the interaction screen (see below), because it avoids background growth while still detecting true interactions.

The aim of this test is to measure the background reporter activity of bait proteins in absence of an interacting prey protein. This measurement is used for choosing the selection conditions for the Y2H screening.

1. Bait strains are arrayed on a single-well Omnitray agar plate; usually standard 96-spot format.
2. The arrayed bait strains are mated with a prey strain carrying the empty prey plasmid, e.g., Y187 strain with pGADT7g. Mating is conducted according to the standard screening protocol as described in Protocol 11.6.8. Note that here an array of baits is tested whereas in a “real” screen (Protocol 11.6.8) an array of preys is tested.

3. After selecting for diploid yeast cells (on –LT agar), the cells are transferred to media selecting for the HIS3p reporter gene activity as described in Protocol 11.6.8. The -LTH transfer may be done to multiple plates with increasing concentrations of 3-AT. Recommended 3-AT concentrations for the –LTH plates are 0, 1, 3, 10, 25, and 50 mM.
4. These –LTH + 3-AT plates are incubated for 4–6 days at 30 °C. The auto-activation level of each of the bait is assessed and the lowest 3-AT concentration that completely prevents colony growth is noted. As this concentration of 3-AT suppresses reporter activation in the absence of an interacting prey, this 3-AT concentration is added to –LTH plates in the actual interaction screens as described in Protocol 11.6.8.

11.6.8 Screening for Protein Interactions Using a Protein Array

The Y2H prey array can be screened for protein interactions by a mating procedure that can be carried out using robotics (Biomek FX work station). A yeast strain expressing a single candidate protein as a DBD fusion is mated to all the colonies in the prey array (Fig. 11.2, step 1). After mating, the colonies are transferred to a diploid-specific medium, and then to the two-hybrid interaction selective medium.

A 384-pin replicating tool (e.g., High-Density Replication Tool; V&P Scientific) can be used to transfer the colonies form one agar plate to another and between the transfer steps, the pinning tool must be sterilized (described below).

Note that not all plastic ware is compatible with robotic devices, although most robots can be reprogrammed to accept different consumables. In the procedure described here, the prey array is gridded on 86 × 128 mm single-well microtiter plates (e.g., OmniTray, Nalge Nunc International) in a 384-colony format (see Fig. 11.2).

1. *Sterilization*: Sterilize a 384-pin replicator by dipping the pins in the sequential order into 20 % bleach for 20 s, sterile water for 1 s, 95 % ethanol for 20 s, and sterile water again for 1 s. Repeat this sterilization after each transfer.

   Note 1: Immersion of the pins into these solutions must be sufficient to ensure complete sterilization. When automatic pinning devices are used, the solutions need to be checked and refilled occasionally (especially ethanol which evaporates faster than the others).

Day 1:

1. *Preparing prey array for screening*: Use the sterile replicator to transfer the yeast prey array from selective plates to single-well microtiter plates containing solid YEPD medium and grow the array overnight in a 30 °C incubator.

   Note 2: Usually in a systematic array-based Y2H screening, duplicate or quadruplicate prey arrays are used. Ideally, the master prey array should be kept on selective agar plates. The master array should only be used to make "working" copies on YEPDA agar plates for mating. The template can be used for 1–2 weeks; after 2 weeks it is recommended to copy the array onto fresh selective agar plates. Preys and bait clones tend to lose the plasmid if stored on YEPDA for longer periods, which may reduce the mating and screening efficiency.

2. *Preparing bait liquid culture (DBD fusion-expressing yeast strain):* Inoculate 20 ml of liquid YEPD medium in a 250-ml conical flask with a bait strain and grow overnight in a 30 °C shaker

   Note 3: If the bait strains are frozen, they are grown on selective agar medium plates and grown for 2–3 days at 30 °C. Baits from this plate are then used to inoculate the liquid YEPD medium. It is important to make a fresh bait culture for Y2H mating, as keeping the bait culture on reach medium (YEPD) for a long time may cause loss of plasmids. Usually we grow baits overnight for the screening.

Day 2:

3. *Mating procedure*: Pour the overnight liquid bait culture into a sterile Omnitray plate. Dip the sterilized pins of the pin replicator (thick pins of ~1.5 mm diameter should be used to pin baits) into the bait liquid culture and place directly onto a fresh single-well microtiter plate containing YEPDA agar media. Repeat with the required number of plates and allow the yeast spots to dry onto the plates.
4. Pick up the fresh prey array yeast colonies with sterilized pins (thin pins of ~1 mm diameter should be used to pin the preys) and transfer them directly onto the baits on the YEPDA plate, so that each of the 384 bait spots per plate receives different prey yeast cells (i.e., a different AD fusion protein). Incubate overnight at 30 °C to allow mating (Fig. 11.2, step 1).

   Note 4: Mating usually take place in <15 h, but a longer period is recommended because some bait strains show poor mating efficiency. Adding adenine into the bait culture before mating increases the mating efficiency of some baits.
5. *Selection of Diploids*: For the selection of diploids, transfer the colonies from YEPDA mating plates to diploids selection minimal media agar plates (–Leu –Trp plates) using the sterilized pinning tool (thick pins should be used in this step). Grow the plates for 2–3 days at 30 °C until the colonies are >1 mm in diameter (Fig. 11.2, step 2).

   Note 5: This step is an essential control step to ensure successful mating because only diploid cells containing the Leu2 and Trp1 markers on the prey and bait vectors, respectively, will grow in this medium. This step also helps the recovery of the colonies and increases the efficiency of the next interaction selection step.
6. *Interaction selection:* Transfer the diploid yeast cells from –Leu –Trp plates to interaction selection minimal media agar plates (–His –Leu –Trp plates), using the sterilized pinning tool. If the baits are auto-activating, they have to be transferred onto –His –Leu –Trp supplemented with a specific concentration of 3-AT plates (Protocol 11.6.1). Incubate the plates at 30 °C for 4–6 days.

7. Score the interactions by looking for growing colonies that are significantly above background by size and are present as duplicate or quadruplicate colonies.
8. Most two-hybrid-positive colonies appear within 3–5 days, but occasionally positive interactions can be observed later. Very small colonies are usually designated as background; however, the real positives signal should be compared with background vector control.
9. Scoring can be done manually or using automated image analysis procedures. When using image analysis, care must be taken not to score contaminated colonies as positives.

11.6.9 Screening for Protein Interactions Using Pooled Libraries

Although the Y2H array screening ensures each pairwise combination in the library will be tested, it may be not feasible of large genomes, as the screening throughput increases exponentially with the genome size. Pooled library screening is an alternative strategy to significantly accelerate the screening. The prey clones are made by systematically cloning the ORFeome into Y2H vectors. The interacting preys in the library are identified by sequencing the Y2H positive yeast colonies (Protocol 11.6.11). Furthermore, to ensure the reproducibility of the interaction, all the interaction pairs will be subjected to Y2H pairwise retest (Protocol 11.6.10)
Day 1:

1. Preparing pooled prey libraries: The prey library strains (yeast) expressing each ORF is grown on a selective liquid medium (in 2 ml deep well plate) for 48 h in a 30 °C shaker. Equal amount of each of the freshly grown preys are combined into to a single pool. Ideally preys should be grown freshly for each batch of screening.
2. Inoculate the empty prey vector in 200 ml selective medium (Y2H negative control)
3. Preparing bait liquid culture (DBD fusion-expressing yeast strain): Inoculate 10 ml of selective medium (medium lacking tryptophan, depending on the selective marker on the bait plasmid) with bait fusion-expressing yeast strain and grow the yeast overnight in a 30 °C incubator.

Day 2:

4. Mating procedure: Mix bait and prey at a 1:1 ratio, for example 4 OD bait (4 ml of OD = 1) and 4 OD prey (4 ml of OD = 1) culture in 15 ml Falcon tubes.

   Note 6: The amount of (OD units) bait and prey used for the screening depends on the complexity of the prey library, in case of E. coli which contains about 4000 ORFs; we use 4 OD units of baits and preys. In case of human cDNA library screening we recommend using 12 OD units of baits and preys each

5. For each of the bait include one negative control, i.e., mix bait and empty prey vector at a 1:1 ratio.
6. Centrifuge the bait and prey solution for 2 min, at 3000 rpm at room temperature, discard supernatant
7. Suspend the yeast pellet in 500 μl YPDA liquid medium, and plate on YEPDA agar plate (60 × 15 mm), and air dry the plates.
8. Incubate the plates at 30 °C for 6 h or overnight at room temperature.
9. After incubation, collect the cells by washing the plate with 2 ml of sterile water.
10. Spin down the cells, remove the supernatant and wash the cells by adding 2 ml of sterile water
11. Suspend the cells in 2 ml of selective medium (media lacking tryptophan, and leucine).
12. Plate 500–1000 μl on the interaction selective agar plates –Leu –Trp -His supplemented with predefined concentration of 3-AT, based on auto-activation test of the bait. The remaining sample can be stored at 4 °C for 4–6 days for further use.

    Note 7: This step is an essential control step to ensure successful Y2H mating procedure, because only diploid cells containing the Leu2 and Trp1 markers on the prey and bait vectors, respectively, will grow on media lacking tryptophan, and leucine medium. This step also helps the recovery of the colonies and increases the efficiency of the next interaction selection step. To measure the diploids make an aliquot of 1:100 dilution of the sample (step 11) and plate the cells on –Leu –Trp plates, the screening depth in millions, should be > 0.1 million, up to 1 million diploids in case of E. coli library screening. This is at least twenty times the number of library size.
13. *Interaction selection*: Incubate the –Leu –Trp -His + 3-AT for 4–6 days at 30 °C until the colonies are ~1 mm in diameter.
14. Two-hybrid positives: The interaction selection plates that show colony growth but no colonies on control plates (bait mated to empty prey vector) are the two-hybrid positive yeast clones. If the control plates show even few colonies the diploids should be plated on selective plates with higher concentration of 3-AT.
15. Identity of interacting preys: The positive yeast colonies are picked either manually or using robotics and subjected to yeast colony PCR (Protocol 11.6.11), followed by DNA sequencing to identify the preys.

11.6.10 Pairwise Y2H Retesting

A major consideration when using the Y2H system is the number of false positives, particularly in the pooled library screening. The major sources for false positives are non-reproducible signals that arise through little-understood mechanisms. Thus, pairwise retesting can identify most of the false positives. We routinely use at least duplicate tests, although quadruplicates should be used if possible (Fig. 11.2).

Retesting is done by mating the interaction pair to be tested and by comparing the activation strength of this pair with the activation strength of a control, usually the bait mated with the strain that contains the empty prey vector. Testing the reproducibility of an interaction greatly increases the reliability of the Y2H data.

1. Re-array bait and prey strains of each interaction pair to be tested into 96-well microtiter plates. Use separate 96-well plates for baits and preys. For each retested interaction, fill one well of the bait plate and one corresponding well of the prey plate with 150 μl selective liquid medium (media lacking Leucine or Tryptophan).
2. For each retested interaction, inoculate the bait strain into a well of the 96-well plate and the prey strain at the corresponding position of the 96-well prey plate, for example, bait at position B2 of the bait plate and prey at position B2 of the prey plate. In addition, inoculate the prey strain with the empty prey vector (e.g., strain Y187 with plasmid pGADT7g) into 20 ml selective liquid medium.
3. Incubate the plates overnight at 30 °C.
4. Mate the baits grown in the bait plate with their corresponding preys in the prey plate. In addition, mate each bait with the prey strain carrying an empty prey vector as a background activation control. The mating is done as described in Protocol 11.6.8, using the bait and prey 96-well plates directly as the source plates.

   Note 8: First the baits are transferred from their 96-well plate to two YEPDA plates (interaction test and control plate) using a 96-well replication tool. Let the plate dry for 10–20 min. Then transfer the prey's from their 96-well plate onto the first YEPDA plate and the empty prey vector control strain onto the second YEPDA plate
5. The transfers to selective plates and incubations are done as described in Protocol 11.6.8. As before, test different baits with different activation strengths on a single plate and pin the diploid cells onto –LTH plates with different concentrations of 3-AT. For choosing the 3-AT range, the activation strengths (Protocol 11.6.7) serve as a guideline.
6. After incubating for 4–6 days at 30 °C on –LTH/3-AT plates, the interactions are scored; positive interactions show a clear colony growth at a certain level of 3-AT, whereas no growth should be seen in the control (bait mated with empty vector strain).

11.6.11 Yeast Colony PCR and Sequencing Sample Preparation

Yeast Colony PCR This protocol is designed to amplify the insert of the preys or baits in the two-hybrid positive yeast clones, using primers that bind to the upstream and downstream region of the insert. The PCR is optimized for 30 μl reaction; the total volume of the reaction can be scaled up and down as required.

1. Pick the yeast colony from interaction selective plate into 100 μl of sterile H_2O, in 96 well plate; store the plate at −80 freezer for longer storage.
2. Take a new 96 well PCR plate and pipette 5 unit of zymolyase (1 μl) enzyme to each well.
3. Add 9 μl of above yeast (step 1), and incubate at 30 °C for 60 min.
4. After incubation, add 20 μl PCR master mix with forward and reverse primers specific to prey or bait vector used in the Y2H screening.
5. Run PCR cycles as recommended by the enzyme (polymerase) provider.
6. After PCR, load 5 μl of PCR reaction into agarose gel to check PCR amplicons.

Purification of the PCR Amplicons for Sequencing To clean up PCR products before sequencing, the PCR reaction is subjected to the exonuclease I, which removes leftover primers while the Shrimp Alkaline Phosphatase (SAP) removes the dNTPs

1. Spin the Yeast colony PCR plate at 2000 rpm for 3 min (to sediment yeast debris).
2. Pipette 8 μl of PCR sample without touching the bottom yeast debris, into new PCR plate.
3. Make the SAP master mix by mixing the following reagents

SAP master mix	
Components	100 samples
10× SAP buffer	50 μl
Water	890 μl
SAP (1 U/μl)	50 μl
Exonuclease I (10 U/μl)	10 μl

4. Add 10 μl of SAP master mix to 8 μl of PCR sample.
5. Incubate in the thermocycler as follows: 37 °C for 60 min, 72 °C for 15 min, then put on hold at 4 °C.
6. Use the sample for DNA sequencing using primers specific to prey or bait vector.

Acknowledgements Work on this chapter was funded by NIH grant RO1GM79710.

References

Bader JS, Chaudhuri A, Rothberg JM, Chant J (2004) Gaining confidence in high-throughput protein interaction networks. Nat Biotechnol 22(1):78–85

Bennett ST, Barnes C, Cox A, Davies L, Brown C (2005) Toward the 1,000 dollars human genome. Pharmacogenomics 6(4):373–382. doi:10.1517/14622416.6.4.373

Braun P, Tasan M, Dreze M, Barrios-Rodiles M, Lemmens I, Yu H, Sahalie JM, Murray RR, Roncari L, de Smet AS, Venkatesan K, Rual JF, Vandenhaute J, Cusick ME, Pawson T, Hill DE, Tavernier J, Wrana JL, Roth FP, Vidal M (2009) An experimentally derived confidence score for binary protein-protein interactions. Nat Methods 6(1):91–97

Brent R, Ptashne M (1985) A eukaryotic transcriptional activator bearing the DNA specificity of a prokaryotic repressor. Cell 43(3 Pt 2):729–736

Cagney G, Uetz P, Fields S (2001) Two-hybrid analysis of the Saccharomyces cerevisiae 26S proteasome. Physiol Genomics 7(1):27–34

Calderwood MA, Venkatesan K, Xing L, Chase MR, Vazquez A, Holthaus AM, Ewence AE, Li N, Hirozane-Kishikawa T, Hill DE, Vidal M, Kieff E, Johannsen E (2007) Epstein-Barr virus and virus human protein interaction maps. Proc Natl Acad Sci U S A 104(18):7606–7611. doi:10.1073/pnas.0702332104

Chen YC, Rajagopala SV, Stellberger T, Uetz P (2010) Exhaustive benchmarking of the yeast two-hybrid system. Nat Methods 7(9):667–668, author reply 8

de Chassey B, Navratil V, Tafforeau L, Hiet MS, Aublin-Gex A, Agaugue S, Meiffren G, Pradezynski F, Faria BF, Chantier T, Le Breton M, Pellet J, Davoust N, Mangeot PE, Chaboud A, Penin F, Jacob Y, Vidalain PO, Vidal M, Andre P, Rabourdin-Combe C, Lotteau V (2008) Hepatitis C virus infection protein network. Mol Syst Biol 4:230. doi:10.1038/msb.2008.66

Ester C, Uetz P (2008) The FF domains of yeast U1 snRNP protein Prp40 mediate interactions with Luc7 and Snu71. BMC Biochem 9:29. doi:10.1186/1471-2091-9-29

Estojak J, Brent R, Golemis EA (1995) Correlation of two-hybrid affinity data with in vitro measurements. Mol Cell Biol 15(10):5820–5829

Fields S, Song O (1989) A novel genetic system to detect protein-protein interactions. Nature 340(6230):245–246

Gavin AC, Bosche M, Krause R, Grandi P, Marzioch M, Bauer A, Schultz J, Rick JM, Michon AM, Cruciat CM, Remor M, Hofert C, Schelder M, Brajenovic M, Ruffner H, Merino A, Klein K, Hudak M, Dickson D, Rudi T, Gnau V, Bauch A, Bastuck S, Huhse B, Leutwein C, Heurtier MA, Copley RR, Edelmann A, Querfurth E, Rybin V, Drewes G, Raida M, Bouwmeester T, Bork P, Seraphin B, Kuster B, Neubauer G, Superti-Furga G (2002) Functional organization of the yeast proteome by systematic analysis of protein complexes. Nature 415(6868):141–147

Giot L, Bader JS, Brouwer C, Chaudhuri A, Kuang B, Li Y, Hao YL, Ooi CE, Godwin B, Vitols E, Vijayadamodar G, Pochart P, Machineni H, Welsh M, Kong Y, Zerhusen B, Malcolm R, Varrone Z, Collis A, Minto M, Burgess S, McDaniel L, Stimpson E, Spriggs F, Williams J, Neurath K, Ioime N, Agee M, Voss E, Furtak K, Renzulli R, Aanensen N, Carrolla S, Bickelhaupt E, Lazovatsky Y, DaSilva A, Zhong J, Stanyon CA, Finley RL Jr, White KP, Braverman M, Jarvie T, Gold S, Leach M, Knight J, Shimkets RA, McKenna MP, Chant J, Rothberg JM (2003) A protein interaction map of Drosophila melanogaster. Science 302(5651):1727–1736

Goll J, Uetz P (2006) The elusive yeast interactome. Genome Biol 7(6):223

Gong W, Shen YP, Ma LG, Pan Y, Du YL, Wang DH, Yang JY, Hu LD, Liu XF, Dong CX, Ma L, Chen YH, Yang XY, Gao Y, Zhu D, Tan X, Mu JY, Zhang DB, Liu YL, Dinesh-Kumar SP, Li Y, Wang XP, Gu HY, Qu LJ, Bai SN, Lu YT, Li JY, Zhao JD, Zuo J, Huang H, Deng XW, Zhu YX (2004) Genome-wide ORFeome cloning and analysis of Arabidopsis transcription factor genes. Plant Physiol 135(2):773–782

Harper JW, Adami GR, Wei N, Keyomarsi K, Elledge SJ (1993) The p21 Cdk-interacting protein Cip1 is a potent inhibitor of G1 cyclin-dependent kinases. Cell 75(4):805–816

Hauser R, Ceol A, Rajagopala SV, Mosca R, Siszler G, Wermke N, Sikorski P, Schwarz F, Schick M, Wuchty S, Aloy P, Uetz P (2014) A second-generation protein-protein interaction network of Helicobacter pylori. Mol Cell Proteomics 13(5):1318–1329. doi:10.1074/mcp.O113.033571

Hegele A, Kamburov A, Grossmann A, Sourlis C, Wowro S, Weimann M, Will CL, Pena V, Luhrmann R, Stelzl U (2012) Dynamic protein-protein interaction wiring of the human spliceosome. Mol Cell 45(4):567–580. doi:10.1016/j.molcel.2011.12.034

Ho Y, Gruhler A, Heilbut A, Bader GD, Moore L, Adams SL, Millar A, Taylor P, Bennett K, Boutilier K, Yang L, Wolting C, Donaldson I, Schandorff S, Shewnarane J, Vo M, Taggart J, Goudreault M, Muskat B, Alfarano C, Dewar D, Lin Z, Michalickova K, Willems AR, Sassi H, Nielsen PA, Rasmussen KJ, Andersen JR, Johansen LE, Hansen LH, Jespersen H, Podtelejnikov A, Nielsen E, Crawford J, Poulsen V, Sorensen BD, Matthiesen J, Hendrickson RC, Gleeson F, Pawson T, Moran MF, Durocher D, Mann M, Hogue CW, Figeys D, Tyers M (2002) Systematic identification of protein complexes in Saccharomyces cerevisiae by mass spectrometry. Nature 415(6868):180–183

Ito T, Chiba T, Ozawa R, Yoshida M, Hattori M, Sakaki Y (2001) A comprehensive two-hybrid analysis to explore the yeast protein interactome. Proc Natl Acad Sci U S A 98(8):4569–4574
James P, Halladay J, Craig EA (1996) Genomic libraries and a host strain designed for highly efficient two-hybrid selection in yeast. Genetics 144(4):1425–1436
Jin F, Avramova L, Huang J, Hazbun T (2007) A yeast two-hybrid smart-pool-array system for protein-interaction mapping. Nat Methods 4(5):405–407
Joung JK, Ramm EI, Pabo CO (2000) A bacterial two-hybrid selection system for studying protein-DNA and protein-protein interactions. Proc Natl Acad Sci U S A 97(13):7382–7387
Kerrien S, Aranda B, Breuza L, Bridge A, Broackes-Carter F, Chen C, Duesbury M, Dumousseau M, Feuermann M, Hinz U, Jandrasits C, Jimenez RC, Khadake J, Mahadevan U, Masson P, Pedruzzi I, Pfeiffenberger E, Porras P, Raghunath A, Roechert B, Orchard S, Hermjakob H (2012) The IntAct molecular interaction database in 2012. Nucleic Acids Res 40(Database issue):D841–D846. doi: 10.1093/nar/gkr1088
Khadka S, Vangeloff AD, Zhang C, Siddavatam P, Heaton NS, Wang L, Sengupta R, Sahasrabudhe S, Randall G, Gribskov M, Kuhn RJ, Perera R, LaCount DJ (2011) A physical interaction network of dengue virus and human proteins. Mol Cell Proteomics 10(12):M111 012187. doi: 10.1074/mcp.M111.012187
Koegl M, Uetz P (2007) Improving yeast two-hybrid screening systems. Brief Funct Genomic Proteomic 6(4):302–312
Lacomble S, Portman N, Gull K (2009) A protein-protein interaction map of the Trypanosoma brucei paraflagellar rod. PLoS One 4(11), e7685. doi:10.1371/journal.pone.0007685
Lamesch P, Milstein S, Hao T, Rosenberg J, Li N, Sequerra R, Bosak S, Doucette-Stamm L, Vandenhaute J, Hill DE, Vidal M (2004) C. elegans ORFeome version 3.1: increasing the coverage of ORFeome resources with improved gene predictions. Genome Res 14(10B):2064–2069
Landy A (1989) Dynamic, structural, and regulatory aspects of lambda site-specific recombination. Annu Rev Biochem 58:913–949
Margulies M, Egholm M, Altman WE, Attiya S, Bader JS, Bemben LA, Berka J, Braverman MS, Chen YJ, Chen Z, Dewell SB, Du L, Fierro JM, Gomes XV, Godwin BC, He W, Helgesen S, Ho CH, Irzyk GP, Jando SC, Alenquer ML, Jarvie TP, Jirage KB, Kim JB, Knight JR, Lanza JR, Leamon JH, Lefkowitz SM, Lei M, Li J, Lohman KL, Lu H, Makhijani VB, McDade KE, McKenna MP, Myers EW, Nickerson E, Nobile JR, Plant R, Puc BP, Ronan MT, Roth GT, Sarkis GJ, Simons JF, Simpson JW, Srinivasan M, Tartaro KR, Tomasz A, Vogt KA, Volkmer GA, Wang SH, Wang Y, Weiner MP, Yu P, Begley RF, Rothberg JM (2005) Genome sequencing in microfabricated high-density picolitre reactors. Nature 437(7057):376–380. doi:10.1038/nature03959
Memisevic V, Zavaljevski N, Pieper R, Rajagopala SV, Kwon K, Townsend K, Yu C, Yu X, DeShazer D, Reifman J, Wallqvist A (2013) Novel Burkholderia mallei virulence factors linked to specific host-pathogen protein interactions. Mol Cell Proteomics 12(11):3036–3051. doi:10.1074/mcp.M113.029041
Parrish JR, Yu J, Liu G, Hines JA, Chan JE, Mangiola BA, Zhang H, Pacifico S, Fotouhi F, DiRita VJ, Ideker T, Andrews P, Finley RL Jr (2007) A proteome-wide protein interaction map for Campylobacter jejuni. Genome Biol 8(7):R130
Rain JC, Selig L, De Reuse H, Battaglia V, Reverdy C, Simon S, Lenzen G, Petel F, Wojcik J, Schachter V, Chemama Y, Labigne A, Legrain P (2001) The protein-protein interaction map of Helicobacter pylori. Nature 409(6817):211–215
Rajagopala SV, Titz B, Goll J, Parrish JR, Wohlbold K, McKevitt MT, Palzkill T, Mori H, Finley RL Jr, Uetz P (2007) The protein network of bacterial motility. Mol Syst Biol 3:128
Rajagopala SV, Hughes KT, Uetz P (2009) Benchmarking yeast two-hybrid systems using the interactions of bacterial motility proteins. Proteomics 9(23):5296–5302
Rajagopala SV, Yamamoto N, Zweifel AE, Nakamichi T, Huang HK, Mendez-Rios JD, Franca-Koh J, Boorgula MP, Fujita K, Suzuki K, Hu JC, Wanner BL, Mori H, Uetz P (2010) The Escherichia coli K-12 ORFeome: a resource for comparative molecular microbiology. BMC Genomics 11:470

Rajagopala SV, Casjens S, Uetz P (2011) The protein interaction map of bacteriophage lambda. BMC Microbiol 11:213. doi:10.1186/1471-2180-11-213

Rajagopala SV, Sikorski P, Caufield JH, Tovchigrechko A, Uetz P (2012) Studying protein complexes by the yeast two-hybrid system. Methods 58(4):392–399. doi:10.1016/j.ymeth.2012.07.015

Rajagopala SV, Sikorski P, Kumar A, Mosca R, Vlasblom J, Arnold R, Franca-Koh J, Pakala SB, Phanse S, Ceol A, Hauser R, Siszler G, Wuchty S, Emili A, Babu M, Aloy P, Pieper R, Uetz P (2014) The binary protein-protein interaction landscape of Escherichia coli. Nat Biotechnol 32(3):285–290. doi:10.1038/nbt.2831

Raquet X, Eckert JH, Muller S, Johnsson N (2001) Detection of altered protein conformations in living cells. J Mol Biol 305(4):927–938

Rolland T, Tasan M, Charloteaux B, Pevzner SJ, Zhong Q, Sahni N, Yi S, Lemmens I, Fontanillo C, Mosca R, Kamburov A, Ghiassian SD, Yang X, Ghamsari L, Balcha D, Begg BE, Braun P, Brehme M, Broly MP, Carvunis AR, Convery-Zupan D, Corominas R, Coulombe-Huntington J, Dann E, Dreze M, Dricot A, Fan C, Franzosa E, Gebreab F, Gutierrez BJ, Hardy MF, Jin M, Kang S, Kiros R, Lin GN, Luck K, MacWilliams A, Menche J, Murray RR, Palagi A, Poulin MM, Rambout X, Rasla J, Reichert P, Romero V, Ruyssinck E, Sahalie JM, Scholz A, Shah AA, Sharma A, Shen Y, Spirohn K, Tam S, Tejeda AO, Trigg SA, Twizere JC, Vega K, Walsh J, Cusick ME, Xia Y, Barabasi AL, Iakoucheva LM, Aloy P, De Las RJ, Tavernier J, Calderwood MA, Hill DE, Hao T, Roth FP, Vidal M (2014) A proteome-scale map of the human interactome network. Cell 159(5):1212–1226. doi:10.1016/j.cell.2014.10.050

Rual JF, Hirozane-Kishikawa T, Hao T, Bertin N, Li S, Dricot A, Li N, Rosenberg J, Lamesch P, Vidalain PO, Clingingsmith TR, Hartley JL, Esposito D, Cheo D, Moore T, Simmons B, Sequerra R, Bosak S, Doucette-Stamm L, Le Peuch C, Vandenhaute J, Cusick ME, Albala JS, Hill DE, Vidal M (2004) Human ORFeome version 1.1: a platform for reverse proteomics. Genome Res 14(10B):2128–2135

Rual JF, Venkatesan K, Hao T, Hirozane-Kishikawa T, Dricot A, Li N, Berriz GF, Gibbons FD, Dreze M, Ayivi-Guedehoussou N, Klitgord N, Simon C, Boxem M, Milstein S, Rosenberg J, Goldberg DS, Zhang LV, Wong SL, Franklin G, Li S, Albala JS, Lim J, Fraughton C, Llamosas E, Cevik S, Bex C, Lamesch P, Sikorski RS, Vandenhaute J, Zoghbi HY, Smolyar A, Bosak S, Sequerra R, Doucette-Stamm L, Cusick ME, Hill DE, Roth FP, Vidal M (2005) Towards a proteome-scale map of the human protein-protein interaction network. Nature 437(7062):1173–1178

Schwartz H, Alvares CP, White MB, Fields S (1998) Mutation detection by a two-hybrid assay. Hum Mol Genet 7(6):1029–1032

SenGupta DJ, Zhang B, Kraemer B, Pochart P, Fields S, Wickens M (1996) A three-hybrid system to detect RNA-protein interactions *in vivo*. Proc Natl Acad Sci U S A 93(16):8496–8501

Shapira SD, Gat-Viks I, Shum BO, Dricot A, de Grace MM, Wu L, Gupta PB, Hao T, Silver SJ, Root DE, Hill DE, Regev A, Hacohen N (2009) A physical and regulatory map of host-influenza interactions reveals pathways in H1N1 infection. Cell 139(7):1255–1267. doi:10.1016/j.cell.2009.12.018

Stanyon CA, Limjindaporn T, Finley RL Jr (2003) Simultaneous cloning of open reading frames into several different expression vectors. Biotechniques 35(3):520–522, 4–6

Stellberger T, Hauser R, Baiker A, Pothineni VR, Haas J, Uetz P (2010) Improving the yeast two-hybrid system with permutated fusions proteins: the Varicella Zoster Virus interactome. Proteome Sci 8:8

Stelzl U, Worm U, Lalowski M, Haenig C, Brembeck FH, Goehler H, Stroedicke M, Zenkner M, Schoenherr A, Koeppen S, Timm J, Mintzlaff S, Abraham C, Bock N, Kietzmann S, Goedde A, Toksoz E, Droege A, Krobitsch S, Korn B, Birchmeier W, Lehrach H, Wanker EE (2005) A human protein-protein interaction network: a resource for annotating the proteome. Cell 122(6):957–968

Titz B, Rajagopala SV, Goll J, Hauser R, McKevitt MT, Palzkill T, Uetz P (2008) The binary protein interactome of Treponema pallidum–the syphilis spirochete. PLoS One 3(5):e2292. doi: 10.1371/journal.pone.0002292

Uetz P, Giot L, Cagney G, Mansfield TA, Judson RS, Knight JR, Lockshon D, Narayan V, Srinivasan M, Pochart P, Qureshi-Emili A, Li Y, Godwin B, Conover D, Kalbfleisch T, Vijayadamodar G, Yang M, Johnston M, Fields S, Rothberg JM (2000) A comprehensive analysis of protein-protein interactions in Saccharomyces cerevisiae. Nature 403(6770): 623–627

Uetz P, Dong YA, Zeretzke C, Atzler C, Baiker A, Berger B, Rajagopala SV, Roupelieva M, Rose D, Fossum E, Haas J (2006) Herpesviral protein networks and their interaction with the human proteome. Science 311(5758):239–242

Vidal M, Endoh H (1999) Prospects for drug screening using the reverse two-hybrid system. Trends Biotechnol 17(9):374–381

Vidal M, Legrain P (1999) Yeast forward and reverse 'n'-hybrid systems. Nucleic Acids Res 27(4):919–929

von Brunn A, Teepe C, Simpson JC, Pepperkok R, Friedel CC, Zimmer R, Roberts R, Baric R, Haas J (2007) Analysis of intraviral protein-protein interactions of the SARS coronavirus ORFeome. PLoS One 2(5), e459. doi:10.1371/journal.pone.0000459

von Mering C, Jensen LJ, Kuhn M, Chaffron S, Doerks T, Kruger B, Snel B, Bork P (2007) STRING 7–recent developments in the integration and prediction of protein interactions. Nucleic Acids Res 35(Database issue):D358–D362

Walhout AJ, Temple GF, Brasch MA, Hartley JL, Lorson MA, van den Heuvel S, Vidal M (2000) GATEWAY recombinational cloning: application to the cloning of large numbers of open reading frames or ORFeomes. Methods Enzymol 328:575–592

Wang X, Wei X, Thijssen B, Das J, Lipkin SM, Yu H (2012) Three-dimensional reconstruction of protein networks provides insight into human genetic disease. Nat Biotechnol 30(2):159–164. doi:10.1038/nbt.2106

Yu H, Braun P, Yildirim MA, Lemmens I, Venkatesan K, Sahalie J, Hirozane-Kishikawa T, Gebreab F, Li N, Simonis N, Hao T, Rual JF, Dricot A, Vazquez A, Murray RR, Simon C, Tardivo L, Tam S, Svrzikapa N, Fan C, de Smet AS, Motyl A, Hudson ME, Park J, Xin X, Cusick ME, Moore T, Boone C, Snyder M, Roth FP, Barabasi AL, Tavernier J, Hill DE, Vidal M (2008) High-quality binary protein interaction map of the yeast interactome network. Science (New York, NY) 322(5898):104–110. doi: 10.1126/science.1158684

Yu H, Tardivo L, Tam S, Weiner E, Gebreab F, Fan C, Svrzikapa N, Hirozane-Kishikawa T, Rietman E, Yang X, Sahalie J, Salehi-Ashtiani K, Hao T, Cusick ME, Hill DE, Roth FP, Braun P, Vidal M (2011) Next-generation sequencing to generate interactome datasets. Nat Methods 8(6):478–480. doi:10.1038/nmeth.1597

Chapter 12
Biogenesis of *Escherichia coli* DMSO Reductase: A Network of Participants for Protein Folding and Complex Enzyme Maturation

Catherine S. Chan and Raymond J. Turner

Abstract Protein folding and structure have been of interest since the dawn of protein chemistry. Following translation from the ribosome, a protein must go through various steps to become a functional member of the cellular society. Every protein has a unique function in the cell and is classified on this basis. Proteins that are involved in cellular respiration are the bioenergetic workhorses of the cell. Bacteria are resilient organisms that can survive in diverse environments by fine tuning these workhorses. One class of proteins that allow survival under anoxic conditions are anaerobic respiratory oxidoreductases, which utilize many different compounds other than oxygen as its final electron acceptor. Dimethyl sulfoxide (DMSO) is one such compound. Respiration using DMSO as a final electron acceptor is performed by DMSO reductase, converting it to dimethyl sulfide in the process. Microbial respiration using DMSO is reviewed in detail by McCrindle et al. (Adv Microb Physiol 50:147–198, 2005). In this chapter, we discuss the biogenesis of DMSO reductase as an example of the participant network for complex iron-sulfur molybdoenzyme maturation pathways.

Keywords *E. coli* • Dimethyl sulfoxide (DMSO) • Reductase

12.1 Introduction

Molybdoenzymes emerged as a superfamily of respiratory oxidoreductases that require a catalytic molybdenum/tungsten-based cofactor to catalyze redox reactions (McCrindle et al. 2005; Hille 2013; Rothery et al. 2012; Iobbi-Nivol and Leimkühler 2012; Romao 2009; Magalon et al. 2011). These enzymes are further classified into three families based on the active site structure that coordinates the molybdenum atom. A key feature that separates the dimethyl sulfoxide (DMSO) reductase family

C.S. Chan • R.J. Turner (✉)
Department of Biological Sciences, University of Calgary, BI156 Biological Sciences Bldg, 2500 University Dr NW, Calgary, AB T2N 1N4, Canada
e-mail: catchan@ucalgary.ca; turnerr@ucalgary.ca

N.J. Krogan, M. Babu (eds.), *Prokaryotic Systems Biology*, Advances in Experimental Medicine and Biology 883, DOI 10.1007/978-3-319-23603-2_12

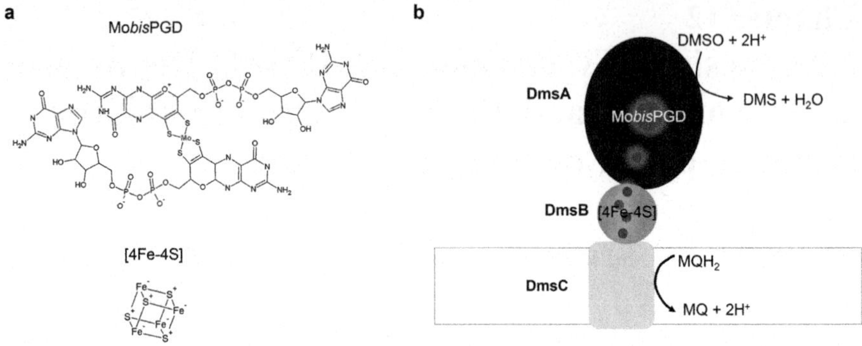

Fig. 12.1 The complex iron sulfur containing molybdoenzyme DMSO reductase. (**a**) Structure of the molybdo-*bis*(pyranopterin guanine dinucleotide) (Mo*bis*PGD) catalytic cofactor and [4Fe-4S] clusters that makeup the electron transfer chain within DMSO reductase. (**b**) Overall architecture and composition of DMSO reductase from *Escherichia coli*, demonstrating the catalytic DmsA, electron conduit DmsB, and membrane anchor DmsC subunits

members from xanthine oxidase and sulfite oxidase families is that it has two pyranopterin groups coordinating the Mo atom, whereas the others have only one. Members of each family have similar structural folds around the catalytic cofactor, and a recent study demonstrated that the protein fold is directly correlated to the pyranopterin conformation (Rothery et al. 2012).

Many members of the DMSO reductase family have also been categorized under another classification system described as the complex iron-sulfur molybdoenzyme (CISM) family (Rothery et al. 2008). CISM family members have similar architecture with one final goal—to provide a modular relay for electron transfer during respiration. The archetypical composition of CISM proteins include a catalytic subunit containing a molybdo-*bis*(pyranopterin guanine dinucleotide) (Mo*bis*PGD) catalytic cofactor and a [4Fe-4S] cluster (Fig. 12.1a), an electron conduit subunit containing four [4Fe-4S] clusters, and a membrane anchor subunit that is imbedded in the cytoplasmic membrane for connecting to the quinone pool. This architecture is not set in stone; some enzymes do not contain all of the archetypical subunits, and some membrane anchor subunits also contain two hemes (note: these subunits follow an all-or-nothing rule with respect to hemes). DMSO reductase is classified as an archetypical CISM family member that contains all the canonical properties of a CISM protein (Fig. 12.1b and Rothery et al. 2008). Other enzymes belonging to both the DMSO reductase molybdoenzyme and CISM families include trimethylamine *N*-oxide (TMAO) reductase and nitrate reductase A (Rothery et al. 2008). The former of the two is an atypical enzyme containing only a catalytic and membrane anchor subunit whereas the latter is an archetypical member.

12.1.1 Twin-Arginine Translocase for Respiratory Enzyme Biogenesis

The twin-arginine translocase (Tat) is known for targeting and translocation of many extra-cytoplasmic proteins in prokaryotes. Proteins that utilize the Tat pathway contain two major defining features—they have a consensus S/T-R-R-x-F-L-K 'twin-arginine' motif in their N-terminal leader/signal peptide that was originally identified in cofactor-containing respiratory oxidoreductases (Berks 1996), and they are fully folded prior to translocation (DeLisa et al. 2003). The discovery of the Tat system has led to a understanding of how cofactor-containing enzymes, such as those in the CISM family, are able to acquire their catalytic cofactors and prosthetic groups to mature into a holoenzyme prior to translocation (Santini et al. 1998; Weiner et al. 1998; Sargent et al. 1998). Since then, the list of Tat substrates has expanded to include non-respiratory proteins and is described in Tullman-Ercek et al. (2007), with the model organism *Escherichia coli* having at least 27 substrates.

In *E. coli*, the translocase itself comprises of the TatABC subunits, and its role in protein translocation has been extensively studied. Subunit composition varies slightly in prokaryotes, and readers are directed to the latest reviews in the literature for a more detailed discussion on the Tat pathway in prokaryotes (Palmer and Berks 2012; Fröbel et al. 2012; Kudva et al. 2013, for example). Figure 12.2 is a highly generalized model of translocation by Tat but shows the overall schematic of the process. The Tat system is also involved in quality control where it rejects improperly folded or mis-assembled substrates (DeLisa et al. 2003; Matos et al. 2008). It has been shown that deletion of some or all of the Tat components eliminates the ability of *E. coli* to grow on media containing substrates dedicated for CISM respiration (Sambasivarao et al. 2001; Sargent et al. 1998; Ray et al. 2003), thus linking the Tat system to CISM maturation. The role of Tat in DMSO reductase maturation will be discussed in further detail in the sections below.

12.1.2 DMSO Reductase for Anaerobic Respiration

DMSO reductase from *E. coli* comprises of the DmsABC heterotrimer (Fig. 12.1b, Bilous and Weiner 1988 and Bilous et al. 1988). DmsA is the largest subunit of the enzyme at $\sim$86 kDa in its mature form. It is translated at 814 amino acids with the first 45 residues forming a twin-arginine motif-containing leader peptide that is proteolytically processed (Sambasivarao et al. 2000; Bilous et al. 1988). The leader peptide of DmsA is essential for production of a fully active enzyme at the cytoplasmic membrane and for anaerobic growth using DMSO as a sole electron acceptor (Sambasivarao et al. 2000). From here on, this sequence will be referred to as the RR-leader and further details on the importance of it for enzyme biogenesis and maturation is discussed in sections below.

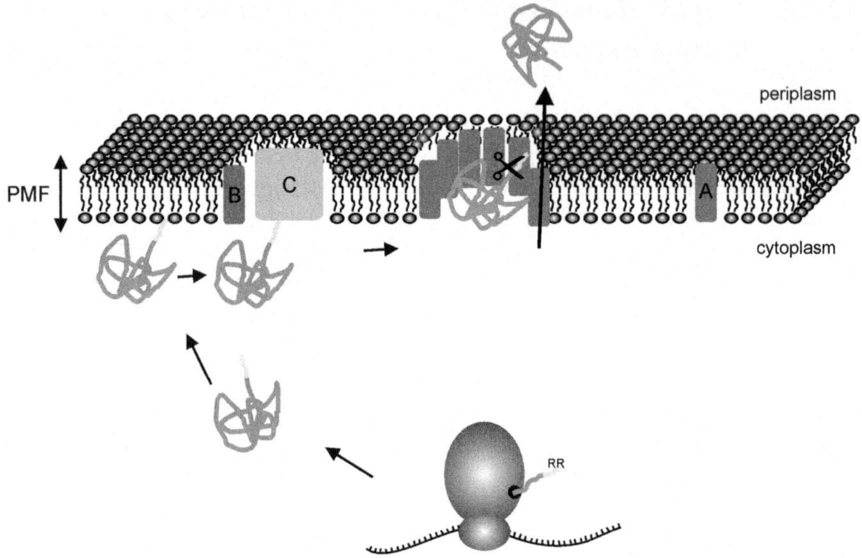

Fig. 12.2 General model for translocation of folded polypeptides by the bacterial Twin-arginine translocation system. Polypeptides are synthesized with a twin-arginine (RR) leader sequence and targeted towards the cytoplasmic membrane after folding. The folded polypeptide is received by the TatBC receptor/quality control complex at the cytoplasmic membrane, and then translocated across the membrane through the homo-oligomeric TatA pore while the leader peptide is cleaved, followed by release into the periplasm. Energy for translocation is provided by the proton motive force (PMF)

DmsA coordinates the Mo*bis*PGD cofactor and one [4Fe-4S] cluster (Cammack and Weiner 1990; Heffron et al. 2001; Tang et al. 2011). Several high resolution crystal structures of the DmsA homologue DorA from *Rhodobacter* sp. have been obtained (Schindelin et al. 1996; Schneider et al. 1996; McAlpine et al. 1998). These structures have led to a further understanding how substrate specificity is enforced by the size of the funnel leading towards the Mo*bis*PGD cofactor (Rothery et al. 2008; Simala-Grant and Weiner 1998). The structures help explain why DMSO reductase is able to reduce a broad range of substrates containing sulfoxide and pyridine *N*-oxide groups (Simala-Grant and Weiner 1996).

DmsB is a 23 kDa protein, serving as the electron conduit subunit. It consists of 205 amino acids, with the last 30 residues essential for anchoring to DmsC (Sambasivarao and Weiner 1991). It coordinates four [4Fe-4S] clusters, each of them through four conserved Cys residues (Cammack and Weiner 1990). The [4Fe-4S] clusters are important for transfer of electrons from menaquinol to DMSO and alternate incorporation of [3Fe-4S] at just one site impairs anaerobic growth on media containing DMSO (Rothery and Weiner 1991). Studies have identified four key residues in DmsB important for electron relay from DmsC. These include Pro80, Ser81, Cys102, and Tyr104, all of which are located at the DmsB-DmsC interface (Cheng et al. 2005).

DmsC is a 287 residue, 31 kDa protein with eight transmembrane helices and both its N- and C-termini localized in the periplasm (Weiner et al. 1993; Sambasivarao et al. 1990). Truncation studies found that the entire length of DmsC is required for DmsAB attachment to the cytoplasmic membrane (Weiner et al. 1993), although attachment to DmsC is not required for activity or accumulation of the DmsAB holoenzyme in the cytoplasm (Sambasivarao and Weiner 1991; Sambasivarao et al. 2001). A previous study showed that proper DMSO reductase activity in *E. coli* membranes requires menaquinone specifically (Wissenbach et al. 1990), with demethylmenaquinone being less efficient (Wissenbach et al. 1992), and that DmsC is the subunit responsible for binding and oxidation (Geijer and Weiner 2004; Zhao and Weiner 1998). It appears to bind at a stoichiometry of 1:1 and involves the residues His65 and Glu87 (Zhao and Weiner 1998; Geijer and Weiner 2004), located in helices 2 and 3 that border the first periplasmic loop. Insertion of DmsC into the cytoplasmic membrane in the absence of DmsAB is lethal (Sambasivarao et al. 2001), explaining similar lethality issues observed in studies overexpressing DmsC (Turner et al. 1997).

DMSO reductase homologues are also found in *Rhodobacter* sp. and *Shewanella oneidensis*. The enzymes from these bacteria differ from the *E. coli* enzyme with the most notable features being that they are soluble, consist of a heterodimer, and the subunits contain a *c*-type cytochrome instead of [Fe-S] clusters [reviewed in McCrindle et al. (2005)].

12.1.2.1 Topological Organization and Controversy

The topological organization of DMSO reductase has been a subject of controversy for quite some time. Studies that suggest that the DmsAB subunits have a cytoplasmic orientation include accessibility by proteases, immunogold labeling, exogenous paramagnetic probes, alkaline phosphatase genetic fusion liability, and immunodetection after cellular fractionation (Sambasivarao et al. 1990, 2001; Rothery and Weiner 1993). Other immunodetection after cellular fractionation studies implicated DmsAB to be localized in the periplasm (Stanley et al. 2002), and argue that the cytoplasmic orientation seen previously is an artifact of overexpression and genetic fusions. Additionally fusing the DmsA RR-leader to GFP resulted in GFP fluorescence in the periplasm, supporting the notion that DmsA is periplasmically localized (Ray et al. 2003).

What added more to the confusion and controversy over the orientation of DMSO reductase was the proposed role that Tat was an alternate translocation pathway to the Sec system, implicating that proteins transported by Tat are solely extra-cytoplasmic. Studies showed that DMSO reductase or its activity was detected in the cytoplasm upon deletion of various Tat subunits (Sargent et al. 1998; Weiner et al. 1998), which led to the debate that deletion of Tat retarded its translocation to the periplasm or that Tat also functions to target cytoplasmically attached enzymes, depending on which 'camp' one was supporting. Further experiments added to the controversy when DmsAB appeared to accumulate in the periplasm in a *tatA* mutant when DmsC was not expressed simultaneously from a plasmid, which the authors attribute that DmsC also has a 'stop-transfer' function to prevent translocation of DmsAB to the periplasm under normal conditions (Weiner et al. 1998).

12.1.2.2 Brief Overview of DMSO Reductase Expression

DMSO reductase of *E. coli* is expressed from the *dmsABC* operon (Bilous et al. 1988). Expression of the *dms* operon is induced under anoxic growth conditions, and is activated by Fnr and the molybdate-responsive regulator ModE (Cotter and Gunsalus 1989; Lamberg and Kiley 2000; McNicholas et al. 1998). Presence of oxygen or nitrate in the growth media suppresses expression of DMSO reductase, where it is repressed by NarXL (Bilous and Weiner 1988; Tseng et al. 1994; Cotter and Gunsalus 1989; Bearson et al. 2002). The topic of Fnr activation by anoxia and nitrate repression is too vast for the purpose of this review, and readers are referred to the numerous reviews such as Kucera Unden et al. (2002) and Green et al. (2014) for example, for more details on these subjects.

12.1.3 DMSO Reductase System-Specific Chaperone

A fourth component of DMSO reductase, a protein that is required for biogenesis and maturation but is not part of the final enzyme, is DmsD. Previously known as YnfI, DmsD is encoded on the paralogous DMSO reductase *ynfFGHI* operon, where *ynfI* was renamed *dmsD* (Oresnik et al. 2001). DmsD was first isolated as DmsA leader-binding protein, and is essential for anaerobic growth on media containing DMSO (Oresnik et al. 2001). These observations led to the hypothesis that DmsD is a system-specific chaperone for DMSO reductase maturation. Phylogenetic analyses found DmsD to be related to system-specific chaperones of other Mo*bis*PGD-containing oxidoreductases that include TorD for TMAO reductase and NarJ for cytoplasmic nitrate reductase A (Turner et al. 2004; Ilbert et al. 2004). Both are CISM enzymes related to DMSO reductase (Rothery et al. 2008). Based on their phylogenetic, structural, and functional relationship, it was proposed that the system-specific chaperones including DmsD be given a collective name of redox enzyme maturation protein (REMP) (Turner et al. 2004, 2010). Further details on DmsD in the participant network for DMSO reductase biogenesis is discussed below.

12.2 System for Maturation into a Functional Holoenzyme

The participant network model for DMSO reductase maturation is presented in Fig. 12.3a. In the following sections, we will walk through the steps that this network is derived from. Expression of *dmsABC* follows the traditional route common for

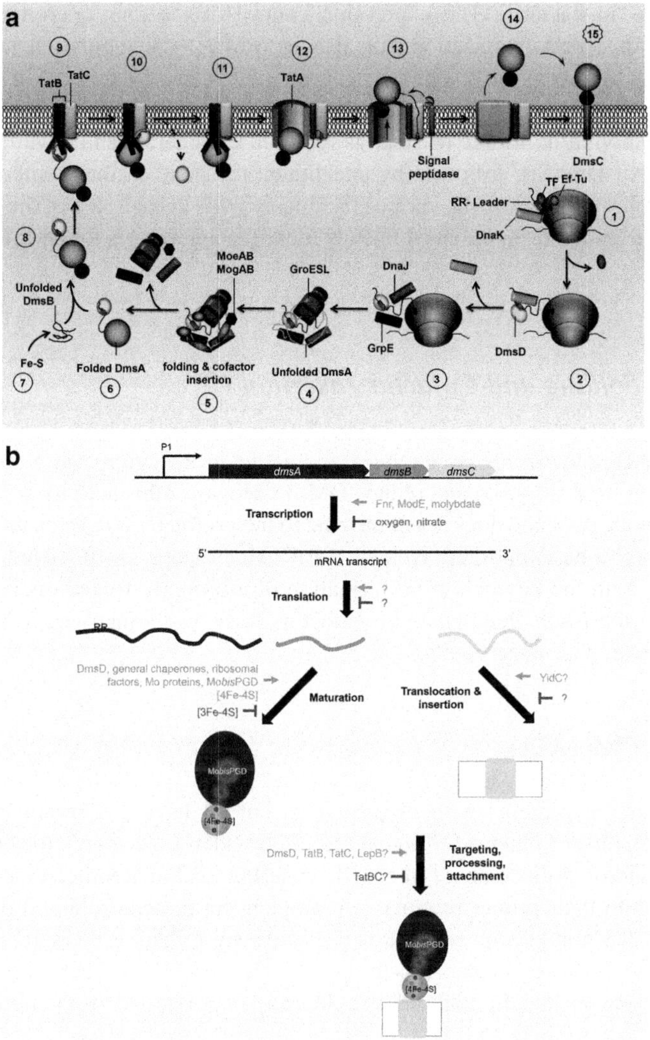

Fig. 12.3 Model of the maturation system pathway for DMSO reductase. (**a**) The cartooned pathway shows stages derived from the interaction web that is hypothesized to occur. (*1*) The nascent chain exiting the ribosome and the RR-leader interacts with DnaK, trigger factor (TF) and elongation factor (Ef-Tu). The REMP chaperone DmsD then joins along with other chaperones (*2–4*). As the protein folds the Mo*bis*PGD cofactor would be inserted through contact with MoeA, MoeB, MogA, and MogB after [4Fe-4S] insertion (*5*). After DmsA is fully folded (*6*), the folded four [4Fe-4S] cluster-containing DmsB interacts with DmsA (*7* and *8*). With DmsAB assembled it is targeted to the Tat system (*9–11*) and translocated (assuming periplasmic, *12–14*) to finally dock to the membrane anchor subunit DmsC to secure a fully functional respiratory enzyme (*15*). (**b**) Participants or factors that regulate production of a mature and functional DMSO reductase at the various expression levels are indicated. Those known or suggested to promote or repress maturation are shown with a (→) and (|—), respectively

all proteins in bacteria. The expression control of the *dms* operon was briefly described in a previous section but also ignores all the regulation mechanisms involved in transcription itself. There have not been any reports of translation control for synthesis of DmsABC in the literature to date. Following transcription and translation, a multitude of steps is required to generate an active holoenzyme consisting of DmsAB, followed by attachment to DmsC at the membrane. In the sections below we discuss the various factors involved in each step of the maturation process and postulate on the role of newly identified interactors for DMSO reductase maturation.

12.2.1 Folding and Cofactor Insertion

The DmsAB holoenzyme is enzymatically active in the cytoplasm with respect to its ability to bind substrate and reduce DMSO. However this activity is not coupled to the quinone pool and does not contribute to the proton motive force until attached to DmsC (Sambasivarao and Weiner 1991). Many steps are involved to achieve this form, with the key being the formation of a properly folded conformation to coordinate the Mo*bis*PGD catalytic cofactor. Participants involved in this step of maturation are discussed below.

12.2.1.1 Chaperones

Chaperones are proteins with diverse roles but all have a common function to assist the folding of other macromolecules. They can be considered the construction workers and/or policemen of the cell, ensuring that macromolecules fold and assemble into their proper productive biological form, or unfold and disassemble those that are non-productive and/or disruptive by targeting them for proteolytic degradation [reviewed in Rodrigo-Brenni and Hegde (2012)]. These quality control systems exist in all kingdoms of life and are important for maintaining protein homeostasis in the cell.

Biogenesis of DMSO reductase involves a plethora of chaperones which are grouped in the sections below to put the entire maturation process into a better perspective.

System-specific chaperone, DmsD. While general molecular chaperones have been described as early as 1978 by the example of heat shock proteins (Kelley and Schlesinger 1978), the concept of public versus private chaperones was also described in the eukaryotic secretory pathway [reviewed in Anelli and Sitia (2008)]. Similarly, this concept is also present in prokaryotes and was described as a family of system-specific chaperones involved in respiratory enzyme biogenesis (Turner et al. 2004; Hatzixanthis et al. 2005; Ilbert et al. 2004). DmsD is one such chaperone.

DmsD is essential for anaerobic growth on media containing DMSO, implicating an essential role in DMSO reductase biogenesis (Ray et al. 2003; Oresnik et al. 2001). Since DmsD was identified as a DmsA RR-leader binding protein (Oresnik et al. 2001), much effort has been placed to understand its role in DMSO reductase maturation. Several studies aimed at understanding the nature of DmsA RR-leader binding were undertaken. It is clear that DmsD binds the DmsA RR-leader at 1:1 stoichiometry (Winstone et al. 2006), and that the entire hydrophobic region in the RR-leader is required for binding, yet the twin-arginine motif only contributes marginally (Winstone et al. 2013). While DmsD binds the RR-leaders of DmsA, TorA, YnfE, and YnfF, it binds the RR-leader of DmsA with higher affinity (Chan et al. 2009), indicating specificity towards its natural biological substrate. N-terminal display of the DmsA RR-leader is also important for the interaction to occur (Winstone et al. 2006), with the weaker interaction towards the TorA RR-leader of TMAO reductase detected under very select conditions (Chan et al. 2009; Oresnik et al. 2001).

DmsD may also participate in the folding of DmsA, since it likely binds the N-terminal RR-leader soon after it is exposed at the ribosome. Protein folding is coordinated with cofactor incorporation, and formation of a catalytically active holoenzyme requires Mo*bis*PGD insertion (Sambasivarao et al. 2002). Therefore, these steps in maturation are intricately connected. Drawing from the example of the homologous TorD chaperone, studies showed that TorD directly participates in Mo*bis*PGD insertion into apoTorA by acting on the enzyme prior to cofactor loading (Ilbert et al. 2003). However, activation of apoTorA in the in vitro system was only 80 %, suggesting the participation of other unidentified factors during *in vivo* maturation. This observation led Li and researchers (2010) to search for potential unidentified factors by targeting the interaction proteome of DmsD using a variety of *in vivo* and in vitro assays. A large number of targets were identified that include general molecular chaperones, ribosomal components, and Mo*bis*PGD cofactor biosynthesis proteins. Table 12.1 lists all the non-substrate proteins that were identified in the DmsD proteome from several studies (Li et al. 2010; Kostecki et al. 2010; Papish et al. 2003). These interactions implicate DmsD in the center of the DMSO reductase maturation pathway for connecting the nascent DmsA polypeptide to proteins that would assist in its path to become an active holoenzyme with DmsB (Fig. 12.3).

A hypothesis connecting these proteins is proposed in Sect. 12.2.2, where DmsD is implicated as a central hub to connect the upstream and downstream processes of DMSO reductase maturation. But first the involvement of other participants for maturation are discussed below.

General Molecular Chaperone, DnaK

The DnaK-DnaJ-GrpE assembly is a well-studied chaperone machine in *E. coli* initially identified for their roles in heat shock induction [reviewed in Genevaux et al. (2007)]. DnaJ and GrpE are often called co-chaperones as they function alongside DnaK. DnaK, also known as Hsp70, has several functions including preventing

Table 12.1 Proteins identified to interact with DmsD through various biochemical studies

Protein	Class/group	Reference
TatB	Translocase	Papish et al. (2003) and Kostecki et al. (2010)
TatC	Translocase	Papish et al. (2003) and Kostecki et al. (2010)
GroEL	General molecular chaperone	Li et al. (2010)
DnaK	General molecular chaperone	Li et al. (2010)
DnaJ[a]	General molecular chaperone	Li et al. (2010)
GrpE[a]	General molecular chaperone	Li et al. (2010)
TufA/Ef-Tu	Ribosome-associated	Li et al. (2010)
Tig	Ribosome-associated	Li et al. (2010)
MoeA	Mo*bis*PGD biosynthesis	Li et al. (2010)
MoeB	Mo*bis*PGD biosynthesis	Li et al. (2010)
MogA	Mo*bis*PGD biosynthesis	Li et al. (2010)
MobB	Mo*bis*PGD biosynthesis	Li et al. (2010)

TufA/Ef-Tu translation elongation factor, *Tig* trigger factor
[a]These co-chaperones work concurrently with DnaK in the DnaK-DnaJ-GrpE chaperone assembly

aggregation of nascent polypeptides, and the refolding and disposal of damaged polypeptides. Its ATPase activity along with assistance from DnaJ/GrpE modulates its ability to participate in such events.

DnaK binds the RR-leader peptide of DmsA (Oresnik et al. 2001), and RR-leader peptides of CueO, TorA, and SufI (Graubner et al. 2007; Pérez-Rodríguez et al. 2007). TorA is the catalytic subunit of TMAO reductase that is also involved in microbial anaerobic respiration (Méjean et al. 1994), CueO is a periplasmic copper oxidase conferring aerobic copper tolerance (Outten et al. 2001), and SufI is a protein involved in cell division during stress conditions (Samaluru et al. 2007); the latter two do not contain complex cofactors like DMSO or TMAO reductase. These observations suggest that the proteins are likely seen and bound by DnaK as they emerge from the ribosome to prevent aggregation. An interesting observation was that the presence of DnaK in the cell was important for the accumulation and/or membrane localization of CueO, TorA, and SufI but not DmsA (Graubner et al. 2007), suggesting that the maturation pathways are not identical.

Chemical chaperone, Mo*bis*PGD. The idea of chemical chaperones was initially proposed to describe osmolytes that influence protein folding (Tatzelt et al. 1996). Further investigations also reveal that chemical chaperones involving methylamines, amino acids, sugars, and polyols can preserve enzyme activity of trypsin under thermal and chemical stress (Levy-Sakin et al. 2014). Following this concept, recent reports support the notion that Mo*bis*PGD is a chemical chaperone for DMSO reductase by stabilizing the tertiary structure of DmsA for cofactor coordination (Tang et al. 2013). This is supported by evaluation of the available structures of CISM proteins where the polypeptide chain appears to be wrapped around the *bis*PGD moiety (Rothery et al. 2012). Previous studies involving another Mo*bis*PGD-containing CISM, TMAO reductase, showed improper localization of the enzyme in mutants deficient in molybdate-uptake and cofactor biosynthesis (Santini et al. 1998). Similar results were observed for the DMSO reductases in *R. sphaeroides* and *R. capsulatus*, where removal of molybdenum from the media or a cofactor biosynthesis mutant inhibited proteolytic processing and accumulation of the mature enzyme in the periplasm (Yoshida et al. 1991; Solomon et al. 1999). Together, these observations implicate Mo*bis*PGD as a chemical chaperone for CISM maturation as this process includes proper coordination of the catalytic cofactor for holoenzyme formation. It should be noted that Mo*bis*PGD is not a chaperone for targeting, as apoDmsA was properly localized and processed in a strain carrying a mutation that cannot produce Mo*bis*PGD (Sambasivarao et al. 2002).

12.2.1.2 Mo*bis*PGD Cofactor Biosynthesis and Coordination

It is without a doubt that the Mo*bis*PGD cofactor is an essential component for DMSO reductase activity. So crucial is the cofactor that DMSO reductase in *R. capsulatus* has its own dedicated cofactor biosynthesis enzymes encoded immediately downstream of its operon (Solomon et al. 1999). Generation of the Mo*bis*PGD cofactor and other cofactor analogues is a long and complicated process [reviewed in Iobbi-Nivol and Leimkühler (2012) and Mendel (2013)]. Briefly, it begins with uptake of molybdate into cells through specific transporters, followed by four key steps that modify molybdate and add it to the organic compounds leading to the final active cofactor form. A process that involves up to 16 proteins. Studies showing impairment of DMSO, TMAO, and nitrate reductase activity by targeting cofactor biosynthesis components directly link this process to CISM maturation (Sambasivarao et al. 2002; Palmer et al. 1996; Genest et al. 2008).

12.2.1.3 [4Fe-4S] Cluster Assembly

Both DmsA and DmsB contain [4Fe-4S] cluster(s) in their subunits for electron relay. These cofactors are synthesized in a complex process that is also regulated in the cell [reviewed in Roche et al. (2013)]. The general scheme starts with donation of sulfur and iron by cysteine desulfurase and an iron donor respectively to a scaffold protein that generates the Fe-S cluster. The Fe-S cluster is then transferred to a carrier protein, which then delivers it to the target protein. Although there is no support in the literature indicating which of the two Fe-S assembly systems (ISC and SUF) are used for DmsAB, it can be presumed that the ISC system through IscA is used since it is more common in anaerobiosis (Vinella et al. 2009). Assembly of the [4Fe-4S] cluster in DmsA must occur prior to insertion of the Mo*bis*PGD cofactor (Tang et al. 2011), so the entire process must be well-coordinated in order to generate an active enzyme. Further, the four [4Fe-4S] clusters in DmsB must assemble to generate a holo folded conformation prior to interacting with DmsA to produce the DmsAB holoenzyme.

12.2.2 Interaction Network of the Participants for DMSO Reductase Maturation

Previously we suggested that the interactions identified for DmsD in Table 12.1 implicates DmsD as a central connector for all the processes required for DMSO reductase maturation. Chaperone networking for protein targeting is a well-studied process for polypeptides destined to the extra-cytoplasmic space (Castanié-Cornet et al. 2013). This idea is not unfounded as such a role has been suggested for DnaK in the *E. coli* chaperone network of protein biogenesis, where it was implicated to connect upstream factors such as the ribosome and trigger factor to downstream ones such as GroEL (Calloni et al. 2012). For example, trigger factor interacts with the signal peptide of emerging nascent TorA and SufI at the ribosome near L23 (Jong et al. 2004). This is presumably to protect it from proteases, which is a known function for trigger factor (Hoffmann et al. 2006). Assuming the same scenario occurs for nascent DmsA, it is hypothesized that DmsD hovers near the ribosome (supported by the interaction with Ef-Tu, Table 12.1). It then binds trigger factor through a hand-off mechanism when DmsA is ready to proceed to the next step in maturation, as part of connecting the upstream process.

DmsD interacting with GroEL and the cofactor biosynthesis proteins is an example of the downstream connections. As a chaperone that provides a protective cavity for newly synthesized or misfolded polypeptides to fold or refold, it is plausible to hypothesize that DmsD having received DmsA from trigger factor, directs it towards GroEL which then assists in the folding of DmsA. Being a polypeptide of ~90 kDa prior to processing, DmsA is clearly larger than the 60 kDa cutoff to which the GroEL cavity can comfortably hold (Houry et al. 1999). A mechanism for GroEL-assisted folding of large proteins involves their binding to the open (*trans*) ring of GroEL-ES and folding in the bulk solution outside of the cavity [reviewed in Marchenkov and Semisotnov (2009)]. If folding does not succeed, the cycle then repeats after the substrate is released by ATP hydrolysis in the *cis* ring. Folding of DmsA outside the cavity of GroEL would also allow for simultaneous insertion of [4Fe-4S] and Mo*bis*PGD during folding. If the cavity of GroEL is not large enough to support DmsA, then it is even less likely to support the entry of the carrier proteins to come within close proximity for cofactor transfer. This also supports the model that DmsD binds the cofactor biosynthesis proteins to connect them to DmsA.

Additional evidence that GroEL participates in respiratory oxidoreductase maturation is supported by the observations that GroEL interacts with the NapD chaperone of the periplasmic nitrate reductase P (Butland et al. 2005), and that it was required for hydrogenase-1 maturation (Rodrigue et al. 1996). Both enzymes are also RR-leader containing Tat substrates. Lastly, GroEL has was implicated for insertion of a molybdenum-iron cofactor into a nitrogenase enzyme from *Azobacter vinelandii* (Ribbe and Burgess 2001), supporting the involvement of GroEL in CISM maturation.

12.3 Holoenzyme Assembly, Targeting, Processing, and Membrane Attachment

The final homestretch of DMSO reductase maturation involves arrival at its permanent location at the cytoplasmic membrane. Regardless of the debate surrounding its membrane orientation, the enzyme must still be targeted and anchored to the cytoplasmic membrane, a process that is dependent on the Tat system (Sargent et al. 1998; Sambasivarao et al. 2001). The individual steps that arrive at this final assembled form of DMSO reductase is discussed below. It should be noted that some of these steps are still poorly understood due to lack of research in those areas.

12.3.1 DmsAB Holoenzyme Assembly

Assembly of the DmsAB catalytically active holoenzyme is one question that remains to be investigated. It is however, suggested that RR-leader bearing oxidoreductases with more than one subunit is targeted to the translocase via a 'piggy-back' mechanism, since all identified multi-subunit oxidoreductases contain a RR-leader in only one of the subunits (Wu et al. 2000). Evidence that DmsAB follows the same mechanism for targeting to the cytoplasmic membrane remains elusive, however, it is assumed this is the case for all CISM enzymes until further research is provided.

12.3.2 Targeting to the Translocase

It is well-accepted that system-specific REMP chaperones, including DmsD, are involved in targeting their oxidreductase substrates to the Tat complex. Evidence that support this include the impaired anaerobic growth of *dmsD* or various *tat* mutants on media containing DMSO (Sambasivarao et al. 2001; Sargent et al. 1998; Ray et al. 2003; Oresnik et al. 2001; Papish et al. 2003). Further, DmsD interacts with the cytoplasmic membrane only under anaerobic growth conditions, an interaction that only occurs in the presence of TatB or TatC (Papish et al. 2003). It was later confirmed that the interaction is directly between DmsD and TatB or TatC using two independent *in vivo* interaction assays (Kostecki et al. 2010). While a transient tripartite interaction is too complicated to analyze by this these assays (and probably most biochemical techniques), an interaction between DmsD, TatB/C, and the DmsA RR-leader suggest that an intermediate complex involving DmsA(B)/DmsD/TatB(C) is highly probable.

The event following targeting likely involves substrate handover and/or dissociation of DmsD, as it is not assembled in the final DMSO reductase complex at the membrane. Clearly DmsA can interact with the translocase independent of DmsD, which suggests the existence of a handover or dissociation step. This is supported by observations that the DmsA RR-leader fused to GFP can still be exported in a *dmsD* mutant (Ray et al. 2003), and that the DmsA RR-leader interacts with TatB and TatC *in vivo* (Kostecki et al. 2010).

12.3.3 Processing of DmsA RR-Leader

The DmsA RR-leader was consistently observed to be removed when isolated from the membrane fraction (Graubner et al. 2007; Bilous et al. 1988; Sambasivarao et al. 2000), implying that it is processed in the final assembled complex. The RR-leader has a typical signal/leader peptidase I recognition sequence at the -1 and -3 positions relative to the cleavage site, suggesting that it is cleaved by this enzyme. However mutation of the -1 and -3 conserved amino acids still led to processing of the DmsA leader, leading to the conclusion that either the residue chosen for substitution had no effect or involvement of a different type of peptidase (Sambasivarao et al. 2000). More recent studies demonstrate that LepB, a type I leader peptidase, is required for processing of three well-known Tat RR-leader substrates (Lüke et al. 2008). Rather than mutating the RR-leader, deliberately repressing *lepB* expression prevented processing of the RR-leaders, indicating direct involvement of LepB. While DmsA was not included in this study, the related CISM catalytic subunit TorA was, and until further research it is assumed that LepB is involved in processing of all Tat RR-leaders. Processing of the RR-leader likely occurs at the translocon since the DmsA RR-leader was not processed in strains devoid of TatB and TatC (Sambasivarao et al. 2001).

12.3.4 Anchoring to DmsC

The last and final step for DMSO reductase assembly involves attachment to the DmsC membrane anchor subunit. Being a typical integral membrane protein, DmsC is likely translocated and integrated by one of the two known pathways in bacteria—Sec- and/or YidC- mediated systems [see Kudva et al. (2013) for a most recent review]. Having more than two transmembrane segments disqualifies DmsC from spontaneous insertion (Engelman and Steitz 1981). And with its topology of both the N- and C- termini in the periplasm (Weiner et al. 1993), this suggests YidC involvement. Recent studies showed that the YidC is involved in cytoplasmic nitrate reductase A biogenesis (Price and Driessen 2008). Being a DMSO reductase-related CISM, this supports the notion that YidC-mediated insertion is a general mechanism for membrane anchor subunit integration.

The direct docking of DmsAB to DmsC is likely solely through interactions with DmsB as truncation studies show that the C-terminus of DmsB is indispensable for anchoring to DmsC (Sambasivarao and Weiner 1991). Much effort has been directed at understanding mechanism of electron transfer between the subunits for respiration (Rothery et al. 2008), but how the final complex is assembled remains to be investigated.

12.4 Summary and Conclusion

Biogenesis of DMSO reductase is a complicated labyrinth involving many participants. Understanding of this process has come a long way since its identification. Although studies have branched off into various areas that have led to significant and sometimes unexpected discoveries, these findings contribute to the final biogenesis process as a whole. Based on the findings discussed in this chapter, we have built an expanded model for DMSO reductase biogenesis shown in Fig. 12.3a. Steps 1–7 of this model follow the scheme of enzyme maturation involving folding and cofactor insertion as discussed in Sect. 12.2. Step 8 depicts the poorly understood holoenzyme assembly step briefly described in Sect. 12.3.1. Lastly, steps 9–15 depicts the targeting, translocation (assuming it is periplasmic localized), and membrane anchoring process described in the remainder of Sect. 12.3. The participants for maturation can also be divided into those that promote or repress maturation to generate the final systems biology model shown in Fig. 12.3b, highlighting the complexity of biogenesis.

Lastly, the model of DMSO reductase biogenesis depicted in Fig. 12.3 is by no means complete. It is clear from the discussion of this chapter that much research is still required for the full understanding of each individual step. This requires a concerted effort from many excellent research groups in the world, but given the significant knowledge advancement that has occurred since the late 1980s the future outlook of this research is encouraging.

References

Anelli T, Sitia R (2008) Protein quality control in the early secretory pathway. EMBO J 27(2):315–327

Bearson SM, Albrecht JA, Gunsalus RP (2002) Oxygen and nitrate-dependent regulation of *dmsABC* operon expression in *Escherichia coli*: sites for Fnr and NarL protein interactions. BMC Microbiol 2:13

Berks BC (1996) A common export pathway for proteins binding complex redox cofactors? Mol Microbiol 22(3):393–404

Bilous PT, Weiner JH (1988) Molecular cloning and expression of the Escherichia coli dimethyl sulfoxide reductase operon. J Bacteriol 170(4):1511–1518

Bilous PT, Cole ST, Anderson WF, Weiner JH (1988) Nucleotide sequence of the dmsABC operon encoding the anaerobic dimethylsulphoxide reductase of Escherichia coli. Mol Microbiol 2(6):785–795. doi:10.1111/j.1365-2958.1988.tb00090.x

Butland G, Peregrin-Alvarez JM, Li J, Yang W, Yang X, Canadien V, Starostine A, Richards D, Beattie B, Krogan N, Davey M, Parkinson J, Greenblatt J, Emili A (2005) Interaction network containing conserved and essential protein complexes in *Escherichia coli*. Nature 433(7025):531–537

Calloni G, Chen T, Schermann Sonya M, Chang H-c, Genevaux P, Agostini F, Tartaglia Gian G, Hayer-Hartl M, Hartl FU (2012) DnaK functions as a central hub in the E. coli chaperone network. Cell Rep 1(3):251–264

Cammack R, Weiner JH (1990) Electron paramagnetic resonance spectroscopic characterization of dimethyl sulfoxide reductase of Escherichia coli. Biochemistry 29(36):8410–8416. doi:10.1021/bi00488a030

Castanié-Cornet M-P, Bruel N, Genevaux P (2013) Chaperone networking facilitates protein targeting to the bacterial cytoplasmic membrane. Biochim Biophys Acta 1843(8):1442–1456. doi:10.1016/j.bbamcr.2013.11.007

Chan CS, Chang L, Rommens KL, Turner RJ (2009) Differential interactions between Tat-specific redox enzyme peptides and their chaperones. J Bacteriol 191(7):2091–2101. doi:10.1128/jb.00949-08

Cheng VWT, Rothery RA, Bertero MG, Strynadka NCJ, Weiner JH (2005) Investigation of the environment surrounding iron – sulfur cluster 4 of Escherichia coli dimethylsulfoxide reductase. Biochemistry 44(22):8068–8077. doi:10.1021/bi050362p

Cotter PA, Gunsalus RP (1989) Oxygen, nitrate, and molybdenum regulation of dmsABC gene expression in Escherichia coli. J Bacteriol 171(7):3817–3823

DeLisa MP, Tullman D, Georgiou G (2003) Folding quality control in the export of proteins by the bacterial twin-arginine translocation pathway. Proc Natl Acad Sci U S A 100(10):6115–6120. doi:10.1073/pnas.0937838100

Engelman DM, Steitz TA (1981) The spontaneous insertion of proteins into and across membranes: the helical hairpin hypothesis. Cell 23(2):411–422

Fröbel J, Rose P, Müller M (2012) Twin-arginine-dependent translocation of folded proteins. Philos Trans R Soc Lond B Biol Sci 367(1592):1029–1046. doi:10.1098/rstb.2011.0202

Geijer P, Weiner JH (2004) Glutamate 87 is important for menaquinol binding in DmsC of the DMSO reductase (DmsABC) from Escherichia coli. Biochim Biophys Acta 1660(1–2):66–74. doi:10.1016/j.bbamem.2003.10.016

Genest O, Neumann M, Seduk F, Stocklein W, Mejean V, Leimkuhler S, Iobbi-Nivol C (2008) Dedicated metallochaperone connects apoenzyme and molybdenum cofactor biosynthesis components. J Biol Chem 283(31):21433–21440. doi:10.1074/jbc.M802954200

Genevaux P, Georgopoulos C, Kelley WL (2007) The Hsp70 chaperone machines of Escherichia coli: a paradigm for the repartition of chaperone functions. Mol Microbiol 66(4):840–857. doi:10.1111/j.1365-2958.2007.05961.x

Graubner W, Schierhorn A, Brüser T (2007) DnaK plays a pivotal role in Tat targeting of CueO and functions beside SlyD as a general Tat signal binding Chaperone. J Biol Chem 282(10):7116–7124. doi:10.1074/jbc.M608235200

Green J, Rolfe MD, Smith LJ (2014) Transcriptional regulation of bacterial virulence gene expression by molecular oxygen and nitric oxide. Virulence 5(4). doi:10.4161/viru.27794

Hatzixanthis K, Richardson DJ, Sargent F (2005) Chaperones involved in assembly and export of N-oxide reductases. Biochem Soc Trans 33(Pt 1):124–126

Heffron K, Léger C, Rothery RA, Weiner JH, Armstrong FA (2001) Determination of an optimal potential window for catalysis by *E. coli* dimethyl sulfoxide reductase and hypothesis on the role of Mo(V) in the reaction pathway. Biochemistry 40(10):3117–3126. doi:10.1021/bi002452u

Hille R (2013) The molybdenum oxotransferases and related enzymes. Dalton Trans 42(9):3029–3042. doi:10.1039/C2DT32376A

Hoffmann A, Merz F, Rutkowska A, Zachmann-Brand B, Deuerling E, Bukau B (2006) Trigger factor forms a protective shield for nascent polypeptides at the ribosome. J Biol Chem 281(10):6539–6545. doi:10.1074/jbc.M512345200

Houry WA, Frishman D, Eckerskorn C, Lottspeich F, Hartl FU (1999) Identification of *in vivo* substrates of the chaperonin GroEL. Nature 402(6758):147–154. doi:10.1038/45977

Ilbert M, Mejean V, Giudici-Orticoni MT, Samama JP, Iobbi-Nivol C (2003) Involvement of a mate chaperone (TorD) in the maturation pathway of molybdoenzyme TorA. J Biol Chem 278(31):28787–28792

Ilbert M, Mejean V, Iobbi-Nivol C (2004) Functional and structural analysis of members of the TorD family, a large chaperone family dedicated to molybdoproteins. Microbiology 150(Pt 4): 935–943

Iobbi-Nivol C, Leimkühler S (2012) Molybdenum enzymes, their maturation and molybdenum cofactor biosynthesis in *Escherichia coli*. Biochim Biophys Acta 1827(8-9):1086–1101. doi:10.1016/j.bbabio.2012.11.007

Jong WS, ten Hagen-Jongman CM, Genevaux P, Brunner J, Oudega B, Luirink J (2004) Trigger factor interacts with the signal peptide of nascent Tat substrates but does not play a critical role in Tat-mediated export. Eur J Biochem 271(23-24):4779–4787

Kelley PM, Schlesinger MJ (1978) The effect of amino acid analogues and heat shock on gene expression in chicken embryo fibroblasts. Cell 15(4):1277–1286. doi:10.1016/0092-8674(78)90053-3

Kostecki JS, Li H, Turner RJ, DeLisa MP (2010) Visualizing interactions along the *Escherichia coli* twin-arginine translocation pathway using protein fragment complementation. PLoS One 5(2), e9225

Kudva R, Denks K, Kuhn P, Vogt A, Müller M, Koch H-G (2013) Protein translocation across the inner membrane of Gram-negative bacteria: the Sec and Tat dependent protein transport pathways. Res Microbiol 164(6):505–534. doi:10.1016/j.resmic.2013.03.016

Lamberg KE, Kiley PJ (2000) FNR-dependent activation of the class II dmsA and narG promoters of *Escherichia coli* requires FNR-activating regions 1 and 3. Mol Microbiol 38(4):817–827. doi:10.1046/j.1365-2958.2000.02172.x

Levy-Sakin M, Berger O, Feibish N, Sharon N, Schnaider L, Shmul G, Amir Y, Buzhansky L, Gazit E (2014) The influence of chemical chaperones on enzymatic activity under thermal and chemical stresses: common features and variation among diverse chemical families. PLoS One 9(2), e88541. doi:10.1371/journal.pone.0088541

Li H, Chang L, Howell JM, Turner RJ (2010) DmsD, a Tat system specific chaperone, interacts with other general chaperones and proteins involved in the molybdenum cofactor biosynthesis. Biochim Biophys Acta 1804(6):1301–1309

Lüke I, Butland G, Moore K, Buchanan G, Lyall V, Fairhurst S, Greenblatt J, Emili A, Palmer T, Sargent F (2008) Biosynthesis of the respiratory formate dehydrogenases from *Escherichia coli*: characterization of the FdhE protein. Arch Microbiol 190(6):685–696

Magalon A, Fedor JG, Walburger A, Weiner JH (2011) Molybdenum enzymes in bacteria and their maturation. Coord Chem Rev 255(9–10):1159–1178. doi:10.1016/j.ccr.2010.12.031

Marchenkov V, Semisotnov G (2009) GroEL-assisted protein folding: does it occur within the chaperonin inner cavity? Int J Mol Sci 10(5):2066–2083

Matos CFRO, Robinson C, Di Cola A (2008) The Tat system proofreads FeS protein substrates and directly initiates the disposal of rejected molecules. EMBO J 27(15):2055–2063

McAlpine AS, McEwan AG, Bailey S (1998) The high resolution crystal structure of DMSO reductase in complex with DMSO. J Mol Biol 275(4):613–623

McCrindle SL, Kappler U, McEwan AG (2005) Microbial dimethylsulfoxide and trimethylamine-N-oxide respiration. Adv Microb Physiol 50:147–198

McNicholas PM, Chiang RC, Gunsalus RP (1998) Anaerobic regulation of the *Escherichia coli* dmsABC operon requires the molybdate-responsive regulator ModE. Mol Microbiol 27(1):197–208. doi:10.1046/j.1365-2958.1998.00675.x

Méjean V, Lobbi-Nivol C, Lepelletier M, Giordano G, Chippaux M, Pascal M-C (1994) TMAO anaerobic respiration in *Escherichia coli*: involvement of the tor operon. Mol Microbiol 11(6):1169–1179. doi:10.1111/j.1365-2958.1994.tb00393.x

Mendel RR (2013) The molybdenum cofactor. J Biol Chem 288(19):13165–13172. doi:10.1074/jbc.R113.455311

Oresnik IJ, Ladner CL, Turner RJ (2001) Identification of a twin-arginine leader-binding protein. Mol Microbiol 40(2):323–331

Outten FW, Huffman DL, Hale JA, O'Halloran TV (2001) The independent cue and cus systems confer copper tolerance during aerobic and anaerobic growth in *Escherichia coli*. J Biol Chem 276(33):30670–30677. doi:10.1074/jbc.M104122200

Palmer T, Berks BC (2012) The twin-arginine translocation (Tat) protein export pathway. Nat Rev Microbiol 10(7):483–496

Palmer T, Santini C-L, Iobbi-Nivol C, Eaves DJ, Boxer DH, Giordano G (1996) Involvement of the *narJ* and *mob* gene products in distinct steps in the biosynthesis of the molybdoenzyme nitrate reductase in *Escherichia coli*. Mol Microbiol 20(4):875–884

Papish AL, Ladner CL, Turner RJ (2003) The twin-arginine leader-binding protein, DmsD, interacts with the TatB and TatC subunits of the *Escherichia coli* twin-arginine translocase. J Biol Chem 278(35):32501–32506

Pérez-Rodríguez R, Fisher AC, Perlmutter JD, Hicks MG, Chanal A, Santini C-L, Wu L-F, Palmer T, DeLisa MP (2007) An essential role for the DnaK molecular chaperone in stabilizing overexpressed substrate proteins of the bacterial twin-arginine translocation pathway. J Mol Biol 367(3):715–730. doi:10.1016/j.jmb.2007.01.027

Price CE, Driessen AJM (2008) YidC is involved in the biogenesis of anaerobic respiratory complexes in the inner membrane of *Escherichia coli*. J Biol Chem 283(40):26921–26927. doi:10.1074/jbc.M804490200

Ray N, Oates J, Turner RJ, Robinson C (2003) DmsD is required for the biogenesis of DMSO reductase in *Escherichia coli* but not for the interaction of the DmsA signal peptide with the Tat apparatus. FEBS Lett 534(1-3):156–160

Ribbe MW, Burgess BK (2001) The chaperone GroEL is required for the final assembly of the molybdenum-iron protein of nitrogenase. Proc Natl Acad Sci U S A 98(10):5521–5525. doi:10.1073/pnas.101119498

Roche B, Aussel L, Ezraty B, Mandin P, Py B, Barras F (2013) Iron/sulfur proteins biogenesis in prokaryotes: formation, regulation and diversity. Biochim Biophys Acta 1827(3):455–469. doi:10.1016/j.bbabio.2012.12.010

Rodrigo-Brenni Monica C, Hegde Ramanujan S (2012) Design principles of protein biosynthesis-coupled quality control. Dev Cell 23(5):896–907. doi:10.1016/j.devcel.2012.10.012

Rodrigue A, Batia N, Müller M, Fayet O, Böhm R, Mandrand-Berthelot MA, Wu LF (1996) Involvement of the GroE chaperonins in the nickel-dependent anaerobic biosynthesis of NiFe-hydrogenases of *Escherichia coli*. J Bacteriol 178(15):4453–4460

Romao MJ (2009) Molybdenum and tungsten enzymes: a crystallographic and mechanistic overview. Dalton Trans 21:4053–4068. doi:10.1039/B821108F

Rothery RA, Weiner JH (1991) Alteration of the iron-sulfur cluster composition of Escherichia coli dimethyl sulfoxide reductase by site-directed mutagenesis. Biochemistry 30(34):8296–8305. doi:10.1021/bi00098a003

Rothery RA, Weiner JH (1993) Topological characterization of Escherichia coli DMSO reductase by electron paramagnetic resonance spectroscopy of an engineered [3Fe-4S] cluster. Biochemistry 32(22):5855–5861

Rothery RA, Workun GJ, Weiner JH (2008) The prokaryotic complex iron–sulfur molybdoenzyme family. Biochim Biophys Acta 1778(9):1897–1929. doi:10.1016/j.bbamem.2007.09.002

Rothery RA, Stein B, Solomonson M, Kirk ML, Weiner JH (2012) Pyranopterin conformation defines the function of molybdenum and tungsten enzymes. Proc Natl Acad Sci U S A 109(37):14773–14778. doi:10.1073/pnas.1200671109

Samaluru H, SaiSree L, Reddy M (2007) Role of SufI (FtsP) in cell division of *Escherichia coli*: evidence for its involvement in stabilizing the assembly of the divisome. J Bacteriol 189(22):8044–8052. doi:10.1128/jb.00773-07

Sambasivarao D, Weiner JH (1991) Dimethyl sulfoxide reductase of *Escherichia coli*: an investigation of function and assembly by use of *in vivo* complementation. J Bacteriol 173(19):5935–5943

Sambasivarao D, Scraba DG, Trieber C, Weiner JH (1990) Organization of dimethyl sulfoxide reductase in the plasma membrane of *Escherichia coli*. J Bacteriol 172(10):5938–5948

Sambasivarao D, Turner RJ, Simala-Grant JL, Shaw G, Hu J, Weiner JH (2000) Multiple roles for the twin arginine leader sequence of dimethyl sulfoxide reductase of *Escherichia coli*. J Biol Chem 275:22526–22531. doi:10.1074/jbc.M909289199

Sambasivarao D, Dawson HA, Zhang G, Shaw G, Hu J, Weiner JH (2001) Investigation of *Escherichia coli* dimethyl sulfoxide reductase assembly and processing in strains defective for the sec-independent protein translocation system membrane targeting and translocation. J Biol Chem 276(23):20167–20174. doi:10.1074/jbc.M010369200

Sambasivarao D, Turner RJ, Bilous PT, Rothery RA, Shaw G, Weiner JH (2002) Differential effects of a molybdopterin synthase sulfurylase (moeB) mutation on *Escherichia coli* molybdoenzyme maturation. Biochem Cell Biol 80(4):435–443

Santini CL, Ize B, Chanal A, Muller M, Giordano G, Wu LF (1998) A novel sec-independent periplasmic protein translocation pathway in *Escherichia coli*. EMBO J 17(1):101–112

Sargent F, Bogsch EG, Stanley NR, Wexler M, Robinson C, Berks BC, Palmer T (1998) Overlapping functions of components of a bacterial Sec-independent protein export pathway. EMBO J 17(13):3640–3650

Schindelin H, Kisker C, Hilton J, Rajagopalan KV, Rees DC (1996) Crystal structure of DMSO reductase: redox-linked changes in molybdopterin coordination. Science 272(5268):1615–1621. doi:10.1126/science.272.5268.1615

Schneider F, Lowe J, Huber R, Schindelin H, Kisker C, Knablein J (1996) Crystal structure of dimethyl sulfoxide reductase from *Rhodobacter capsulatus* at 1.88 Å resolution. J Mol Biol 263(1):53–69

Simala-Grant JL, Weiner JH (1996) Kinetic analysis and substrate specificity of Escherichia coli dimethyl sulfoxide reductase. Microbiology 142(11):3231–3239. doi:10.1099/13500872-142-11-3231

Simala-Grant JL, Weiner JH (1998) Modulation of the substrate specificity of *Escherichia coli* dimethylsulfoxide reductase. Eur J Biochem 251(1-2):510–515. doi:10.1046/j.1432-1327.1998.2510510.x

Solomon PS, Shaw AL, Lane I, Hanson GR, Palmer T, McEwan AG (1999) Characterization of a molybdenum cofactor biosynthetic gene cluster in Rhodobacter capsulatus which is specific for the biogenesis of dimethylsulfoxide reductase. Microbiology 145(6):1421–1429. doi:10.1099/13500872-145-6-1421

Stanley NR, Sargent F, Buchanan G, Shi J, Stewart V, Palmer T, Berks BC (2002) Behaviour of topological marker proteins targeted to the Tat protein transport pathway. Mol Microbiol 43(4):1005–1021

Tang H, Rothery RA, Voss JE, Weiner JH (2011) Correct assembly of iron-sulfur cluster FS0 into *Escherichia coli* dimethyl sulfoxide reductase (DmsABC) is a prerequisite for molybdenum cofactor insertion. J Biol Chem 286(17):15147–15154. doi:10.1074/jbc.M110.213306

Tang H, Rothery RA, Weiner JH (2013) A variant conferring cofactor-dependent assembly of Escherichia coli dimethylsulfoxide reductase. Biochim Biophys Acta 1827(6):730–737. doi:10.1016/j.bbabio.2013.02.009

Tatzelt J, Prusiner SB, Welch WJ (1996) Chemical chaperones interfere with the formation of scrapie prion protein. EMBO J 15(23):6363–6373

Tseng CP, Hansen AK, Cotter P, Gunsalus RP (1994) Effect of cell growth rate on expression of the anaerobic respiratory pathway operons *frdABCD*, *dmsABC*, and *narGHJI* of *Escherichia coli*. J Bacteriol 176(21):6599–6605

Tullman-Ercek D, DeLisa MP, Kawarasaki Y, Iranpour P, Ribnicky B, Palmer T, Georgiou G (2007) Export pathway selectivity of *Escherichia coli* twin-arginine translocation signal peptides. J Biol Chem 282(11), M610507200. doi:10.1074/jbc.M610507200

Turner RJ, Busaan JL, Lee JH, Michalak M, Weiner JH (1997) Expression and epitope tagging of the membrane anchor subunit (DmsC) of *Escherichia coli* dimethyl sulfoxide reductase. Protein Eng 10(3):285–290. doi:10.1093/protein/10.3.285

Turner RJ, Papish AL, Sargent F (2004) Sequence analysis of bacterial redox enzyme maturation proteins (REMPs). Can J Microbiol 50(4):225–238

Turner RJ, Winstone TL, Tran VA, Chan CS (2010) System specific chaperones for membrane redox enzymes maturation in bacteria. In: Durante P, Colucci L (eds) Molecular chaperones: roles, structures and mechanisms. Nova Science Publishers Inc., New York, pp 179–207

Unden G, Achebach S, Holighaus G, Tran HG, Wackwitz B, Zeuner Y (2002) Control of FNR function of *Escherichia coli* by O2 and reducing conditions. J Mol Microbiol Biotechnol 4(3):263–268

Vinella D, Brochier-Armanet C, Loiseau L, Talla E, Barras F (2009) Iron-sulfur (Fe/S) protein biogenesis: phylogenomic and genetic studies of A-type carriers. PLoS Genet 5(5), e1000497. doi:10.1371/journal.pgen.1000497

Weiner JH, Shaw G, Turner RJ, Trieber CA (1993) The topology of the anchor subunit of dimethyl sulfoxide reductase of *Escherichia coli*. J Biol Chem 268(5):3238–3244

Weiner JH, Bilous PT, Shaw GM, Lubitz SP, Frost L, Thomas GH, Cole JA, Turner RJ (1998) A novel and ubiquitous system for membrane targeting and secretion of cofactor-containing proteins. Cell 93(1):93–101

Winstone TL, Workentine ML, Sarfo KJ, Binding AJ, Haslam BD, Turner RJ (2006) Physical nature of signal peptide binding to DmsD. Arch Biochem Biophys 455(1):89–97

Winstone TML, Tran VA, Turner RJ (2013) The hydrophobic region of the DmsA twin-arginine leader peptide determines specificity with chaperone DmsD. Biochemistry. doi:10.1021/bi4009374

Wissenbach U, Kröger A, Unden G (1990) The specific functions of menaquinone and demethylmenaquinone in anaerobic respiration with fumarate, dimethylsulfoxide, trimethylamine N-oxide and nitrate by Escherichia coli. Arch Microbiol 154(1):60–66. doi:10.1007/BF00249179

Wissenbach U, Ternes D, Unden G (1992) An Escherichia coli mutant containing only demethylmenaquinone, but no menaquinone: effects on fumarate, dimethylsulfoxide, trimethylamine N-oxide and nitrate respiration. Arch Microbiol 158(1):68–73. doi:10.1007/BF00249068

Wu L-F, Chanal A, Rodrigue A (2000) Membrane targeting and translocation of bacterial hydrogenases. Arch Microbiol 173(5):319–324

Yoshida Y, Takai M, Satoh T, Takami S (1991) Molybdenum requirement for translocation of dimethyl sulfoxide reductase to the periplasmic space in a photodenitrifier, *Rhodobacter sphaeroides* f. sp. denitrificans. J Bacteriol 173(11):3277–3281

Zhao Z, Weiner JH (1998) Interaction of 2-n-heptyl-4-hydroxyquinoline-N-oxide with dimethyl sulfoxide reductase of *Escherichia coli*. J Biol Chem 273(33):20758–20763. doi:10.1074/jbc.273.33.20758

Chapter 13
Microbial Proteome Profiling and Systems Biology: Applications to *Mycobacterium tuberculosis*

Olga T. Schubert and Ruedi Aebersold

Abstract Each year, 1.3 million people die from tuberculosis, an infectious disease caused by *Mycobacterium tuberculosis*. Systems biology-based strategies might significantly contribute to the knowledge-guided development of more effective vaccines and drugs to prevent and cure infectious diseases. To build models simulating the behaviour of a system in response to internal or external stimuli and to identify potential targets for therapeutic intervention, systems biology approaches require the acquisition of quantitative molecular profiles on many perturbed states. Here we review the current state of proteomic analyses in *Mycobacterium tuberculosis* and discuss the potential of recently emerging targeting mass spectrometry-based techniques which enable fast, sensitive and accurate protein measurements.

Keywords Systems biology • *Mycobacterium tuberculosis* • Proteomics • Proteome mapping • Mass spectrometry • SWATH-MS • Selected reaction monitoring (SRM)

Prokaryotic systems biology seeks an understanding of how the constituents of molecular networks function together and how they give rise to a microbe's phenotype. Nowadays, the sequence of a prokaryotic genome can be obtained within a few hours. However, even though the genome encodes directly or indirectly all other biomolecules of a cell, such as mRNAs, proteins, and metabolites, neither the subset of genes expressed on mRNA or protein level at a particular cellular state nor the dynamic change of these profiles during state changes are presently

O.T. Schubert (✉)
Institute of Molecular Systems Biology, ETH Zurich, Zurich, CH-8093, Switzerland

Systems Biology Graduate School, Zurich, CH-8057, Switzerland
e-mail: schubert@imsb.biol.ethz.ch

R. Aebersold
Institute of Molecular Systems Biology, ETH Zurich, Zurich, CH-8093, Switzerland

Faculty of Science, University of Zurich, Zurich, CH-8057, Switzerland
e-mail: aebersold@imsb.biol.ethz.ch

N.J. Krogan, M. Babu (eds.), *Prokaryotic Systems Biology*, Advances in Experimental Medicine and Biology 883, DOI 10.1007/978-3-319-23603-2_13

predictable from the genomic sequence alone. Investigating the structure, function and regulation of such molecular systems requires techniques to measure all its components at different states in a quantitatively accurate and reproducible fashion. Such measurements have become routine on mRNA level, but, due to technical limitations, on protein level they still lag behind in sensitivity, coverage and consistency.

Here we will discuss the current state of knowledge and recent technological advances in prokaryotic proteomics on the example of Mycobacteria. Due to their high clinical importance, the best-known mycobacterial strains are the human pathogens *Mycobacterium tuberculosis*, *Mycobacterium leprae* and *Mycobacterium ulcerans*. *M. tuberculosis*, the causative agent of tuberculosis, alone infects about one third of the world's population and caused 1.3 million deaths in 2012 (World Health Organization 2013). Currently, 120 mycobacterial species are recognised. Most of these have been isolated from clinical specimens and were also associated with pathogenicity in humans or animals (Tortoli 2006). Here we will focus on *M. tuberculosis*, because it is the best-studied mycobacterial strain, and notably, from the proteomic point of view, even one of the best characterised prokaryotes in general.

In the following pages, we discuss the topic of proteome mapping and its importance towards the goal of measuring the proteome of *M. tuberculosis* at different states (both, originating from *in vitro* cultures or clinical samples) fast and with high sensitivity, accuracy and reproducibility. We will (1) introduce microbial proteomics and review the current state of MS-based proteome mapping in mycobacteria, (2) discuss improved genome annotation by proteogenomics, the state of functional annotation of mycobacterial proteins, and provide an overview of key databases for protein-based research, (3) review the mapping of subcellular localisation, post-translational modifications, protein interactions, and the immunoproteome, (4) show how proteomics has been applied to understand proteomic differences of mycobacterial strains as well as their protein-level adaptation to stress conditions and during infection, and (5) conclude the chapter with an outlook on the future role of proteomic profiling in basic, systems biological and applied clinical research on mycobacteria.

13.1 Introduction to Microbial Proteomics

The first complete genome sequence of *M. tuberculosis* became available in 1998 (Cole et al. 1998). Few years after, the genome sequences of other mycobacteria followed (Cole et al. 2001; Fleischmann et al. 2002; Garnier et al. 2003; Li et al. 2005; Stinear et al. 2007; Brosch et al. 2007; Stinear et al. 2008; Ripoll et al. 2009) and today more than hundred mycobacterial genomes are published. The genomic variability of clinical isolates of the *M. tuberculosis* complex, a group of closely related mycobacteria causing tuberculosis in humans and various animal species, has been studied in much detail (Gagneux et al. 2006; Hershberg et al. 2008; Comas and Gagneux 2009; Comas et al. 2011). However, it remains to be elucidated to which extent the genotypic diversity among these strains influence phenotypic traits, such as virulence and drug-resistance.

mRNAs provide a direct readout for gene expression. Thanks to techniques for the amplification of nucleic acid sequences, mRNA measurements, earlier by microarrays, and more currently by RNA-sequencing, can reach high sensitivity and complete coverage. However, over the past years numerous studies have shown that, generally, mRNA levels correlate poorly with protein levels and that dynamic changes in transcript and protein levels also significantly diverge (Gygi et al. 1999; de Godoy et al. 2008; De Sousa Abreu et al. 2009; Maier et al. 2011; Marguerat et al. 2012; Cortes et al. 2013). These modest correlations highlight that mRNA abundances are only a rough proxy for protein levels and that they cannot reliably predict protein abundances. Furthermore, neither genomics nor transcriptomics can capture effects of post-translational processes, such as phosphorylation, protein-protein interactions or protein stability. To measure protein abundances and protein modifications directly can therefore be important, as proteins are the most critical class of biomolecules in the cell constituting essential structural components, controlling gene expression as transcription factors, catalysing various reactions as enzymes, and transducing signals in response to various stimuli.

Traditionally, protein measurements were based on specific affinity reagents, such as antibodies, which allow protein detection by Western blot or ELISA, or directly in intact cells and tissues by immunocytochemistry and immunohistochemistry, respectively. On a proteome-wide scale, mass spectrometry (MS)-based techniques are most powerful. Proteome coverage achievable with MS-based proteomic techniques has greatly increased over the past years. Early studies using two-dimensional gel electrophoresis (2D-GE) for protein separation identified and quantified below 100 proteins in a mycobacterial sample. With the advent of liquid chromatography-coupled tandem MS (LC-MS/MS), 100s to 1000s of proteins can be identified and quantified in a single MS injection. These bottom-up proteomic techniques have in common that proteins are first proteolytically digested into peptides, which are then separated by liquid chromatography, ionised and injected into the mass spectrometer. The variety of implementations on the MS-level can be roughly divided into two groups: (1) discovery proteomic techniques (e.g. shotgun proteomics), typically aimed at maximal numbers of protein identifications and (2) targeting proteomics techniques (e.g. selected reaction monitoring, SRM), typically aimed at monitoring a smaller, pre-defined subset of proteins with highest possible sensitivity, quantitative accuracy, reproducibility, and throughput across a large number of samples. Novel MS-based proteomic methods (e.g. SWATH-MS) combine untargeted, data-independent acquisition with targeted data extraction, thereby alleviating limitations of either the discovery or SRM method (Gillet et al. 2012). Importantly, both SRM and SWATH-MS, require prior knowledge to specifically target the proteins of interest. The different MS modes of the most commonly used proteomic techniques are summarised in Fig. 13.1 (Leitner and Aebersold 2013).

Each of the above-mentioned proteomic techniques has its advantages: Shotgun proteomics is optimally suited to discover large numbers of proteins but, due to

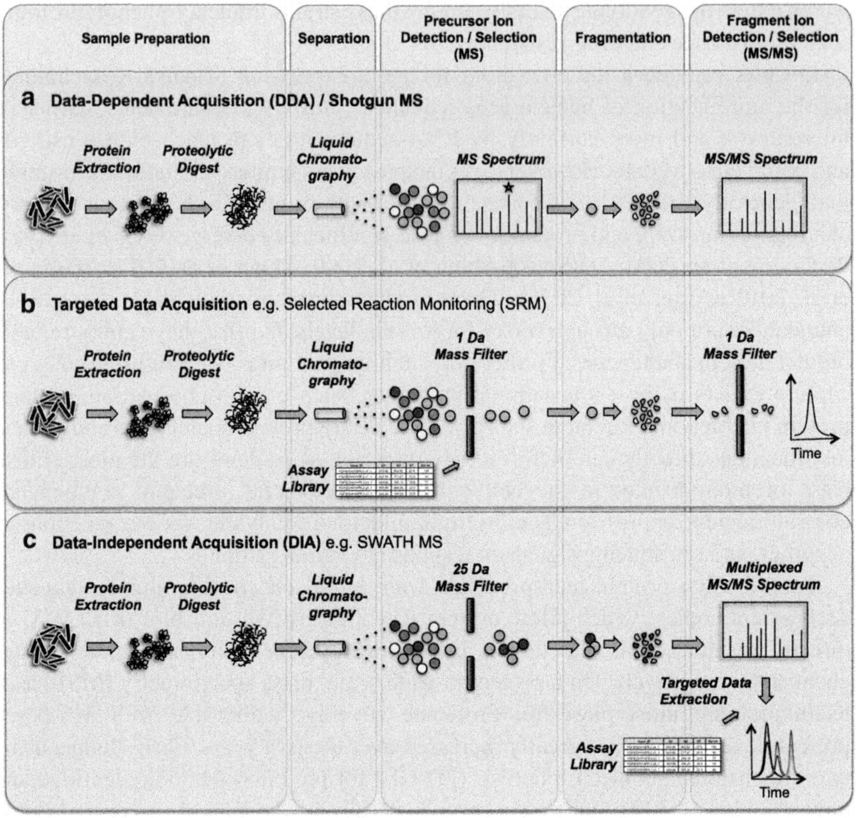

Fig. 13.1 Most commonly used mass spectrometric techniques**.** In a typical bottom-up proteomic approach, extracted proteins are first digested into peptides using a specific proteolytic enzyme, such as trypsin. The resulting peptides are then separated by liquid chromatography and ionised by electrospray ionisation before entering the mass spectrometer. In simple terms, a mass spectrometer determines the mass-to-charge ratio (m/z) of ions and counts them. The m/z of a peptide ion is, however, usually not sufficient for its identification. To infer its sequence the peptide ions are thus fragmented into smaller parts, whose m/z is also recorded by the same or a second mass analyser; this two-step approach is referred to as tandem mass spectrometry (MS/MS). MS/MS methods have been implemented in different flavours: (**a**) **Shotgun MS/Data-dependent acquisition (DDA).** Peptide ions are recorded in a survey MS spectrum from which the most intense peptide ion is selected for fragmentation. The resulting fragment ions are recorded in a characteristic MS/MS spectrum which can then be matched to the corresponding peptide sequence. (**b**) **Targeted data acquisition, e.g. Selected Reaction Monitoring (SRM).** The mass spectrometer, typically a triple quadrupole mass spectrometer, selects a specific peptide ion for fragmentation and then filters for a pre-determined fragment ion. The optimal peptide-fragment ion pairs (transitions) for each peptide have to be determined *a priori* and stored in an SRM assay library. Typically, several transitions are being measured per peptide over time, resulting in a specific, quantitative peak group for each peptide. In a standard SRM experiment, the mass spectrometer cycles through tens to several hundred transitions. (**c**) **Data-independent acquisition (DIA), e.g. SWATH-MS.** The mass spectrometer, typically a new generation quadrupole-time-of-flight/TripleTOF instrument, recursively cycles through a large m/z range (400–1200 m/z) and co-fragments all peptide ions in bins of 25 m/z. The resulting fragment ions are recorded in a composite MS/MS spectrum. MS intensities are then extracted from these highly multiplexed MS/MS spectra in a targeted manner using the MS coordinates of the peptides of interest, which were determined *a priori* and stored in a SWATH assay library. The resulting data resembles the data of targeted data acquisition

irreproducible precursor selection and undersampling, especially in the case of lower abundant proteins, it is less suited for quantification across large sample sets (Domon and Aebersold 2010). SRM excels at consistent and accurate quantification over large sample sets with coefficients of variation (CV) below 15 % and offers the largest dynamic range of all MS techniques available today (Picotti et al. 2009). However, SRM measurements are limited to a few dozens of target proteins per sample injection. In contrast, SWATH-MS analyses are not limited in the number of target proteins, as long as they are present in the assay library. In terms of CVs SWATH-MS performs similarly to SRM and offers a dynamic range of at least three orders of magnitude which is close to the abundance range of a prokaryotic proteome, making it a very promising technology for complete and high-quality microbial proteome measurements (Gillet et al. 2012; Liu et al. 2013b).

13.2 Proteome Mapping: Determining the MS Coordinates of a Protein

Maps of any kind typically record the spatial distribution of objects and capture associations or links between them. It would usually take an initial effort to create a map, but once established, the map helps to find places or retrieve particular information in a targeted manner and thus much faster and more reliably than through pure discovery without the benefit of a map. Life sciences have been revolutionised by the possibility to compile a complete genome map of an organism by sequencing its entire DNA content. The availability of such genome maps fuelled the development of technologies to measure gene expression, for instance microarrays or RNA-sequencing to quantify mRNA abundances. Also the field of proteomics is strongly dependent on high-quality genome maps, because almost all proteomic strategies rely on a database of all possible protein sequences encoded by a genome (Fig. 13.2). Proteome maps, however, have so far not yet been fully exploited in MS-based proteomics (Ahrens et al. 2010). On the contrary, most proteomic studies perpetually re-discover the proteome and are ignorant to information that has been collected previously. Targeting proteomic strategies, such as SRM and SWATH-MS, have succeeded in leveraging the use of prior knowledge in a proteome map to quantify proteins with higher sensitivity, quantitative accuracy and reproducibility at high throughput (Fig. 13.2). For targeted MS-based analysis, a protein assay consists of a set of MS coordinates uniquely describing the protein of interest in the proteome map. The complete set of coordinates describes (1) which peptides of a protein are most representative, i.e. unique and well detectable, (2) the elution time of these peptides from the LC, (3) their pre-dominant charge state, (4) the manner they fragment during collision induced dissociation, thus indicating the most abundant fragment ions formed and their relative intensity.

The MS coordinates of a protein are determined in proteome mapping experiments, which are usually performed using discovery proteomics. To increase proteome coverage in these mapping experiments, the complexity of the samples

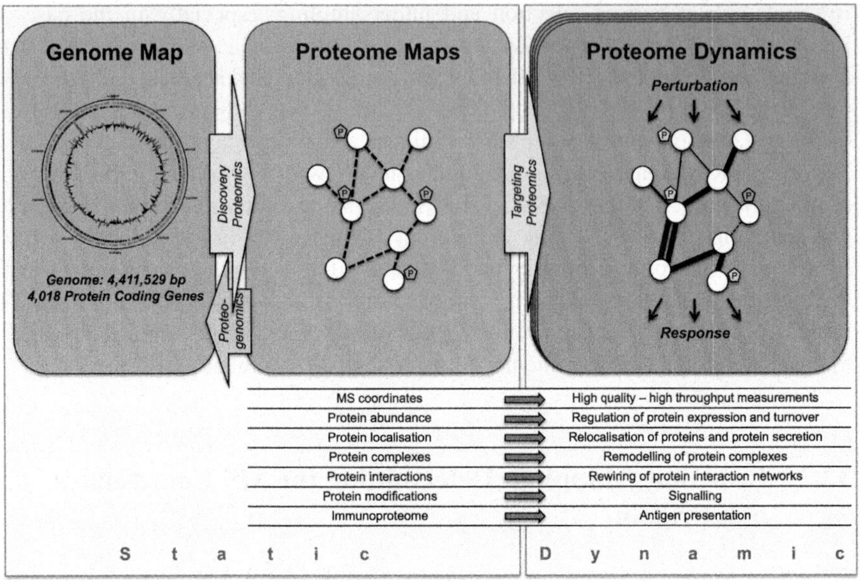

Fig. 13.2 Proteome maps are the basis for high-quality targeted proteomic studies. The genome of *M. tuberculosis* encodes for roughly 4000 genes. The annotated genome sequence defines the space of all possible transcripts and proteins. By discovery proteomics the expressed proteome of a cell can be mapped out and different approaches enable the mapping of potential protein localisations, interactions and post-translational modifications. These proteome maps provide a protein-level inventory of the cell and are aimed at maximal coverage. In contrast, targeting proteomic approaches typically focus on subsets of proteins to be profiled dynamically over a large number of differentially perturbed states fast and reproducibly. To specifically target the proteins of interest, targeting proteomic techniques require prior knowledge in form of MS coordinates/assays contained in a proteome map

can be reduced by fractionation before MS analysis, for instance on protein level by SDS-PAGE, or on peptide level by isoelectric focussing of peptides using off-gel electrophoresis (Heller et al. 2005). The most recent mapping experiments in mycobacteria all cover more than 3000 proteins corresponding to over 75 % of all annotated proteins and have shown that the protein abundances span at least four orders of magnitude (Kelkar et al. 2011; Zheng et al. 2012; Schubert et al. 2013). Using synthetic peptides, the remaining fraction of the proteome was mapped as well, thereby reaching a proteome coverage of 97 % (Schubert et al. 2013). Once a microbial proteome is mapped out, fractionation is typically not required anymore as targeting proteomic techniques, such as SRM or SWATH-MS, are capable to cover the full dynamic range of protein expression in microbial cells in a single analysis (Picotti et al. 2009; Gillet et al. 2012).

13.2.1 Proteogenomics Improves Genome Annotation

Most MS-based proteomic techniques rely on a correctly annotated protein sequence database. For mycobacteria, comparisons of genome annotation efforts have however revealed that genome annotation is incomplete and erroneous (de Souza et al. 2008; de Souza et al. 2009). High-coverage proteomic datasets can help refining genome annotation by providing experimental evidence for genes not previously annotated or wrongly assigned translational start sites, but also by simply corroborating existing open reading frames. This strategy is called proteogenomics and is based on the full translation of the genome in all six reading frames, resulting in a very large amino acid sequence database. The fragment ion spectra of extensive shotgun experiments are then searched against this database and may provide evidence to correct for missing or erroneous gene annotation (Armengaud 2009; Renuse et al. 2011). Over the past few years, several large proteogenomic analyses of *M. tuberculosis* have each reported between 20 and 40 so far not annotated proteins highlighting the shortcomings of current genome annotations (de Souza et al. 2008; de Souza et al. 2011; Kelkar et al. 2011; Schubert et al. 2013).

13.2.2 Functional Annotation of Mycobacterial Proteins

To draw conclusions from proteome measurements, proteins need to be functionally annotated. For *M. tuberculosis* H37Rv 3924 protein-coding genes were originally annotated, but to less than two thirds of these an explicit or putative function could be assigned (Cole et al. 1998). The field still mostly uses the 10 functional classes defined in the original publication (Cole et al. 1998) and protein functions according to this system are being updated in the TubercuList database (Lew et al. 2013). The current annotation of *M. tuberculosis* H37Rv in TubercuList (R27) contains 4018 protein-coding genes of which 26 % belong to the classes "conserved hypothetical" and "unknown" functions. Recent efforts reduced this number to below 12 % based on orthology, integrated genomic context analysis, and literature mining (Doerks et al. 2012). An alternative functional annotation system for *M. tuberculosis* has been provided by Sanger (http://www.sanger.ac.uk/), but has not been updated since 2010. The Sanger system consists of 6 main classes that further split into 29 subclasses; 40 % of the *M. tuberculosis* proteins belong to the classes "conserved hypothetical" and "unknown". The most generic and species-independent functional annotation scheme was initiated by the Gene Ontology (GO) consortium with the aim of standardising the representation of gene and gene product attributes across species and databases (Ashburner et al. 2000). In the UniProt-GOA database, currently (as of February 2014) over 75 % of all *M. tuberculosis* proteins are annotated with at least one GO term (UniProt Consortium 2014). The GO annotation coverage for other mycobacterial proteomes is also relatively high with 60–70 %. Proteins with known function can be visualised on a metabolic or pathway map

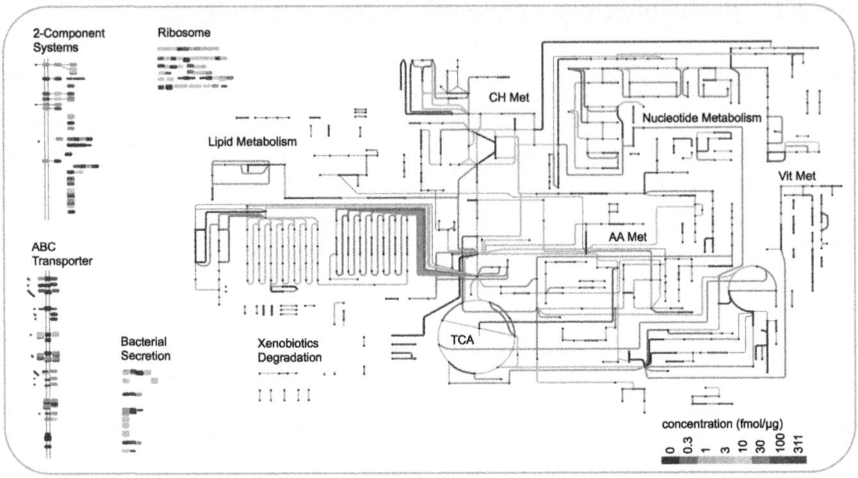

Fig. 13.3 Visualisation of large-scale data on a pathway map. Absolute protein concentrations of *M. tuberculosis* have been mapped using iTuby (http://pathways.embl.de/iTuby). Reprinted from (Schubert et al. 2013) with permission from Elsevier

to enable intuitive understanding of large-scale information; a web-based, *M. tuberculosis*-specific visualisation tool has been recently developed and is called iTuby (http://pathways.embl.de/iTuby) (Fig. 13.3).

13.2.3 Databases Providing Protein-Level Information for Mycobacteria

Databases to collect and organise information and to allow its fast and easy retrieval through web interfaces have become indispensable in today's research. The most comprehensive database to retrieve protein sequence information and functional annotation of proteins in general is *UniProt* (www.UniProt.org) (UniProt Consortium 2014). UniProt integrates, interprets and standardises data from the literature and various resources and along with the protein sequence provides functional annotation, subcellular localisation and a large number of cross-references to external resources. Because of its importance, the proteome of *M. tuberculosis* belongs to the 600 reference proteomes within UniProt, with over 2000 manually curated ("reviewed") proteins.

TubercuList (TubercuList.epfl.ch) is a database for the genome sequence annotation of *M. tuberculosis* H37Rv (Lew et al. 2013). TubercuList provides gene and protein sequences, functional annotation, subcellular localisation, and cross-references similarly to UniProt through a fast and easy to use web interface. TubercuList is more tailored to *M. tuberculosis* research than UniProt and is therefore more straightforward to use if a researcher is interested in *M. tuberculosis* H37Rv only. Its greatest strength lies in the fact that it has been subject to continuous manual curation since its inception in 1998.

Another important database for tuberculosis research is the *Tuberculosis Database* (TBDB, www.tbdb.org), which contains detailed structural and functional annotation of all genes of *M. tuberculosis* (Reddy et al. 2009). The focus of TBDB is on genomic data, as well as microarray and RT-PCR data, but it is nevertheless a useful resource to proteomic researchers.

To browse and disseminate MS-based proteomic data for *M. tuberculosis*, the following databases are available: The *M. tuberculosis* build in the *PeptideAtlas* (www.PeptideAtlas.org) database (Deutsch 2010) covers over 3370 proteins identified from data of several large shotgun MS studies on *M. tuberculosis* (Kelkar et al. 2011; Albrethsen et al. 2013; Schubert et al. 2013; Galagan et al. 2013). Through a web interface, PeptideAtlas offers interactive browsing of all peptide and protein identifications, as well as the corresponding fragment ion spectra. The *SRMAtlas* (www.SRMAtlas.org) database (Picotti et al. 2008) provides downloadable high-quality SRM assays, which had been generated from synthetic peptides, for 97 % of all annotated protein coding genes of *M. tuberculosis* (Schubert et al. 2013). These SRM assays were tested for the detection of proteins in unfractionated mycobacterial lysates by SRM and the assays of 2884 proteins could thus be validated. The SRM traces and statistical scores of the validation measurements can be viewed in the *PASSEL* (www.PeptideAtlas.org/passel) database (Farrah et al. 2012; Schubert et al. 2013). The SWATHAtlas (www.SWATHAtlas.org) database provides for all annotated M. tuberculosis proteins assays to support proteome-wide SWATH-MS analysis (Schubert et al. 2015).

13.2.4 Proteome-Wide Mapping of Subcellular Localisations

In microbes, and in particular in pathogenic microbes, secreted proteins and proteins of the cell envelope (i.e. cell membrane and/or cell wall) have caught attention due to their potential role as antigens, drug targets, biomarkers, and in mediating host-pathogen interactions. In mycobacteria, the assignment of subcellular localisations of proteins has therefore been of great interest for many years. Several early studies compared the secretome of *M. tuberculosis* and the non-pathogenic vaccine strain *M. bovis* BCG and identified, among others, the major secreted virulence factors ESAT-6 and CFP-10 (Mattow et al. 2001; Mattow et al. 2003). Bell and co-workers provide a comprehensive overview of mycobacterial proteomic studies on subcellular localisations and in general over the past 15 years (Bell et al. 2012). In *M. tuberculosis*, using different extraction, fractionation, and enrichment protocols the proteome of the cell wall, membrane, cytosol, and culture filtrate could be determined, in total covering over 1000 proteins (Bell et al. 2012). Latest studies report the identification and quantification of 1176 secreted proteins in culture filtrates of exponentially growing and nutrient starved *M. tuberculosis* cultures

(Albrethsen et al. 2013) and 2203 membrane-associated proteins in exponentially growing *M. tuberculosis* and *M. bovis* BCG (Gunawardena et al. 2013). Overall, these studies provide a comprehensive catalogue of subcellular localisations for many mycobacterial proteins. The dynamic re-localisation of proteins in response to different stimuli and in clinical isolates however warrants further investigation.

13.2.5 Proteome-Wide Mapping of Post-Translational Modifications

Much of the diversity of protein functions not only in eukaryotes but also in prokaryotes, can be attributed to post-translational modifications, altering for instance enzyme activity and protein complex formation (Cain et al. 2014). To study post-translationally modified proteins, enrichment protocols are often applied prior to MS analysis to increase sensitivity by enriching for the usually sub-stoichiometrically modified proteins or peptides. Proteome maps of post-translationally modified proteins in mycobacteria are currently scarce. A remarkable study by Prisic and colleagues reports 516 Serine/Threonine phosphorylation sites in 301 phosphorylated proteins over a large number of different stress conditions (Prisic et al. 2010). The phosphorylation state of the 11 two-component systems in *M. tuberculosis* has so far not been amenable to proteomic analysis due to the acid-labile nature of Histidine and Aspartate phosphorylations. Glycosylation, more specifically O-mannosylation has been shown to attenuate pathogenicity of *M. tuberculosis* in mice (Liu et al. 2013a) and proteomic approaches reported over 40 O-glycosylated proteins in the culture supernatant of *M. tuberculosis* (González-Zamorano et al. 2009; Smith et al. 2014). Pupylation is a recently discovered post-translational modification of lysine residues with Pup, the *p*rokaryotic *u*biquitin-like *p*rotein. Since its discovery in 2008 (Pearce et al. 2008), several groups used mycobacterial cells expressing a tagged-version of pup to affinity purify and identify potential pupylation targets, collectively called the pupylome (Festa et al. 2010; Watrous et al. 2010; Poulsen et al. 2010). To date, in *M. tuberculosis* 602 proteins have been associated with Pup, but only for 55 the pupylation site has been experimentally confirmed (Tung 2012). Other post-translational modifications such as lipidation, methylation and acetylation have yet to be investigated systematically in mycobacteria.

13.2.6 Connecting the Nodes in the Proteome Map: Protein Complexes and Protein Interactions

Most cellular functions and phenotypes are not regulated by individual molecules, but arise from interactions among many molecular components of the cell. Several experimental approaches, such as co-immunoprecipitation as well as yeast two- or three-hybrid and similar systems have been used to study protein interactions in mycobacteria on small scale (Steyn et al. 2003; Veyron-Churlet et al. 2004; Singh et al. 2006; O'Hare et al. 2008; Tharad et al. 2011; Parikh et al. 2013). To study protein complexes and protein-protein interactions on a larger scale, the method of choice is affinity purification coupled to MS analysis (Gingras et al. 2007). However, affinity purification approaches require genetic manipulation of the cells to introduce an affinity tag in the bait protein. This is a particularly time-consuming and demanding undertaking in the slow-growing and difficult-to-handle mycobacteria. An alternative technique to query the protein complexome of a cell, overcoming these challenges, is MS-based protein correlation profiling. Therefore, soluble protein complexes are fractionated, e.g. by size-exclusion chromatography, and subsequently, proteins in each fraction are quantified by MS and assigned to complexes based on the correlation of their fractionation profiles (Ranish et al. 2003; Havugimana et al. 2012; Kristensen et al. 2012). Challenging as they are, both these techniques have so far not been applied to experimentally study protein interactions and complexes in mycobacteria and global protein interaction networks remain limited to computational predictions (Raman and Chandra 2008; Cui et al. 2009; Wang et al. 2010; Liu et al. 2012; Vashisht et al. 2012).

13.2.7 Mapping the Mycobacterial Immunoproteome

The immunoproteome of an organism is its antigenic repertoire as recognised by the human immune system. Antigens are used in vaccines to trigger a specific immune response and it is therefore of high clinical interest to map out the *M. tuberculosis* immunoproteome (Kunnath-Velayudhan and Porcelli 2013). To interrogate the entire *M. tuberculosis* proteome for antigens, two approaches have been applied. One was based on a library of peptides predicted to bind with high affinity to commonly expressed MHC class II alleles; this library was then used to screen for $CD4^+$ T cell responses (Lindestam Arlehamn et al. 2013). The other approach is based on purified proteins or protein microarrays covering almost all annotated *M. tuberculosis* proteins to test patient serum antibody responses as a surrogate for $CD4^+$ T cell responses (Li et al. 2010; Kunnath-Velayudhan et al. 2012). From the results of these screens it becomes apparent that immune responses target a small sub-proteome, which is enriched for membrane-associated and secreted proteins, and that members of the PE/PPE and ESX protein families are dominant in the immunoproteome.

13.3 Proteomic Profiling: Studying Responses to Genetic and Environmental Stimuli

13.3.1 Comparative Proteomics of Mycobacterial Strains

To find novel protective antigens, early studies compared the proteomes or sub-proteomes of different mycobacterial strains by 2D-GE (Jungblut et al. 1999; Betts et al. 2000; Mattow et al. 2001; Mattow et al. 2003; Schmidt et al. 2004; Pheiffer et al. 2005). More recently, De Souza and colleagues used untargeted LC-MS/MS to compare the proteomes of hypo- and hypervirulent *M. tuberculosis* Beijing strains and found about 50 proteins over-represented in either strain (de Souza et al. 2010). Other studies compared membrane associated proteins between the virulent *M. tuberculosis* strain H37Rv and its avirulent counterpart H37Ra or the *M. bovis* BCG and reported dozens to few hundreds of proteins that are of significant differential abundance (Målen et al. 2011; Gunawardena et al. 2013). These studies indicate that genomic variation of mycobacterial strains indeed manifests on the proteome level and comparative proteomic studies may thus help to understand differences in phenotypic properties, such as virulence. However, more extensive comparisons, both on transcriptome-level, such as the one by Rose and colleagues (Rose et al. 2013), as well as on proteome-level, will be required to elucidate how genetic variability between mycobacterial isolates is translated into their distinct phenotypes.

13.3.2 Proteomics of Stress Responses in M. tuberculosis

To study proteomic adaptation of *M. tuberculosis* during infection and persistence in the human host, various *in vitro* culture models have been used, mimicking conditions thought to play a role during infection of macrophages or within tuberculous granuloma. In the face of an active host immune response, a subpopulation of the bacilli is thought to enter a metabolically highly reduced state called dormancy where they replicate rarely, if at all (Chao and Rubin 2010). To model these latent infections *in vitro*, the most commonly used stress condition is hypoxia, which has been shown to be prevalent in tuberculous granuloma (Via et al. 2008). There are different models of hypoxic stress, one of which is the Wayne model in which oxygen is gradually self-depleted by growing bacteria in sealed culture vessels (Wayne and Hayes 1996). Early proteomic studies showed the strong up-regulation of a number of proteins, including HspX, Hrp1, DosR, Ald, GroEL-2, and Tuf, in response to the hypoxic conditions (Cunningham and Spreadbury 1998; Boon et al. 2001; Rosenkrands et al. 2002; Starck et al. 2004). Several of the induced proteins belong to the roughly 50 members of the DosR regulon, which are regulated by the two component system DosS/DosT-DosR (Boon and Dick 2002; Park et al. 2003). The DosR regulon is thought to be involved in the metabolic adaptation to the anaerobic state, as well as survival and resuscitation from dormancy (Rustad et al. 2009; Leistikow et al. 2010). A chemostat-adapted version of the Wayne model analysed by ICAT, a chemical labelling-based MS approach, identified over 200 differentially expressed proteins in the two distinct phases of non-replicative persistence induced by gradual oxygen depletion (Cho et al. 2006). The most significantly increased functional classes were small molecule degradation and energy metabolism along with the induction of 13 proteins of the DosR regulon. Further cell culture models for dormancy also showing strong induction of HspX and other members of the DosR regulon on protein level are for instance standing cultures (Florczyk et al. 2001; Purkayastha et al. 2002; Schubert et al. 2013) or long-term stationary phase cultures (Yuan et al. 1996).

Besides hypoxia, another frequently applied *in vitro* stress is starvation in phosphate-buffered saline (Loebel et al. 1933). Using this setup increased protein levels for HspX, Rv2557 and Rv2558 and decreased protein levels for Tig, MPT64 and MPT32 (a.k.a. apa/modD/45 kD antigen) were found (Betts et al. 2002). A recent proteomic study investigated changes in the culture filtrate of cultures subjected to this type of nutrient starvation and found 230 proteins to be increased and 208 proteins to be decreased of the 1176 proteins identified in total; among the strongly increased proteins were many members of toxin-antitoxin systems (Albrethsen et al. 2013). Overall, these *in vitro* studies indicate a significant remodelling of the *M. tuberculosis* proteome in response to various stresses.

13.3.3 Proteomics of M. tuberculosis Infection

In vitro cultures of bacteria are usually simple to conduct and the analysis of samples does not suffer from interfering human proteins. A human background proteome, such as would be present in the case of infected human cell or tissue samples, drastically reduces selectivity and sensitivity of proteomic techniques aimed at identifying and quantifying bacterial proteins. However, *in vitro* cultures obviously fall short in mimicking all aspects of an *in vivo* environment and therefore the findings derived from such cultures will eventually have to be validated in more *in vivo*-like experimental setups.

Proteomic experiments of mycobacteria from infected macrophages were conducted already almost 20 years ago using 2D-GE analysis, revealing between 40 and 70 differentially regulated protein spots compared to broth culture (Lee and Horwitz 1995; McDonough et al. 2000). Later 2D-GE-based studies involved MS analysis to identify proteins with increased expression levels in *M. bovis* BCG during macrophage infection (HspX, GroEL-1, GroEL-2, Rv2623, InhA and Tuf) (Monahan et al. 2001), and proteins unique to *M. tuberculosis* during macrophage infection (Mattow et al. 2006), such as the methyl citrate pathway enzyme PrpD.

Technically even more challenging is the investigation of *M. tuberculosis* in infected tissue, for instance the granulomatous tissue structures typically forming after infection of the lung. Due to the challenges associated with the low number of bacteria in relation to host cells, to date, only a single study analysed the mycobacterial proteome in infected lung tissue. Here the guinea pig model of aerosol infection was used to examine the mycobacterial proteome during early (30 days) and chronic (90 days) stages of the disease resulting in the identification of over 300 mycobacterial proteins at each of the two stages (Kruh et al. 2010). The two major functional classes of proteins identified in the infected lung were cell wall and cell processes as well as intermediate metabolism. Furthermore were PE/PPE proteins found to be consistently expressed during infection.

Even though the studies mentioned above have provided new insight into mycobacterial proteome reorganisation during infection, it remains challenging to robustly and sensitively quantify mycobacterial proteins in the host environment by discovery-based proteomics. Targeted proteomic techniques, such as SRM, together with the corresponding *M. tuberculosis*-specific assays, for instance provided through the SRMAtlas database (Schubert et al. 2013), will help to overcome some of these challenges thanks to superior selectivity, sensitivity and dynamic range compared to conventional untargeted proteomic techniques. They can thus provide data on which models of bacterial physiology during infection and persistence in the human host can be built. These models, in turn, can provide assistance in the rational design of new preventive or curative measures for tuberculosis.

13.4 Outlook and Conclusions

Proteomics is at the transition from exploratory, discovery-driven approaches to sophisticated, dynamic and quantitative analyses of biological systems and to applied clinical applications. Proteomic research on a particular organism typically starts with the discovery and cataloguing of the components of a system, i.e. with the development of a map that would form the basis for future experiments. Once such a comprehensive proteome map is established, the focus typically shifts from qualitative to quantitative measurements to investigate molecular mechanisms of cell physiology by understanding how proteins change abundance, post-translational modification state, and interactions with other proteins in response to genetic and environmental perturbations. In the context of systems biology, where the goal is to mathematically describe and model a cellular process or even an entire organism, the system's adaptations in response to many different conditions need to be queried. For such studies, high quality and complete datasets are a must and therefore well-established proteome maps and corresponding tools to navigate them are indispensable. For clinical studies, consistent and accurate proteome measurements are of even higher importance. Clinical studies typically focus on few proteins, but large patient cohorts and require very robust techniques to ensure the required reproducibility at high throughput, for instance in the field of biomarker research. For mycobacteria, in particular *M. tuberculosis*, the proteome in its various facets has been mapped already relatively well, allowing us to transit from exploratory proteome mapping to querying the systems of interest.

References

Ahrens CH, Brunner E, Qeli E et al (2010) Generating and navigating proteome maps using mass spectrometry. Nat Rev Mol Cell Biol 11:789–801. doi:10.1038/nrm2973

Albrethsen J, Agner J, Piersma SR et al (2013) Proteomic profiling of the Mycobacterium tuberculosis identifies nutrient starvation responsive toxin-antitoxin systems. Mol Cell Proteomics. doi:10.1074/mcp.M112.018846

Armengaud J (2009) A perfect genome annotation is within reach with the proteomics and genomics alliance. Curr Opin Microbiol 12:292–300. doi:10.1016/j.mib.2009.03.005

Ashburner M, Ball CA, Blake JA et al (2000) Gene ontology: tool for the unification of biology. The Gene Ontology Consortium. Nat Genet 25:25–29. doi:10.1038/75556

Bell C, Smith GT, Sweredoski MJ, Hess S (2012) Characterization of the Mycobacterium tuberculosis proteome by liquid chromatography mass spectrometry-based proteomics techniques: a comprehensive resource for tuberculosis research. J Proteome Res 11:119–130. doi:10.1021/pr2007939

Betts JC, Dodson P, Quan S et al (2000) Comparison of the proteome of Mycobacterium tuberculosis strain H37Rv with clinical isolate CDC 1551. Microbiology 146(Pt 12): 3205–3216

Betts JC, Lukey PT, Robb LC et al (2002) Evaluation of a nutrient starvation model of Mycobacterium tuberculosis persistence by gene and protein expression profiling. Mol Microbiol 43:717–731

Boon C, Dick T (2002) Mycobacterium bovis BCG response regulator essential for hypoxic dormancy. J Bacteriol 184:6760–6767

Boon C, Li R, Qi R, Dick T (2001) Proteins of Mycobacterium bovis BCG induced in the Wayne dormancy model. J Bacteriol 183:2672–2676. doi:10.1128/JB.183.8.2672-2676.2001

Brosch R, Gordon SV, Garnier T et al (2007) Genome plasticity of BCG and impact on vaccine efficacy. Proc Natl Acad Sci U S A 104:5596–5601. doi:10.1073/pnas.0700869104

Cain JA, Solis N, Cordwell SJ (2014) Beyond gene expression: the impact of protein post-translational modifications in bacteria. Proteomics 97:265–286. doi:10.1016/j.jprot.2013.08.012

Chao MC, Rubin EJ (2010) Letting sleeping dos lie: does dormancy play a role in tuberculosis? Annu Rev Microbiol 64:293–311. doi:10.1146/annurev.micro.112408.134043

Cho SH, Goodlett D, Franzblau S (2006) ICAT-based comparative proteomic analysis of non-replicating persistent Mycobacterium tuberculosis. Tuberculosis (Edinb) 86:445–460. doi:10.1016/j.tube.2005.10.002

Cole ST, Brosch R, Parkhill J et al (1998) Deciphering the biology of Mycobacterium tuberculosis from the complete genome sequence. Nature 393:537–544. doi:10.1038/31159

Cole ST, Eiglmeier K, Parkhill J et al (2001) Massive gene decay in the leprosy bacillus. Nature 409:1007–1011. doi:10.1038/35059006

Comas I, Gagneux S (2009) The past and future of tuberculosis research. PLoS Pathog 5, e1000600. doi:10.1371/journal.ppat.1000600

Comas I, Borrell S, Roetzer A et al (2011) Whole-genome sequencing of rifampicin-resistant Mycobacterium tuberculosis strains identifies compensatory mutations in RNA polymerase genes. Nat Genet 44:106–110. doi:10.1038/ng.1038

Cortes T, Schubert OT, Rose G et al (2013) Genome-wide mapping of transcriptional start sites defines an extensive leaderless transcriptome in Mycobacterium tuberculosis. Cell Rep. doi:10.1016/j.celrep.2013.10.031

Cui T, Zhang L, Wang X, He Z-G (2009) Uncovering new signaling proteins and potential drug targets through the interactome analysis of Mycobacterium tuberculosis. BMC Genomics 10:118. doi:10.1186/1471-2164-10-118

Cunningham AF, Spreadbury CL (1998) Mycobacterial stationary phase induced by low oxygen tension: cell wall thickening and localization of the 16-kilodalton alpha-crystallin homolog. J Bacteriol 180:801–808 de Godoy LMF, Olsen JV, Cox J et al (2008) Comprehensive mass-spectrometry-based proteome quantification of haploid versus diploid yeast. Nature 455:1251–1254. doi:10.1038/nature07341

De Sousa Abreu R, Penalva LO, Marcotte EM, Vogel C (2009) Global signatures of protein and mRNA expression levels. Mol Biosyst 5:1512–1526. doi:10.1039/b908315d

de Souza GA, Målen H, Søfteland T et al (2008) High accuracy mass spectrometry analysis as a tool to verify and improve gene annotation using Mycobacterium tuberculosis as an example. BMC Genomics 9:316. doi:10.1186/1471-2164-9-316

de Souza GA, Søfteland T, Koehler CJ et al (2009) Validating divergent ORF annotation of the Mycobacterium leprae genome through a full translation data set and peptide identification by tandem mass spectrometry. Proteomics 9:3233–3243. doi:10.1002/pmic.200800955

de Souza GA, Fortuin S, Aguilar D et al (2010) Using a label-free proteomics method to identify differentially abundant proteins in closely related hypo- and hypervirulent clinical Mycobacterium tuberculosis Beijing isolates. Mol Cell Proteomics 9:2414–2423. doi:10.1074/mcp.M900422-MCP200

de Souza GA, Arntzen MØ, Fortuin S et al (2011) Proteogenomic analysis of polymorphisms and gene annotation divergences in prokaryotes using a clustered mass spectrometry-friendly database. Mol Cell Proteomics 10:M110.002527. doi:10.1074/mcp.M110.002527

Deutsch EW (2010) The PeptideAtlas Project. Methods Mol Biol 604:285–296. doi:10.1007/978-1-60761-444-9_19

Doerks T, van Noort V, Minguez P, Bork P (2012) Annotation of the M. tuberculosis Hypothetical Orfeome: adding functional information to more than half of the uncharacterized proteins. PLoS One 7, e34302. doi:10.1371/journal.pone.0034302

Domon B, Aebersold R (2010) Options and considerations when selecting a quantitative proteomics strategy. Nat Biotechnol 28:710–721. doi:10.1038/nbt.1661

Farrah T, Deutsch EW, Kreisberg R et al (2012) PASSEL: the PeptideAtlas SRMexperiment library. Proteomics 12:1170–1175. doi:10.1002/pmic.201100515

Festa RA, McAllister F, Pearce MJ et al (2010) Prokaryotic ubiquitin-like protein (Pup) proteome of Mycobacterium tuberculosis. PLoS One 5, e8589. doi:10.1371/journal.pone.0008589

Fleischmann RD, Alland D, Eisen JA et al (2002) Whole-genome comparison of Mycobacterium tuberculosis clinical and laboratory strains. J Bacteriol 184:5479–5490

Florczyk MA, McCue LA, Stack RF et al (2001) Identification and characterization of mycobacterial proteins differentially expressed understanding and shaking culture conditions, including Rv2623 from a novel class of putative ATP-binding proteins. Infect Immun 69:5777–5785

Gagneux S, DeRiemer K, Van T et al (2006) Variable host-pathogen compatibility in Mycobacterium tuberculosis. Proc Natl Acad Sci U S A 103:2869–2873. doi:10.1073/pnas.0511240103

Galagan JE, Minch K, Peterson M et al (2013) The Mycobacterium tuberculosis regulatory network and hypoxia. Nature. doi:10.1038/nature12337

Garnier T, Eiglmeier K, Camus J-C et al (2003) The complete genome sequence of Mycobacterium bovis. Proc Natl Acad Sci U S A 100:7877–7882. doi:10.1073/pnas.1130426100

Gillet LC, Navarro P, Tate S et al (2012) Targeted data extraction of the MS/MS spectra generated by data-independent acquisition: a new concept for consistent and accurate proteome analysis. Mol Cell Proteomics 11:O111.016717. doi:10.1074/mcp.O111.016717

Gingras A-C, Gstaiger M, Raught B, Aebersold R (2007) Analysis of protein complexes using mass spectrometry. Nat Rev Mol Cell Biol 8:645–654. doi:10.1038/nrm2208

González-Zamorano M, Mendoza-Hernández G, Xolalpa W et al (2009) Mycobacterium tuberculosis glycoproteomics based on ConA-lectin affinity capture of mannosylated proteins. J Proteome Res 8:721–733. doi:10.1021/pr800756a

Gunawardena HP, Feltcher ME, Wrobel JA et al (2013) Comparison of the membrane proteome of virulent Mycobacterium tuberculosis and the attenuated Mycobacterium bovis BCG vaccine strain by label-free quantitative proteomics. J Proteome Res. doi:10.1021/pr400334k

Gygi SP, Rochon Y, Franza BR, Aebersold R (1999) Correlation between protein and mRNA abundance in yeast. Mol Cell Biol 19:1720–1730

Havugimana PC, Hart GT, Nepusz T et al (2012) A census of human soluble protein complexes. Cell 150:1068–1081. doi:10.1016/j.cell.2012.08.011

Heller M, Ye M, Michel PE et al (2005) Added value for tandem mass spectrometry shotgun proteomics data validation through isoelectric focusing of peptides. J Proteome Res 4: 2273–2282. doi:10.1021/pr050193v

Hershberg R, Lipatov M, Small PM et al (2008) High functional diversity in Mycobacterium tuberculosis driven by genetic drift and human demography. PLoS Biol 6, e311. doi:10.1371/journal.pbio.0060311

Jungblut PR, Schaible UE, Mollenkopf HJ et al (1999) Comparative proteome analysis of Mycobacterium tuberculosis and Mycobacterium bovis BCG strains: towards functional genomics of microbial pathogens. Mol Microbiol 33:1103–1117

Kelkar DS, Kumar D, Kumar P et al (2011) Proteogenomic analysis of Mycobacterium tuberculosis by high resolution mass spectrometry. Mol Cell Proteomics 10:M111.011445. doi:10.1074/mcp.M111.011445

Kristensen AR, Gsponer J, Foster LJ (2012) A high-throughput approach for measuring temporal changes in the interactome. Nat Methods 9:907–909. doi:10.1038/nmeth.2131

Kruh NA, Troudt J, Izzo A et al (2010) Portrait of a pathogen: the Mycobacterium tuberculosis proteome *in vivo*. PLoS One 5, e13938. doi:10.1371/journal.pone.0013938

Kunnath-Velayudhan S, Porcelli SA (2013) Recent advances in defining the immunoproteome of Mycobacterium tuberculosis. Front Immunol 4:335. doi:10.3389/fimmu.2013.00335

Kunnath-Velayudhan S, Davidow AL, Wang H-Y et al (2012) Proteome-scale antibody responses and outcome of Mycobacterium tuberculosis infection in nonhuman primates and in tuberculosis patients. J Infect Dis 206:697–705. doi:10.1093/infdis/jis421

Lee BY, Horwitz MA (1995) Identification of macrophage and stress-induced proteins of Mycobacterium tuberculosis. J Clin Invest 96:245–249. doi:10.1172/JCI118028

Leistikow RL, Morton RA, Bartek IL et al (2010) The Mycobacterium tuberculosis DosR regulon assists in metabolic homeostasis and enables rapid recovery from nonrespiring dormancy. J Bacteriol 192:1662–1670. doi:10.1128/JB.00926-09

Leitner A, Aebersold R (2013) SnapShot: mass spectrometry for protein and proteome analyses. Cell 154:252–252.e1. doi: 10.1016/j.cell.2013.06.025

Lew JM, Mao C, Shukla M et al (2013) Database resources for the tuberculosis community. Tuberculosis (Edinb) 93:12–17. doi:10.1016/j.tube.2012.11.003

Li L, Bannantine JP, Zhang Q et al (2005) The complete genome sequence of Mycobacterium avium subspecies paratuberculosis. Proc Natl Acad Sci U S A 102:12344–12349. doi:10.1073/pnas.0505662102

Li Y, Zeng J, Shi J et al (2010) A proteome-scale identification of novel antigenic proteins in Mycobacterium tuberculosis toward diagnostic and vaccine development. J Proteome Res. doi:10.1021/pr1005108

Lindestam Arlehamn CS, Gerasimova A, Mele F et al (2013) Memory T cells in latent Mycobacterium tuberculosis infection are directed against three antigenic islands and largely contained in a CXCR3+CCR6+ Th1 subset. PLoS Pathog 9, e1003130. doi:10.1371/journal.ppat.1003130

Liu Z-P, Wang J, Qiu Y-Q et al (2012) Inferring a protein interaction map of Mycobacterium tuberculosis based on sequences and interologs. BMC Bioinformatics 13:S6. doi:10.1186/1471-2105-8-475

Liu C-F, Tonini L, Malaga W et al (2013a) Bacterial protein-O-mannosylating enzyme is crucial for virulence of Mycobacterium tuberculosis. Proc Natl Acad Sci. doi:10.1073/pnas.1219704110

Liu Y, Hüttenhain R, Surinova S et al (2013b) Quantitative measurements of N-linked glycoproteins in human plasma by SWATH-MS. Proteomics. doi:10.1002/pmic.201200417

Loebel RO, Shorr E, Richardson HB (1933) The influence of adverse conditions upon the respiratory metabolism and growth of human tubercle bacilli. J Bacteriol 26:167–200

Maier T, Schmidt A, Güell M et al (2011) Quantification of mRNA and protein and integration with protein turnover in a bacterium. Mol Syst Biol 7:511. doi:10.1038/msb.2011.38

Målen H, de Souza GA, Pathak S et al (2011) Comparison of membrane proteins of Mycobacterium tuberculosis H37Rv and H37Ra strains. BMC Microbiol 11:18. doi:10.1186/1471-2180-11-18

Marguerat S, Schmidt A, Codlin S et al (2012) Quantitative analysis of fission yeast transcriptomes and proteomes in proliferating and quiescent cells. Cell 151:671–683. doi:10.1016/j.cell.2012.09.019

Mattow J, Jungblut PR, Schaible UE et al (2001) Identification of proteins from Mycobacterium tuberculosis missing in attenuated Mycobacterium bovis BCG strains. Electrophoresis 22:2936–2946. doi:10.1002/1522-2683(200108)22:14<2936::AID-ELPS2936>3.0.CO;2-S

Mattow J, Schaible UE, Schmidt F et al (2003) Comparative proteome analysis of culture supernatant proteins from virulent Mycobacterium tuberculosis H37Rv and attenuated M. bovis BCG Copenhagen. Electrophoresis 24:3405–3420. doi:10.1002/elps.200305601

Mattow J, Siejak F, Hagens K et al (2006) Proteins unique to intraphagosomally grown Mycobacterium tuberculosis. Proteomics 6:2485–2494. doi:10.1002/pmic.200500547

McDonough KA, Florczyk MA, Kress Y (2000) Intracellular passage within macrophages affects the trafficking of virulent tubercle bacilli upon reinfection of other macrophages in a serum-dependent manner. Tuber Lung Dis 80:259–271. doi:10.1054/tuld.2000.0268

Monahan IM, Betts J, Banerjee DK, Butcher PD (2001) Differential expression of mycobacterial proteins following phagocytosis by macrophages. Microbiology 147:459–471

O'Hare H, Juillerat A, Dianisková P, Johnsson K (2008) A split-protein sensor for studying protein-protein interaction in mycobacteria. J Microbiol Methods 73:79–84. doi:10.1016/j.mimet.2008.02.008

Parikh A, Kumar D, Chawla Y et al (2013) Development of a new generation of vectors for gene expression, gene replacement, and protein-protein interaction studies in mycobacteria. Appl Environ Microbiol 79:1718–1729. doi:10.1128/AEM.03695-12

Park H-D, Guinn KM, Harrell MI et al (2003) Rv3133c/dosR is a transcription factor that mediates the hypoxic response of Mycobacterium tuberculosis. Mol Microbiol 48:833–843

Pearce MJ, Mintseris J, Ferreyra J et al (2008) Ubiquitin-like protein involved in the proteasome pathway of Mycobacterium tuberculosis. Science 322:1104–1107. doi:10.1126/science.1163885

Pheiffer C, Betts JC, Flynn HR et al (2005) Protein expression by a Beijing strain differs from that of another clinical isolate and Mycobacterium tuberculosis H37Rv. Microbiology 151:1139–1150. doi:10.1099/mic.0.27518-0

Picotti P, Lam H, Campbell DS et al (2008) A database of mass spectrometric assays for the yeast proteome. Nat Methods 5:913–914. doi:10.1038/nmeth1108-913

Picotti P, Bodenmiller B, Mueller LN et al (2009) Full dynamic range proteome analysis of S. cerevisiae by targeted proteomics. Cell 138:795–806. doi:10.1016/j.cell.2009.05.051

Picotti P, Bodenmiller B, Aebersold R (2013) Proteomics meets the scientific method. Nat Methods 10:24–27. doi:10.1038/nmeth.2291

Poulsen C, Akhter Y, Jeon AH-W et al (2010) Proteome-wide identification of mycobacterial pupylation targets. Mol Syst Biol 6:386. doi:10.1038/msb.2010.39

Prisic S, Dankwa S, Schwartz D et al (2010) Extensive phosphorylation with overlapping specificity by Mycobacterium tuberculosis serine/threonine protein kinases. Proc Natl Acad Sci U S A 107:7521–7526. doi:10.1073/pnas.0913482107

Purkayastha A, McCue LA, McDonough KA (2002) Identification of a Mycobacterium tuberculosis putative classical nitroreductase gene whose expression is coregulated with that of the acr aene within macrophages, in standing versus shaking cultures, and under low oxygen conditions. Infect Immun 70:1518–1529

Raman K, Chandra N (2008) Mycobacterium tuberculosis interactome analysis unravels potential pathways to drug resistance. BMC Microbiol 8:234. doi:10.1186/1471-2180-8-234

Ranish JA, Yi EC, Leslie DM et al (2003) The study of macromolecular complexes by quantitative proteomics. Nat Genet 33:349–355. doi:10.1038/ng1101

Reddy TBK, Riley R, Wymore F et al (2009) TB database: an integrated platform for tuberculosis research. Nucleic Acids Res 37:D499–D508. doi:10.1093/nar/gkn652

Renuse S, Chaerkady R, Pandey A (2011) Proteogenomics. Proteomics 11:620–630. doi:10.1002/pmic.201000615

Ripoll F, Pasek S, Schenowitz C et al (2009) Non mycobacterial virulence genes in the genome of the emerging pathogen Mycobacterium abscessus. PLoS One 4, e5660. doi:10.1371/journal.pone.0005660

Rose G, Cortes T, Comas I et al (2013) Mapping of genotype-phenotype diversity among clinical isolates of Mycobacterium tuberculosis by sequence-based transcriptional profiling. Genome Biol Evol 5:1849–1862. doi:10.1093/gbe/evt138

Rosenkrands I, Slayden RA, Crawford J et al (2002) Hypoxic response of Mycobacterium tuberculosis studied by metabolic labeling and proteome analysis of cellular and extracellular proteins. J Bacteriol 184:3485–3491

Rustad TR, Sherrid AM, Minch KJ, Sherman DR (2009) Hypoxia: a window into Mycobacterium tuberculosis latency. Cell Microbiol 11:1151–1159. doi:10.1111/j.1462-5822.2009.01325.x

Schmidt F, Donahoe S, Hagens K et al (2004) Complementary analysis of the Mycobacterium tuberculosis proteome by two-dimensional electrophoresis and isotope-coded affinity tag technology. Mol Cell Proteomics 3:24–42. doi:10.1074/mcp.M300074-MCP200

Schubert OT, Mouritsen J, Ludwig C et al (2013) The Mtb proteome library: a resource of assays to quantify the complete proteome of Mycobacterium tuberculosis. Cell Host Microbe 13: 602–612. doi:10.1016/j.chom.2013.04.008

Schubert OT, Ludwig C, Kogadeeva M et al (2015) Absolute proteome composition and dynamics during dormancy and resuscitation of Mycobacterium tuberculosis. Cell Host Microbe 18: 96–108. doi:10.1016/j.chom.2015.06.001

Singh A, Mai D, Kumar A, Steyn AJC (2006) Dissecting virulence pathways of Mycobacterium tuberculosis through protein-protein association. Proc Natl Acad Sci U S A 103:11346–11351. doi:10.1073/pnas.0602817103

Smith GT, Sweredoski MJ, Hess S (2014) O-linked glycosylation sites profiling in Mycobacterium tuberculosis culture filtrate proteins. Proteomics 97:296–306. doi:10.1016/j.jprot.2013.05.011

Starck J, Källenius G, Marklund B-I et al (2004) Comparative proteome analysis of Mycobacterium tuberculosis grown under aerobic and anaerobic conditions. Microbiology 150:3821–3829. doi:10.1099/mic.0.27284-0

Steyn AJC, Joseph J, Bloom BR (2003) Interaction of the sensor module of Mycobacterium tuberculosis H37Rv KdpD with members of the Lpr family. Mol Microbiol 47:1075–1089

Stinear TP, Seemann T, Pidot S et al (2007) Reductive evolution and niche adaptation inferred from the genome of Mycobacterium ulcerans, the causative agent of Buruli ulcer. Genome Res 17:192–200. doi:10.1101/gr.5942807

Stinear TP, Seemann T, Harrison PF et al (2008) Insights from the complete genome sequence of Mycobacterium marinum on the evolution of Mycobacterium tuberculosis. Genome Res 18:729–741. doi:10.1101/gr.075069.107

Tharad M, Samuchiwal SK, Bhalla K et al (2011) A three-hybrid system to probe *in vivo* protein-protein interactions: application to the essential proteins of the RD1 complex of M. tuberculosis. PLoS One 6, e27503. doi:10.1371/journal.pone.0027503

Tortoli E (2006) The new mycobacteria: an update. FEMS Immunol Med Microbiol 48:159–178. doi:10.1111/j.1574-695X.2006.00123.x

Tung C-W (2012) PupDB: a database of pupylated proteins. BMC Bioinformatics 13:40. doi:10.1186/1471-2105-13-40

UniProt Consortium (2014) Activities at the Universal Protein Resource (UniProt). Nucleic Acids Res 42:D191–D198. doi:10.1093/nar/gkt1140

Vashisht R, Mondal AK, Jain A et al (2012) Crowd sourcing a new paradigm for interactome driven drug target identification in Mycobacterium tuberculosis. PLoS One 7, e39808. doi:10.1371/journal.pone.0039808

Veyron-Churlet R, Guerrini O, Mourey L et al (2004) Protein-protein interactions within the fatty acid synthase-II system of Mycobacterium tuberculosis are essential for mycobacterial viability. Mol Microbiol 54:1161–1172. doi:10.1111/j.1365-2958.2004.04334.x

Via LE, Lin PL, Ray SM et al (2008) Tuberculous granulomas are hypoxic in guinea pigs, rabbits, and nonhuman primates. Infect Immun 76:2333–2340. doi:10.1128/IAI.01515-07

Wang Y, Cui T, Zhang C et al (2010) Global protein-protein interaction network in the human pathogen Mycobacterium tuberculosis H37Rv. J Proteome Res 9:6665–6677. doi:10.1021/pr100808n

Watrous J, Burns K, Liu W-T et al (2010) Expansion of the mycobacterial "PUPylome". Mol Biosyst 6:376–385. doi:10.1039/b916104j

Wayne LG, Hayes LG (1996) An in vitro model for sequential study of shiftdown of Mycobacterium tuberculosis through two stages of nonreplicating persistence. Infect Immun 64: 2062–2069

World Health Organization (2013) Global tuberculosis report 2013. World Health Organization

Yuan Y, Crane DD, Barry CE (1996) Stationary phase-associated protein expression in Mycobacterium tuberculosis: function of the mycobacterial alpha-crystallin homolog. J Bacteriol 178:4484–4492

Zheng J, Liu L, Wei C et al (2012) A comprehensive proteomic analysis of Mycobacterium bovis bacillus Calmette-Guérin using high resolution Fourier transform mass spectrometry. Proteomics 77:357–371. doi:10.1016/j.jprot.2012.09.010

Chapter 14
Structural Aspects of Bacterial Outer Membrane Protein Assembly

Charles Calmettes, Andrew Judd, and Trevor F. Moraes

Abstract The outer membrane of Gram-negative bacteria is predominantly populated by β-Barrel proteins and lipid anchored proteins that serve a variety of biological functions. The proper folding and assembly of these proteins is essential for bacterial viability and often plays a critical role in virulence and pathogenesis. The β-barrel assembly machinery (Bam) complex is responsible for the proper assembly of β-barrels into the outer membrane of Gram-negative bacteria, whereas the localization of lipoproteins (Lol) system is required for proper targeting of lipoproteins to the outer membrane.

Keywords Outer membrane biogenesis • Bam machinery • Omp85 family • Lol pathway • Protein trafficking

14.1 OMPs and Lipoproteins Require Translocation Across the Inner Membrane

The proper assembly of outer membrane proteins (OMP) in Gram-negative bacteria requires the targeting and chaperone functions of a number of cytoplasmic, inner membrane, periplasmic and outer membrane proteins. The majority of proteins that are destined to reside in either the periplasm or outer membrane are expressed with a cleavable N-terminal signal peptide (SP). In Gram-negative bacteria, the SP consists of a hydrophobic stretch of 15+ amino acids sequence that is recognized by SecB. SecB binds the peptide and prevents folding, leaving the SP exposed and directing the nascent polypeptide chain to the inner membrane (Bechtluft et al. 2010). The peripheral protein SecA recognizes the SP and allows for its release to the SecYEG

C. Calmettes • A. Judd • T.F. Moraes (✉)
Department of Biochemistry, University of Toronto, Medical Science Building, Rm. 5366, 1 King's College Circle, Toronto, ON Canada, M5S 1A8
e-mail: charles.calmettes@utoronto.ca; andrew.judd@mail.utoronto.ca; trevor.moraes@utoronto.ca

N.J. Krogan, M. Babu (eds.), *Prokaryotic Systems Biology*, Advances in Experimental Medicine and Biology 883, DOI 10.1007/978-3-319-23603-2_14

translocase. SecYEG translocates the peptide across the inner membrane and into the periplasm in an ATP-dependent manner (Kudva et al. 2013). Here a signal peptidase cleaves the SP, releasing the protein into the periplasm. It is worth mentioning that while the Sec pathway is the most common secretion pathway in Gram-negative bacteria, the twin-arginine translocation (TAT) translocon is also used to translocate folded proteins across the inner membrane (Kudva et al. 2013). While very few proteins seem to use the TAT system to enter the periplasm, with only 27 TAT substrates predicted in *E. coli* (Palmer and Berks 2012), it appears to be important in the export of virulence factors of some Gram-negative pathogens such as *Pseudomonas aeruginosa* (De Buck et al. 2008). To date, β-barrel OMPs have not been shown to use the TAT system to access the periplasm, although some outer membrane lipoproteins appear to do so (Palmer and Berks 2012). If the protein is destined to remain in the periplasm, its journey would end here. However, β-barrel OMPs and lipoproteins must undergo further steps before they can carry out their function in the outer membrane.

14.2 Passage Across the Periplasm

Two periplasmic pathways have been reported to assist the trafficking of OMPs from the inner membrane to the outer membrane. These pathways require the three important chaperone SurA and Skp/DegP to respectively constitute the major and minor OMP-transiting routes across the periplasm.

14.2.1 The Major SurA Pathway

Upon entry into the periplasm in an unfolded conformation, the lipophilic β-barrel OMPs are prone to aggregation in this aqueous compartment and require dedicated chaperones for proper trafficking. The OMP clients are escorted through the periplasm by chaperone proteins to prevent any premature misfolding while they transit to the outer membrane. As there is no ATP in periplasm, these holdase chaperones function in an ATP-independent manner in contrast with cytoplasmic chaperonin. Instead of directly facilitating OMPs folding, they protect the substrates from aggregation and target them to the Bam machinery, a downstream folding system localized in the outer membrane. In *E. coli*, SurA binds the amphipathic OMP polypeptides and is described as the main OMP-carrier across the periplasm (Hennecke et al. 2005).

SurA is composed of four different segments, two of which are peptidyl-prolyl isomerases (PPIase) domains (Fig. 14.1). Its crystallographic structure reveals a bi-lobed organization: the first lobe constitutes the core structural and functional

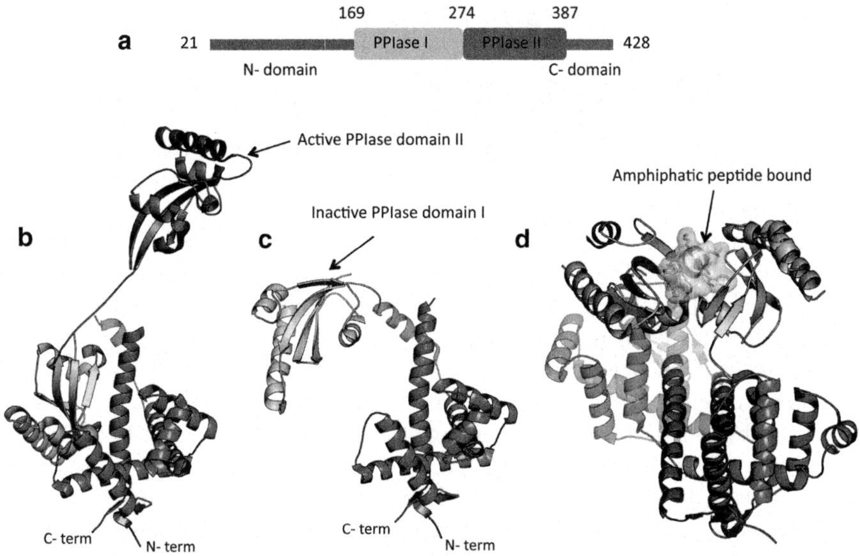

Fig. 14.1 Structural representations of the holdase/chaperone SurA from *E. coli*. Panel (**a**) illustrates the domains annotation of the mature SurA protein, an 45 kDa periplasmic and holdase chaperone with peptidyl-prolyl isomerase function located within its active PPIase-II domain colored in *dark green*. The OMP holdase chaperone activity resides in a functional core comprising the N- and C- terminal domains docked onto the inactive PPIase-I domain colored in *blue*, *red* and *yellow*, respectively. The full length SurA structure from the pdb entry 1M5Y is illustrated in panel (**b**) in cartoon representation. The panels (**c**) and (**d**) are cartoon visualizations of a dimeric SurA holdase core functional domain crystallized in bound conformation to an amphiphatic polypeptide mimicking the OMP substrates (pdb entry 2PV3). Panel (**c**) represents a single monomer from the SurA dimer drawn in panel (**d**). The structures from panels (**b**) and (**c**) are aligned on their N-terminal domain to illustrate a possible domain rearrangement of SurA upon binding of its substrates

module including the N- and C- terminal fragments folded onto the first PPIase domain, while the dispensable second PPIase domain forms a satellite module flexibly anchored to the larger core unit (Bitto and McKay 2002). In vitro, the core functional module has demonstrated the ability to organize alternative domain architecture in order to adapt binding to diverse OMP-mimicking polypeptides (Xu et al. 2007). Domain deletion studies and mutant characterization illustrated that PPIase activity is not require to sustain SurA function; consequently, SurA chaperoning functions mostly consist of a sequestering of OMP polypeptides during trafficking as a mechanism to both increase its solubility in the aqueous periplasm and prevents aggregation from occurring due to its hydrophobic content (Rigel and Silhavy 2012).

14.2.2 *The Minor Skp & DegP Pathway*

14.2.2.1 The Holdase Chaperone Skp

An alternative pathway to SurA involves the Skp and DegP chaperones, which have been identified to escort OMP clients across the periplasm as well. The Skp and DegP chaperones function together to form a redundant system for β-barrel OMPs that are unable to move forward on the initial SurA pathway (Sklar et al. 2007). In *E. coli*, these two pathways may complement each-other with at least one of them being required for cell viability as the *surA/skp* or *surA/degP* double knockouts have a synthetic lethality phenotype (Rizzitello et al. 2001).

The structure of Skp, a 18 kDa protein, oligomerizes to resembles a "jellyfish" architecture held together in a homotrimeric arrangement creating a central cavity delimited by α-helical tentacles protruding from a β-barrel body (Fig. 14.2) (Walton and Sousa 2004). The Skp holdase functions specifically with OMP; it binds and protects the hydrophobic β-barrel from aggregation within its inner-cavity, whereas any potential periplasmic domain from its OMP substrate folds into its native conformation outside of the cavity (Walton et al. 2009).

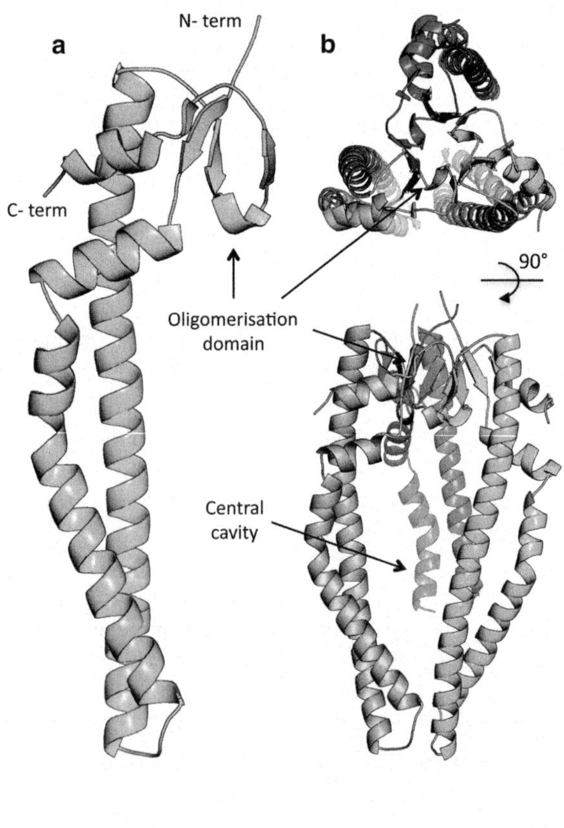

Fig. 14.2 Structure of the holdase/chaperone Skp from *E. coli*. The mature Skp chaperone, a 143 residue long protein, adopts a homotrimeric layout resembling a jellyfish structure. Panel (**a**) illustrates the Skp monomer while panel (**b**) consists of a top and side view of the functional trimeric chaperone. The monomers assemble each other creating a central cavity to protect unfolded OMP substrates while trafficking across the periplasmic compartment

14.2.2.2 The Holdase and Protease Chaperone DegP

DegP exhibits dual protease and general chaperone activities regulated in a temperature-dependant fashion impacting its oligomeric equilibrium (Spiess et al. 1999). The mature DegP segment contains a N-terminal chymotrypsin like serine protease domain, and two C-terminal PDZ domains mediating oligomeric arrangement of DegP trimeric units (Fig. 14.3). As a consequence, the DegP trimeric blocks have the capability to adopt diverse quaternary structures that have specific

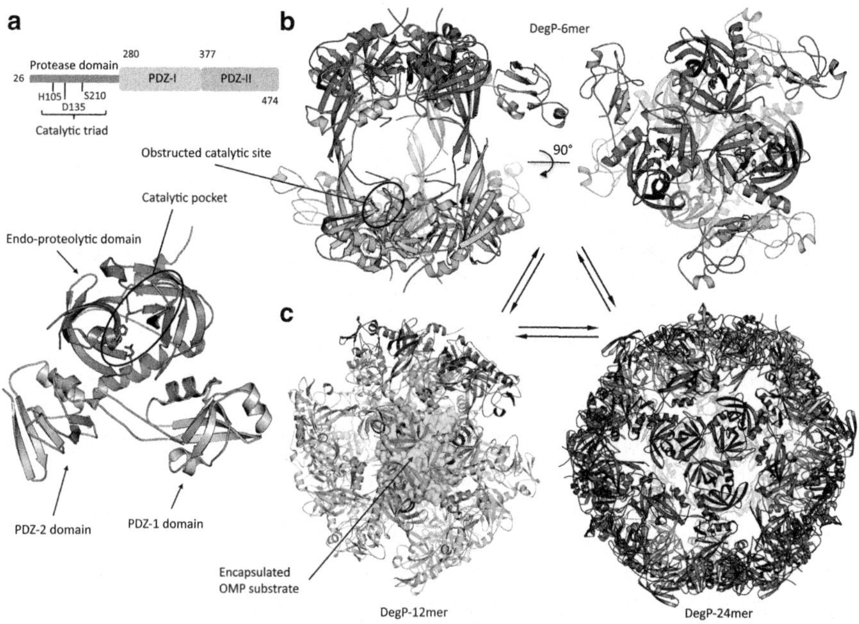

Fig. 14.3 Structure of the active and inactive DegP protease-chaperone complexes from *E. coli*. The 47 kDa mature DegP is a 449 amino acid protein subdivided in three domains represented in panel (**a**). The DegP monomer contains an endopeptidase domain (colored *blue*) containing the conserved His105-Asp135-Ser210 catalytic triad regulated by the surrounding loops, and two C-terminal PDZ domains (colored *pink* and *yellow*) important for DegP oligomerization. DegP forms homotrimeric subunits that are partitioned into the Deg-6mer, Deg-12mer and Deg-24mer of high molecular weight complexes. Panel (**b**) illustrates the DegP resting conformation resulting from the dimerization of two trimer subunits facing each other on their degradative sites. The trimer–trimer interactions observed in the resting Deg-6mer induce severe obstructions and distortions inhibiting proteolytic activity. None of the six PDZ-II domains are resolved in the crystallographic DegP-6mer structure (pdb entry 1KY9). Panel (**c**); upon binding to its substrates, the hexameric DegP particles dissociate and convert into the molecular cages Deg-12mer (4 trimer subunits) and Deg-24mer (8 trimer subunits) cemented through diverse PDZ-PDZ interactions. DegP-12mer (quasi-atomic model, pdb entry 4A8D) and Deg-24mer (pdb code 3CS0) are illustrated in panel (**c**); they self-associate around their cognate substrates that get encapsulated. All trimer subunits are drawn in cartoon representation and are identified using a unique color code; the DegP-12mer encapsulated ompA substrate, colored in *yellow*, is in surface representation

functions in protein quality control. In the absence of substrates, DegP adopts a rested proteolytically inactive conformation consisting of a hexamer arrangement that disorders the catalytic site. Upon binding of its substrates, DegP trimers reorganize into a dodecamer ($DegP_{12}$) or 24-mer ($DegP_{24}$) based molecular cage assemblies enclosing the protein substrates into large inner-cavities of 78 and 110 Å, respectively (Krojer et al. 2008). Both the periplasmic and the outer membrane proteins are potential DegP-encapsulated clients. The inner-cavity exhibits dual protease and chaperone antagonistic functions in both the $DegP_{12}$ and $DegP_{24}$ oligomeric states. A favorite model for the DegP OMP-chaperone suggests that the central compartment provides a protective environment to prevent OMPs from aggregating, where it actively selects for protein clients prone to adopt native or partially folded conformations to escape its degradative activity (Krojer et al. 2008). Indeed, DegP endoproteolytic activity is restricted to unfolded peptides.

14.3 Uncatalyzed Versus Catalyzed Insertion of Native OMPs

A subset of OMPs have been described to spontaneously insert and fold within "artificial" liposome membrane. However, spontaneous events are dependent on the phospholipid headgroups composition, such as phosphoethanolamine and phosphoglycerol, which have recently been reported to impose a kinetic barrier to prevent OMPs insertion and folding (Gessmann et al. 2014). Interestingly, these two phospholipid headgroups are enriched onto the periplasmic membrane surface of both the inner- and outer- membrane where they inhibit random and uncontrolled folding of native OMPs. Indeed, this kinetic retardation of spontaneous porin insertions—that would kill the cell by disrupting the protomotive force across inner membrane—allows the bacteria to deploy specialized mechanisms to compete again uncatalyzed OMP partitioning. Such bacterial mechanisms consist of a sequestering system involving SurA, Skp and DegP holdase-chaperones that prevent unfolded OMPs from being promiscuously inserted into the inner membrane (Wu et al. 2011). These bacteria challenges against unfavorable spontaneous insertion into the inner membrane is further supported by the establishment of a dedicated β-barrel assembly machinery (Bam) catalyzing efficient foldase and insertase functions at the outer membrane (Voulhoux et al. 2003).

14.4 The Bam Complex

The Carboxy-terminus of all OMPs contains a species-specific β-signal motif responsible of its final trafficking destination to the Bam machinery, an outer membrane complex consisting of OMP chaperone/foldase and insertase function (Robert et al. 2006). In *E. coli*, the Bam complex consists of five proteins

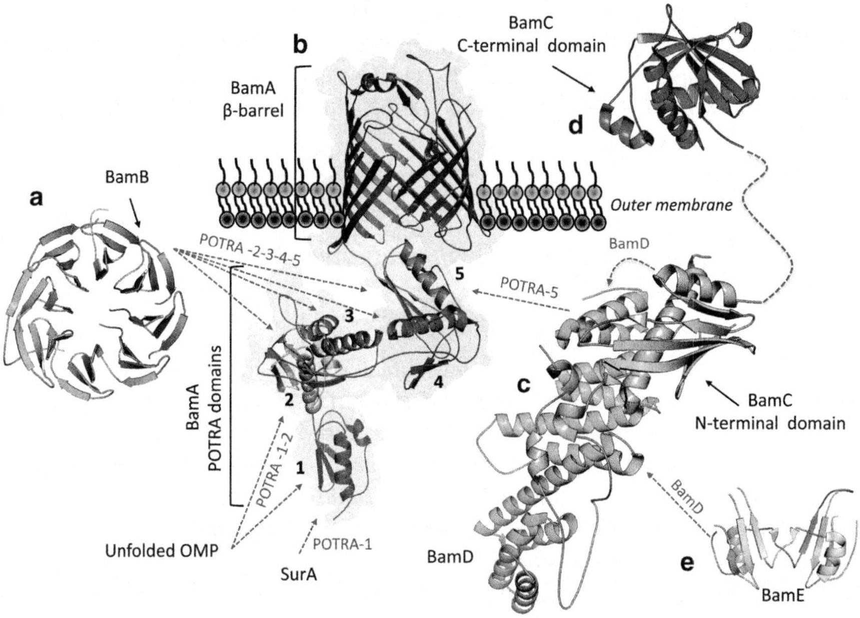

Fig. 14.4 Bam complex protein–protein mapping interactions in *E. coli*. (**a**, **b**, **c**, **d** and **e**) illustrate the crystal structures of BamB (pdb code 3Q7M), BamA (pdb code 4K3B), BamCD complex (pdb code 3TGO), the BamC C-carboxy-terminal domain (pdb code 2HY5) and the BamE dimer (pdb code 2YH9), respectively. All proteins are labeled with *black arrow* while their binding partners identified in *E. coli* are indicated with *red line*. All five periplasmic POTRA domains from bamA are identified with their respective number; the amino terminal POTRA being POTRA-1 and the most carboxy terminal domain POTRA-5

named BamA (88 kDa), BamB (40 kDa), BamC (34 kDa), BamD (26 kDa), and BamE (10 kDa) (Fig. 14.4) (Ricci and Silhavy 2012), from which BamA and BamD have been identified as the two core components essential to OMP biogenesis. All four BamBCDE lipoproteins are anchored in the inner leaflet of the outer membrane, however, the carboxy terminal domain of BamC is intriguingly exposed at the cell surface through an unknown mechanism (Webb et al. 2012). BamA is itself a β-barrel OMP thought to catalyze the final assembly of OMPs into the outer membrane, while its four associated BamBCDE components are lipoproteins thought to influence BamA function (Rigel et al. 2013). BamA belongs to the Omp85 superfamily and possesses five periplasmic POTRA (polypeptide translocation associated) domains that constitute a scaffold architecture recruiting OMP chaperones and the other Bam components (Ricci et al. 2012). The association consists of two subcomplexes, BamAB and BamCDE, that sequester each other to form the functional Bam machinery. Domain interaction studies in *E. coli* have shown that BamA recruits BamB through interactions with POTRA 2-3-4-5 domains, while the POTRA 5 is required to bind the BamCDE subcomplex

(Kim et al. 2007). The exact role of the POTRA domains is not clear but they demonstrate an ability to associate with unfolded OMP clients and also recruit the major OMP chaperone SurA specifically via the POTRA 1 domain (Knowles et al. 2008; Bennion et al. 2010), making the POTRA domains potential candidates for chaperone-like functions.

14.4.1 Bam Machinery in Gram-negative Bacteria: Variation on a Theme

Most of the Bam complex experimental characterizations reported so far have been performed in the *E. coli* and *N. meningitidis* bacterial systems. However, phylogenic study of the Bam components illustrates the Bam architectural diversity among Gram-negative bacteria-with a wide discrepancy of subunit compositions and domain arrangements. With the exceptions of BamA and BamD, which were identified in the genomes of all sequenced proteobacteria, the other lipoproteins BamB BamC and BamE were sporadically missing from particular lineages (Webb et al. 2012); an additional subunit BamF was also discovered in the α-proteobacterium *Caulobacter cresentus* where the unrelated BamC component is missing (Anwari et al. 2012). In addition, POTRA-deletion mutants have revealed inconsistent dependency patterns between species, such that the POTRAs 3-4-5 were identified as essential domains to maintain outer membrane integrity in *E. coli* while the BamA ortholog from *N. meningitidis* remains functional with a single POTRA 5 domain (Kim et al. 2007; Bos et al. 2007). These differences reflect various species-specific organization and subunit arrangement around the BamA catalytic core, which is likely to employ a conserved mechanism to insert and assist the folding of OMP clients.

Furthermore, the Omp85 superfamily is subdivided into 10 protein subfamilies including BamA and BamA-paralogs that may have conserved BamA-related functions (Heinz and Lithgow 2014). Particular attention should be focused on TamA, a BamA paralog that has been proposed to assist in the OMP biogenesis of the autotransporter family. Interestingly, the TamA and BamA proteins reveal striking structural resemblance with TamA exhibiting the same BamA key-features thought to contribute in the insertase and chaperone/foldase catalytic function (Gruss et al. 2013). Consequently, it was recently proposed that some of these BamA paralog subfamilies acquired in diverse bacterial taxa, such as TamA, might assist the ubiquitous Bam machinery to assemble subsets of OMPs (Heinz and Lithgow 2014). Nonetheless, it is worth mentioning that the role of TamA during the autotransporter biogenesis remains debated (Sauri et al. 2009; Roman-Hernandez et al. 2014), and such insertase and foldase functions have not been clearly demonstrated for TamA or any of the other BamA paralogs.

14.4.2 Protein Insertion and Folding Models

While some researchers have speculated how the Bam complex inserts OMPs into the outer membrane, the mechanism remains a mystery. Recently, the crystal structure of BamA has been solved in two different confirmations, which provides insight into the potential assembly mechanism of the Bam complex (Noinaj et al. 2013). The two structures show that the inner cavity of the barrel can either be in a closed or open conformation. Also, BamA appears to be able to open laterally towards the inside of the membrane, possibly allowing for the insertion of OMPs into the outer membrane. The structures also reveal an asymmetrical hydrophobic belt along the exterior rim of the membrane embedded barrel domain, indicating that BamA may be able to perturb the membrane at this location and could destabilize the outer membrane, facilitating subsequent insertion or folding of OMPs (Fig. 14.5). Interestingly, the OMP assembly pathway has been recreated in vitro by adding an unfolded OMP to SurA and proteoliposomes containing the reconstitute Bam complex, and the rate of membrane insertion and activity of the target OMP has been measured (Roman-Hernandez et al. 2014; Hagan et al. 2010). Combined with the recent structures of BamA, this in vitro assay could allow researchers to probe the pathway in ways they previously could not, which could bring us closer to understanding exactly how the Bam complex is performing this task.

Nonetheless, the structure of the open 16-stranded BamA barrel provides useful insight to decipher the mechanisms driving OMP folding. Indeed, the loose interaction observed in between unzipped strands 1 and 16 of BamA suggests it uses a lateral opening of its barrel domain to sequentially incorporate β-strands from the nascent OMP: the exposed strands of BamA providing a template interface for the insertion of additional strands from the OMP clients by β-augmentation, through which a client would finally bud off from the hydrid BamA-OMP barrel into the bilayer membrane (Fig. 14.5). Another possible scenario has been proposed where the OMP clients fold within the cavity of BamA prior to its release through a lateral event; however, the volume of the inner-chamber from BamA does not fit the 8-stranded OMP dimensions, which disfavors the latter model. The BamA structure provided researchers with a template to designed pair-wise cysteine mutants to challenge the lateral opening model using constitutive BamA locked mutants with bridged strands 1 and 16: these experiments have clearly illustrated that the lateral opening is an essential feature required for BamA functionality (Noinaj et al. 2014).

The asymmetric hydrophobic thickness along the exterior rim of BamA is also suggested to contribute to the insertion of OMP clients. Indeed, molecular dynamic simulations have predicted membrane disturbance around the BamA barrel junction of the unzipped strands 1 and 16 (Noinaj et al. 2013). This membrane stress, experimentally observed in BamA proteoliposomes using electron microscopy (Sinnige et al. 2014), has been proposed to prime the membrane bilayer to synergistically facilitate the bilayer insertion of the OMP substrates. Furthermore, it is reported in vitro that thinner lipid interfaces are prone to facilitate OMP self-assembly and membrane insertion by lowering the kinetic barrier to intrinsic OMP folding

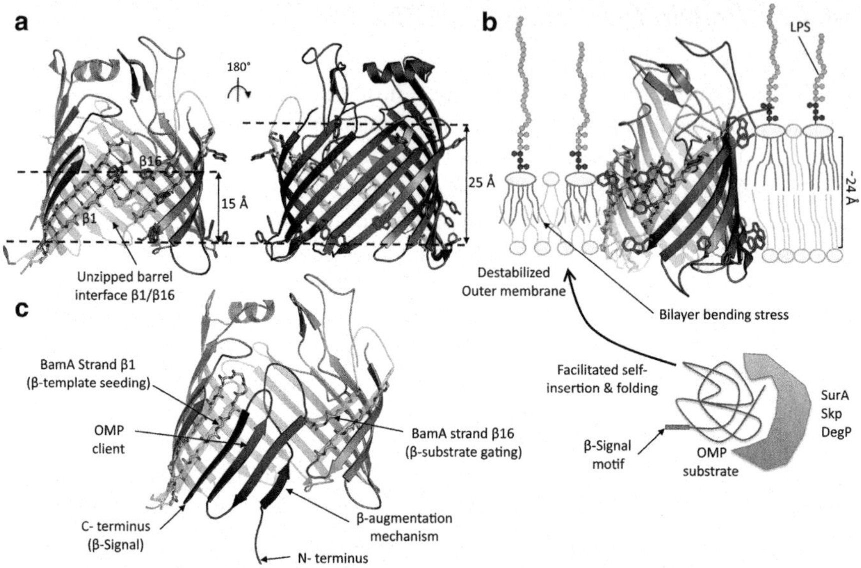

Fig. 14.5 BamA foldase and insertase models. Panel (**a**) illustrates the unzipped β-barrel interface from BamA, and the hydrophobic asymmetry around the outer-rim of the barrel. The BamA β-barrel is drawn in cartoon representation and colored in *dark green*. The two extremes N-terminus and C-terminus β-strands are colored in *light green*, and their hydrogen bond networking revealed in *blue dotted lines*. The unzipped β-strands 1 and 16 are labeled for clarity, and the aromatic residues exposed in the lipid phase are highlighted in *purple sticks*. Hydrophobic thicknesses are measured 15 and 25 Å at the two extreme asymmetry limits, with the unzipped strand 1 and 16 corresponding to the thinner interface with the lipidic phase. Panel (**b**) illustrates the membrane destabilization resulted from the bilayer adaptation to the low hydrophobic thickness of BamA in close proximity of the β-strands 1 and 16. Thus lipid distortions would lower the kinetic barrier to favor the self-insertion and spontaneous folding of simple OMP substrates brought in close proximity of BamA; presence of β-signal motif and SurA–BamA interactions contribute to traffic OMPs to their final Bam destination. Panel (**c**) illustrates another model consisting of the formation of a BamA-OMP hybrid β-barrel. In this model, the β-strand 1 from BamA gates the sequential insertion of additional strands from the OMP substrate that initially use the β-strand 16 as a template to seed the folding by β-augmentation. Each newly inserted strand will serve as a template to nucleate folding of the following strand while they get inserted into the hybrid BamA-OMP barrel. The new OMP would ultimately buds off from the hybrid pore to populate the outer membrane

reaction (Kleinschmidt and Tamm 2002). One other inclusive hypothesis suggests this membrane destabilization could be fully sufficient to promote the spontaneous folding of simple OMP substrates that have been brought in close proximity to BamA and the destabilized outer membrane by the SurA and Skp/DegP trafficking pathways. This model is supported by the critical and unusually thin 15 Å hydrophobic thickness measured at the vicinity of the BamA β-strand 1 (Fig. 14.5) that should impose a severe bilayer bending stress on the lipid acyl-chain; indeed, BamA minimal membrane thickness measures 9 Å compared to an average 24 Å hydrophobic thickness being calculated from all reported OMP structures, which is believed to match the outer membrane lipid-phase thickness (Pogozheva et al. 2013).

14.5 Lol System for Lipoprotein Trafficking

The second major class of proteins in the outer membrane is lipoproteins. Only 90 lipoproteins are predicted to be expressed in *E. coli* and the function of most of these is still unknown (Tokuda and Matsuyama 2004). However, it is known that some lipoproteins carry out essential functions in the cell, such as BamD in the Bam complex. They begin their road to maturation exactly the same way as a β-barrel OMP does, with expression occurring in the cytoplasm followed by translocation across the inner membrane through the Sec translocon (Bechtluft et al. 2010). Some lipoproteins lack a traditional SP and are translocated into the periplasm through the TAT pathway (Valente et al. 2007) where the signal is a twin arginine repeat near the N-terminus of the protein, although this appears to be rare. While the SEC translocon machinery actively transports proteins in an unfolded manner, the Tat pathway translocates folded protein clients associated with their cofactors as a way to allow secreted proteins to acquire periplasmic-deprived cofactors such as molybdenum ion or metal sulfur cluster (Berks et al. 2014). The SP of lipoproteins contains a consensus sequence that targets the protein for lipidation known as a lipobox (Leu-(Ala-Ser)-(Gly/Ala)-Cys) (Hayashi and Wu 1990; Tokuda and Matsuyama 2004). Three enzymes are required for the maturation of lipoproteins and they are essential for growth in *E. coli* (Wu and Tokunaga 1986). Upon translocation into the periplasm, the lipobox is recognized by phosphatidylglycerol/prolipoprotein diacylglycerol transferase (Lgt), which adds a diacylglycerol group to the sulfur atom of the cysteine residue (Fig. 14.6). Then the SP is removed either by signal peptidase II (SpII) followed by the acetylation of the N-terminus of the cysteine residue by phospholipid/apolipoprotein transacylase (Lnt). After these processing steps, the mature lipoprotein now enters the localization of lipoprotein (Lol) pathway (Okuda and Tokuda 2011).

The Lol pathway is composed of five proteins, the inner membrane ABC transporter LolCDE, the periplasmic carrier LolA and the outer membrane receptor LolB (Okuda and Tokuda 2011). This begins by the lipoprotein either being recognized or avoided by the LolCDE inner membrane complex. If a retention signal is present, the lipoprotein is ignored by the Lol system and remains anchored to the inner membrane. This is dependent on the amino acid sequence following the lipidated cysteine and the retention signal varies between organisms (Narita and Tokuda 2007; Yamaguchi and Inouye 1988). Lacking this signal, the lipoprotein will be bound by LolCDE that actively transfer its client to the periplasmic carrier LolA in an ATP-dependant manner. The LolCDE complex is divided into a cytosolic ATPase domain LolD bound to the membrane-embedded heterodimer LolCE. The LolD ATPase activity is coupled with the LolCE transporter subunit, which sorts lipoproteins

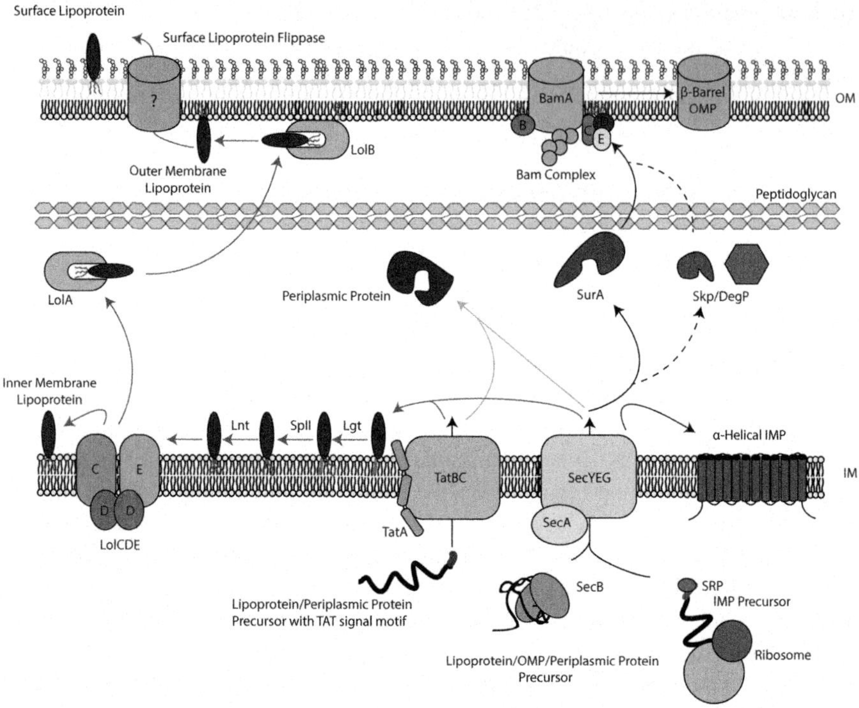

Fig. 14.6 Protein biogenesis pathways of the inner membrane, periplasm and outer membrane. Inner membrane proteins, outer membrane proteins, and lipoproteins all follow different pathways to reach their destination of function once out the cytosol. α-Helical inner membrane proteins (IMPs) are moved across the inner membrane (IM) and inserted within it by the Sec translocon. Outer membrane proteins (OMPs) are also brought into the periplasm this way but are bound to SurA for transport to the outer membrane and to prevent aggregation along the way. Outer membrane proteins are inserted into the outer membrane (OM) by the Bam complex. Lipoproteins can be moved across the inner membrane by either the Tat or Sec translocons. Once in the periplasm, three enzymes process the lipoprotein by removing its signal peptide and replacing it with a lipid anchor. The lipoprotein is then either recognized or avoided by the LolCDE complex. Lipoproteins that are not bound by LolCDE remain anchored to the inner membrane. The ones that are recognized are passed over to the periplasmic carrier, LolA for transport to the outer membrane. The outer membrane lipoprotein receptor LolB takes the lipoprotein from LolA and inserts it into the outer membrane. Some organisms have some lipoproteins translocated to the cell surface through an unknown mechanism

based on residues adjacent to the lipidated Cys, and transfers the lipid anchors of its substrate into the hydrophobic cavity of LolA (Fig. 14.7). By sequestering the hydrophobic acyl-chains from its lipoprotein client LolA is then able to traffic it to the outer membrane. LolA transfers the lipoprotein to LolB on the inner leaflet of the outer membrane. The structures of LolA and LolB are very similar (Fig. 14.7) but researchers were able to locate a single 3_{10}-helix in LolA that prevents the retrograde transfer of lipoproteins within the inner membrane, making the Lol system a one

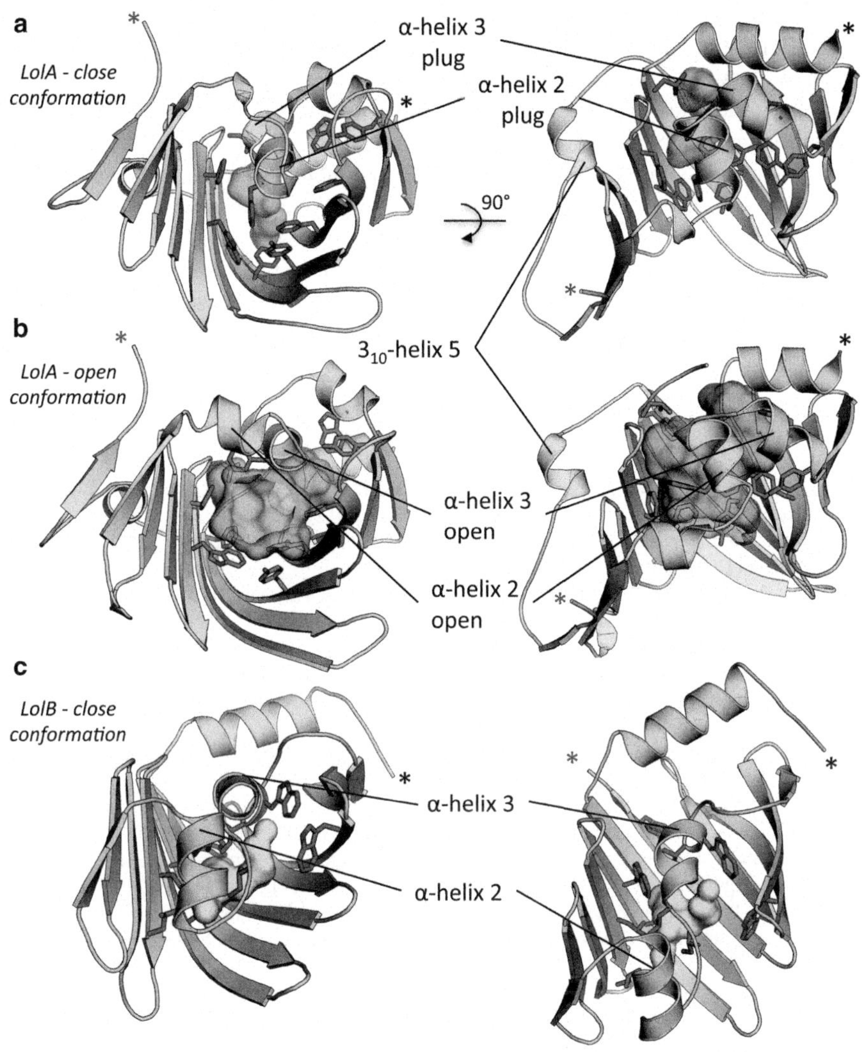

Fig. 14.7 LolA and LolB structures from *E. coli*. (**a**, **b** and **c**) panels represent orthogonal views of the crystallographic structure of LolA in closed and open conformations (pdb entries 2ZPC and 2ZPD), and the closed conformation of LolB (pdb code 1IWN), respectively. LolA and LolB are very similar; they both enclose a hydrophobic cavity sealed by an unclosed β-barrel and a α-helical lid. Their main differences reside in an extra helix-strand motif at the carboxy-terminus of LolA, and the presence of a lipid anchor at he LolB amino-terminus. The extra helix-strand motif unique to LolA contains the 3_{10}-helix 5, that is important to lock the lipid anchor of its substrate and subsequently prevent abortive membrane localization. Access to the hydrophobic cavity, illustrated using an internal surface representation (colored in *orange* and *yellow* in LolA and LolB, respectively), requires the displacement of the α-helices 2 and 3 (plug) that are obstructing the lipid anchor binding site within the LolA and LolB closed conformations (**a** and **c**). Hydrophobic residues covering the lipid-binding cavities are in *purple stick* representation. The N- and C- terminus are indicated with a *black* and *red asterisk*, respectively

way pathway (Okuda et al. 2008; Takeda et al. 2003). Once the lipoprotein is bound to LolB, the lipid anchor is inserted into the outer membrane, where the lipoprotein can now carry out its intended function. However, the molecular details allowing the lipid substrate to swap from LolB to the outer membrane, or the mechanisms driving the exchange of the lipid anchors in between the Lol components remain elusive to date. Finally, in some Gram-negative organisms including *Neisseria meningitidis, Haemophilis influenzae* ... etc, lipoproteins can be secreted or flipped to the outer surface of the cell. The protein(s) involved in this flippase function are currently not known but have a significant role in presenting surface lipoproteins and may behave like type 5 secretion system proteins or two partner secretion systems.

References

Anwari K, Webb CT, Poggio S, Perry AJ, Belousoff M, Celik N, Ramm G, Lovering A, Sockett RE, Smit J, Jacobs-Wagner C, Lithgow T (2012) The evolution of new lipoprotein subunits of the bacterial outer membrane BAM complex. Mol Microbiol 84(5):832–844. doi:10.1111/j.1365-2958.2012.08059.x

Bechtluft P, Kedrov A, Slotboom DJ, Nouwen N, Tans SJ, Driessen AJ (2010) Tight hydrophobic contacts with the SecB chaperone prevent folding of substrate proteins. Biochemistry 49(11):2380–2388. doi:10.1021/bi902051e

Bennion D, Charlson ES, Coon E, Misra R (2010) Dissection of beta-barrel outer membrane protein assembly pathways through characterizing BamA POTRA 1 mutants of Escherichia coli. Mol Microbiol 77(5):1153–1171. doi:MMI7280

Berks BC, Lea SM, Stansfeld PJ (2014) Structural biology of Tat protein transport. Curr Opin Struct Biol 27C:32–37. doi:S0959-440X(14)00025-6

Bitto E, McKay DB (2002) Crystallographic structure of SurA, a molecular chaperone that facilitates folding of outer membrane porins. Structure 10(11):1489–1498. doi:S0969212602008778

Bos MP, Robert V, Tommassen J (2007) Functioning of outer membrane protein assembly factor Omp85 requires a single POTRA domain. EMBO Rep 8(12):1149–1154. doi:7401092

De Buck E, Lammertyn E, Anne J (2008) The importance of the twin-arginine translocation pathway for bacterial virulence. Trends Microbiol 16(9):442–453. doi:S0966-842X(08)00166-2

Gessmann D, Chung YH, Danoff EJ, Plummer AM, Sandlin CW, Zaccai NR, Fleming KG (2014) Outer membrane beta-barrel protein folding is physically controlled by periplasmic lipid head groups and BamA. Proc Natl Acad Sci U S A 111(16):5878–5883. doi:1322473111

Gruss F, Zahringer F, Jakob RP, Burmann BM, Hiller S, Maier T (2013) The structural basis of autotransporter translocation by TamA. Nat Struct Mol Biol 20(11):1318–1320. doi:nsmb.2689

Hagan CL, Kim S, Kahne D (2010) Reconstitution of outer membrane protein assembly from purified components. Science 328(5980):890–892. doi:science.1188919

Hayashi S, Wu HC (1990) Lipoproteins in bacteria. J Bioenerg Biomembr 22(3):451–471

Heinz E, Lithgow T (2014) A comprehensive analysis of the Omp85/TpsB protein superfamily structural diversity, taxonomic occurrence, and evolution. Front Microbiol 5:370. doi:10.3389/fmicb.2014.00370

Hennecke G, Nolte J, Volkmer-Engert R, Schneider-Mergener J, Behrens S (2005) The periplasmic chaperone SurA exploits two features characteristic of integral outer membrane proteins for selective substrate recognition. J Biol Chem 280(25):23540–23548. doi:M413742200

Kim S, Malinverni JC, Sliz P, Silhavy TJ, Harrison SC, Kahne D (2007) Structure and function of an essential component of the outer membrane protein assembly machine. Science 317(5840):961–964. doi:317/5840/961

Kleinschmidt JH, Tamm LK (2002) Secondary and tertiary structure formation of the beta-barrel membrane protein OmpA is synchronized and depends on membrane thickness. J Mol Biol 324(2):319–330. doi:S0022283602010719
Knowles TJ, Jeeves M, Bobat S, Dancea F, McClelland D, Palmer T, Overduin M, Henderson IR (2008) Fold and function of polypeptide transport-associated domains responsible for delivering unfolded proteins to membranes. Mol Microbiol 68(5):1216–1227. doi:MMI6225
Krojer T, Sawa J, Schafer E, Saibil HR, Ehrmann M, Clausen T (2008) Structural basis for the regulated protease and chaperone function of DegP. Nature 453(7197):885–890. doi:nature07004
Kudva R, Denks K, Kuhn P, Vogt A, Muller M, Koch HG (2013) Protein translocation across the inner membrane of Gram-negative bacteria: the Sec and Tat dependent protein transport pathways. Res Microbiol 164(6):505–534. doi:S0923-2508(13)00059-4
Narita S, Tokuda H (2007) Amino acids at positions 3 and 4 determine the membrane specificity of Pseudomonas aeruginosa lipoproteins. J Biol Chem 282(18):13372–13378. doi:M611839200
Noinaj N, Kuszak AJ, Gumbart JC, Lukacik P, Chang H, Easley NC, Lithgow T, Buchanan SK (2013) Structural insight into the biogenesis of beta-barrel membrane proteins. Nature 501(7467):385–390. doi:nature12521
Noinaj N, Kuszak AJ, Balusek C, Gumbart JC, Buchanan SK (2014) Lateral opening and exit pore formation are required for BamA function. Structure 22(7):1055–1062. doi:S0969-2126(14)00150-6
Okuda S, Tokuda H (2011) Lipoprotein sorting in bacteria. Annu Rev Microbiol 65:239–259. doi:10.1146/annurev-micro-090110-102859
Okuda S, Watanabe S, Tokuda H (2008) A short helix in the C-terminal region of LolA is important for the specific membrane localization of lipoproteins. FEBS Lett 582(15): 2247–2251. doi:S0014-5793(08)00424-9
Palmer T, Berks BC (2012) The twin-arginine translocation (Tat) protein export pathway. Nat Rev Microbiol 10(7):483–496. doi:nrmicro281410.1038/nrmicro2814
Pogozheva ID, Tristram-Nagle S, Mosberg HI, Lomize AL (2013) Structural adaptations of proteins to different biological membranes. Biochim Biophys Acta 1828(11):2592–2608. doi:S0005-2736(13)00212-5
Ricci DP, Silhavy TJ (2012) The Bam machine: a molecular cooper. Biochim Biophys Acta 1818(4):1067–1084. doi:S0005-2736(11)00282-3
Ricci DP, Hagan CL, Kahne D, Silhavy TJ (2012) Activation of the Escherichia coli beta-barrel assembly machine (Bam) is required for essential components to interact properly with substrate. Proc Natl Acad Sci U S A 109(9):3487–3491. doi:1201362109
Rigel NW, Silhavy TJ (2012) Making a beta-barrel: assembly of outer membrane proteins in Gram-negative bacteria. Curr Opin Microbiol 15(2):189–193. doi:S1369-5274(11)00216-5
Rigel NW, Ricci DP, Silhavy TJ (2013) Conformation-specific labeling of BamA and suppressor analysis suggest a cyclic mechanism for beta-barrel assembly in Escherichia coli. Proc Natl Acad Sci U S A 110(13):5151–5156. doi:1302662110
Rizzitello AE, Harper JR, Silhavy TJ (2001) Genetic evidence for parallel pathways of chaperone activity in the periplasm of Escherichia coli. J Bacteriol 183(23):6794–6800. doi:10.1128/JB.183.23.6794-6800.2001
Robert V, Volokhina EB, Senf F, Bos MP, Van Gelder P, Tommassen J (2006) Assembly factor Omp85 recognizes its outer membrane protein substrates by a species-specific C-terminal motif. PLoS Biol 4(11):e377. doi:06-PLBI-RA-0371R1
Roman-Hernandez G, Peterson JH, Bernstein HD (2014) Reconstitution of bacterial autotransporter assembly using purified components. Elife:e04234. doi:10.7554/eLife.04234
Sauri A, Soprova Z, Wickstrom D, de Gier JW, Van der Schors RC, Smit AB, Jong WS, Luirink J (2009) The Bam (Omp85) complex is involved in secretion of the autotransporter haemoglobin protease. Microbiology 155(Pt 12):3982–3991. doi:mic.0.034991-0
Sinnige T, Weingarth M, Renault M, Baker L, Tommassen J, Baldus M (2014) Solid-state NMR studies of full-length BamA in lipid bilayers suggest limited overall POTRA mobility. J Mol Biol 426(9):2009–2021. doi:S0022-2836(14)00073-4

Sklar JG, Wu T, Kahne D, Silhavy TJ (2007) Defining the roles of the periplasmic chaperones SurA, Skp, and DegP in Escherichia coli. Genes Dev 21(19):2473–2484. doi:21/19/2473

Spiess C, Beil A, Ehrmann M (1999) A temperature-dependent switch from chaperone to protease in a widely conserved heat shock protein. Cell 97(3):339–347. doi:S0092-8674(00)80743-6

Takeda K, Miyatake H, Yokota N, Matsuyama S, Tokuda H, Miki K (2003) Crystal structures of bacterial lipoprotein localization factors, LolA and LolB. EMBO J 22(13):3199–3209. doi:10.1093/emboj/cdg324

Tokuda H, Matsuyama S (2004) Sorting of lipoproteins to the outer membrane in E. coli. Biochim Biophys Acta 1694(1–3):IN1–9

Valente FM, Pereira PM, Venceslau SS, Regalla M, Coelho AV, Pereira IA (2007) The [NiFeSe] hydrogenase from Desulfovibrio vulgaris Hildenborough is a bacterial lipoprotein lacking a typical lipoprotein signal peptide. FEBS Lett 581(18):3341–3344. doi:S0014-5793(07)00669-2

Voulhoux R, Bos MP, Geurtsen J, Mols M, Tommassen J (2003) Role of a highly conserved bacterial protein in outer membrane protein assembly. Science 299(5604):262–265. doi:10.1126/science.1078973299/5604/262

Walton TA, Sousa MC (2004) Crystal structure of Skp, a prefoldin-like chaperone that protects soluble and membrane proteins from aggregation. Mol Cell 15(3):367–374. doi:10.1016/j.molcel.2004.07.023

Walton TA, Sandoval CM, Fowler CA, Pardi A, Sousa MC (2009) The cavity-chaperone Skp protects its substrate from aggregation but allows independent folding of substrate domains. Proc Natl Acad Sci U S A 106(6):1772–1777. doi:0809275106

Webb CT, Heinz E, Lithgow T (2012) Evolution of the beta-barrel assembly machinery. Trends Microbiol 20(12):612–620. doi:S0966-842X(12)00149-7

Webb CT, Selkrig J, Perry AJ, Noinaj N, Buchanan SK, Lithgow T (2012) Dynamic association of BAM complexes modules includes surface exposure of the lipoprotein BamC. J Mol Biol 422:545–555

Wu HC, Tokunaga M (1986) Biogenesis of lipoproteins in bacteria. Curr Top Microbiol Immunol 125:127–157

Wu S, Ge X, Lv Z, Zhi Z, Chang Z, Zhao XS (2011) Interaction between bacterial outer membrane proteins and periplasmic quality control factors: a kinetic partitioning mechanism. Biochem J 438(3):505–511. doi:BJ20110264

Xu X, Wang S, Hu YX, McKay DB (2007) The periplasmic bacterial molecular chaperone SurA adapts its structure to bind peptides in different conformations to assert a sequence preference for aromatic residues. J Mol Biol 373(2):367–381. doi:S0022-2836(07)01048-0

Yamaguchi K, Inouye M (1988) Lipoprotein 28, an inner membrane protein of Escherichia coli encoded by nlpA, is not essential for growth. J Bacteriol 170(8):3747–3749

Chapter 15
Substrate Interaction Networks of the *Escherichia coli* Chaperones: Trigger Factor, DnaK and GroEL

Vaibhav Bhandari and Walid A. Houry

Abstract In the dense cellular environment, protein misfolding and inter-molecular protein aggregation compete with protein folding. Chaperones associate with proteins to prevent misfolding and to assist in folding to the native state. In *Escherichia coli*, the chaperones trigger factor, DnaK/DnaJ/GrpE, and GroEL/ES are the major chaperones responsible for insuring proper de novo protein folding. With multitudes of proteins produced by the bacterium, the chaperones have to be selective for their substrates. Yet, chaperone selectivity cannot be too specific. Recent biochemical and high-throughput studies have provided important insights highlighting the strategies used by chaperones in maintaining proteostasis in the cell. Here, we discuss the substrate networks and cooperation among these protein folding chaperones.

Keywords Molecular chaperones • Trigger factor • GroEL/GroES • DnaK/DnaJ/GrpE • Protein folding • Protein aggregation • Chaperone interaction network

15.1 Introduction

Amongst the most renowned tenets in the study of protein folding, Anfinsen's thermodynamic principle states that the native conformation of a protein is achieved to attain the structure with minimum free energy for the respective polypeptide sequence (Anfinsen 1973). This is found to be true for many small proteins that have been experimentally studied and which have been observed to have a funnel-like energy landscape folding pathway to reach the lowest free energy native state (Mayor et al. 2003; Hartl et al. 2011). Burial of hydrophobic residues in the interior of the protein is a major driving force for the folding of soluble proteins. However, large proteins often seem to have a ragged pathway where protein folding can

V. Bhandari • W.A. Houry (✉)
Department of Biochemistry, University of Toronto, 1 King's College Circle, Medical Sciences Building, Room 5308, Toronto, ON M5S 1A8, Canada
e-mail: walid.houry@utoronto.ca

N.J. Krogan, M. Babu (eds.), *Prokaryotic Systems Biology*, Advances in Experimental Medicine and Biology 883, DOI 10.1007/978-3-319-23603-2_15

be inefficient and error prone. Folding intermediates in this situation can fall into kinetic traps and may expose hydrophobic residues or unstructured elements, which lead to misfolding of the protein and/or its aggregation (Dill and Chan 1997; Ellis 2006).

Molecular chaperones are a class of proteins that function to prevent such misfolding and aggregation that may occur in the chaotically dense medium that is the cellular environment. Chaperones bind, stabilize, fold and remodel proteins in healthy and stressed cells (Hartl et al. 2011). Many chaperones are present in the cell. De novo protein folding in a model bacteria such as *Escherichia coli* is mainly performed by three highly conserved chaperone systems: trigger factor (TF), DnaK/DnaJ/GrpE (Hsp70/Hsp40/nucleotide exchange factor) system and the GroEL/GroES (Hsp60/Hsp10) system (Horwich et al. 1993; Hesterkamp et al. 1996; Hartl and Hayer-Hartl 2002; Mayer and Bukau 2005).

The number of proteins that require chaperone assistance in the cell, the extent of the required assistance and the structural properties allowing for this assistance are actively being studied by many groups. Furthermore, the chaperone interaction networks of these different chaperones are also being investigated as this can provide answers to many pertinent questions about protein biogenesis, post-translational protein regulation and protein evolution.

15.2 Trigger Factor

15.2.1 Trigger Factor Structure and Substrate Recognition

Trigger factor (TF) is the only ribosome-associated bacterial chaperone (Hesterkamp et al. 1996). It is a 48 kDa protein that can be divided into three structural domains: namely, the N-terminal domain (NTD), the C-terminal domain (CTD) and the PPIase domain (Fig. 15.1a) (Ferbitz et al. 2004). The NTD and PPIase domains are connected by an extended linker, which allows the TF to have an elongated three-dimensional structure where the CTD is in the middle with the NTD and PPIase domain at opposite ends (Martinez-Hackert and Hendrickson 2009). The CTD contains helical-extensions that mimic protruding arms. CTD and PPIase domain form a cleft-like concave binding pocket for potential substrates (Merz et al. 2006; Mashaghi et al. 2013) (Fig. 15.1a).

Within the cell, TF can be found freely in the cytosol or attached to ribosomes. It is estimated that there is two to threefold molar excess of TF over ribosomes (Lill et al. 1988). TF associates transiently in a 1:1 stoichiometry with the ribosome, binding and acting on most nascent polypeptides emerging from the ribosome polypeptide exit tunnel (Kramer et al. 2002; Ferbitz et al. 2004; Raine et al. 2006; Rutkowska et al. 2008). TF associates with a vacant ribosome with a K_d of about 1–2 μM and a mean residence time of 10 s (Patzelt et al. 2001; Maier et al. 2003; Kaiser et al. 2006; Hoffmann et al. 2010). Nascent polypeptides increase TF's

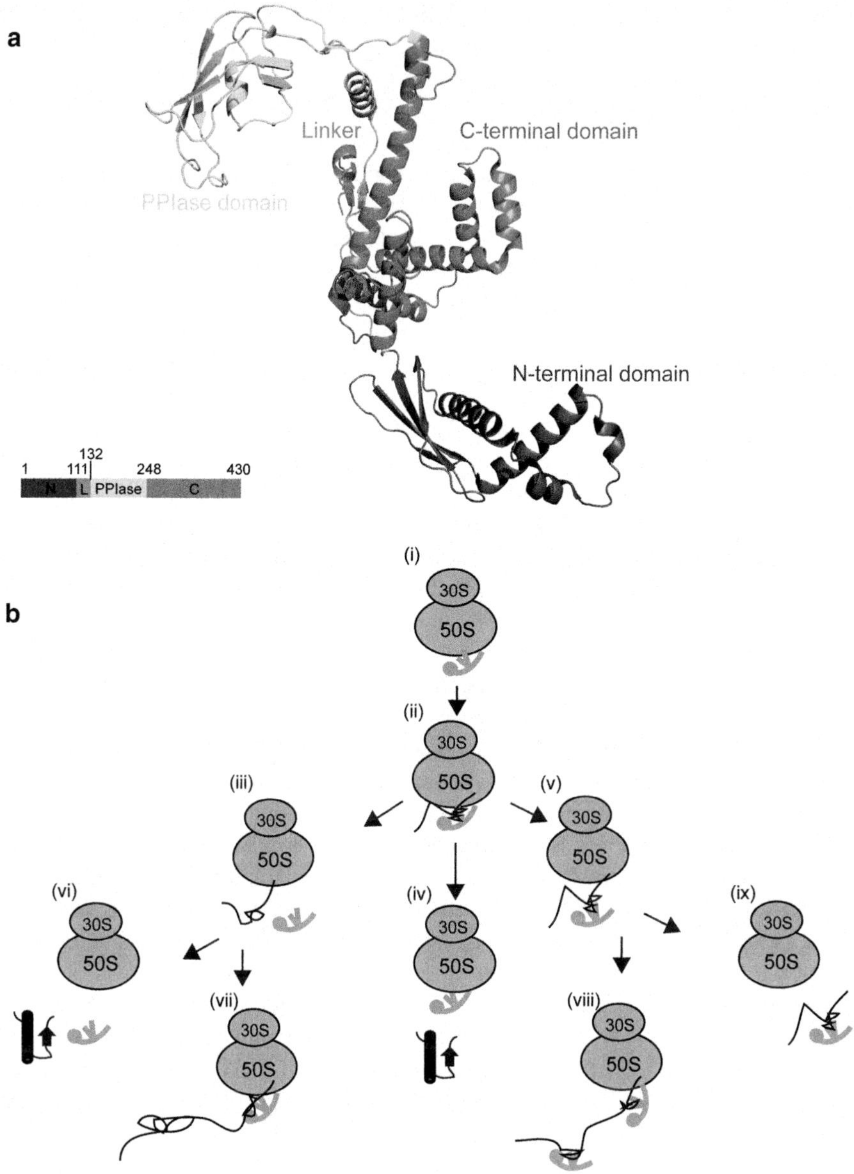

Fig. 15.1 Structure and function of trigger factor. (**a**) Structure of TF [PDB ID 1W26 (Ferbitz et al. 2004)] with N-terminal domain (*violet*), C-terminal domain (*blue*), PPIase domain (*yellow*), and linker (*green*) highlighted. Structures were drawn using the PyMOL molecular graphics system (DeLano 2002). A bar graph of TF domain arrangement is shown below the structure. (**b**) Mechanism of TF (*in green*) function is shown at various states of substrate interaction on and off the ribosome (*blue*)

affinity for ribosomes by 2–30-fold, based on their size, folded state and amino acid composition (Raine et al. 2006; Hoffmann et al. 2010). This enables the chaperone to differentiate between translating ribosomes and vacant ones (Rutkowska et al. 2008). Binding polypeptides on ribosomes also increases the half-life of the TF-ribosome association (Rutkowska et al. 2008).

TF associates promiscuously with polypeptides as they exit the ribosome during translation, protecting hydrophobic elements of emerging polypeptides from the hydrophilic environment of the cytoplasm through direct interactions with these elements (Hesterkamp et al. 1996; Hoffmann et al. 2006; Kaiser et al. 2006; Lakshmipathy et al. 2007; Rutkowska et al. 2008). Approximately eight amino acid-long sequences rich in hydrophobic and aromatic residues with a positive net charge are thought to be responsible for TF-substrate recognition (Patzelt et al. 2001; Saio et al. 2014). These sites occur regularly in most polypeptides, approximately once every 32 residues (Bukau et al. 2000; Patzelt et al. 2001).

The X-ray crystal structure for TF from *Thermatoga maritima* in complex with the ribosomal small subunit protein S7 has been solved (Martinez-Hackert and Hendrickson 2009) and provides some clues as to the basis of substrate recognition by this chaperone. The TF–S7 interaction was found to be a non-specific interaction, as would be expected for an interaction between a promiscuous chaperone and one of its many substrates. The interaction interface was very large, poorly packed, dominantly polar and sharing low shape complementarity. The interaction between these two proteins depicts a non-specific association and offers insights into the promiscuity displayed by the chaperone, which is necessary for TF function.

More recently, NMR-based techniques were used to map the interaction of TF with an unfolded substrate, *E. coli* alkaline phosphatase (PhoA) (Saio et al. 2014). The authors show that three TF molecules bind to one unfolded PhoA molecule. At least four substrate binding sites were identified in TF: one in the PPIase domain and three in the CTD. TF was found to use these four sites to bind to several regions of PhoA primarily through hydrophobic contacts. The TF–PhoA interaction was found to be highly dynamic, however, a more stable complex was formed as the length of the substrate protein and the number of regions recognized by TF increased.

Approximately 70 % of proteins are thought to fold to their native structures after association with TF. Other proteins can be passed onto downstream chaperone systems, DnaK and GroEL for further folding. Indeed, DnaK can compensate for the loss of TF in the cell (Deuerling et al. 1999; Teter et al. 1999) (discussed further below).

15.2.2 Trigger Factor Functional Cycle

The mechanism of TF action has been described as dynamic, consisting of a series of substrate binding and release events on and off the ribosome (Fig. 15.1b) (Kaiser et al. 2006; Hoffmann et al. 2010; Saio et al. 2014). TF is assumed to contact most polypeptides upon their exit from the ribosome, but many of these interactions are transient and weak (Valent et al. 1995). TF can bind the vacant ribosome (i) but the association is enhanced upon interaction of TF with an emerging nascent polypeptide (ii). Once on the ribosome, TF remains bound for a minimum of about 10 s, which is enough time for the ribosome to translate a polypeptide chain of up to 200 residues. Following this, TF can be released from the nascent chain and ribosome (iii, vi) or the completed nascent chain might be released to fold to the native state with TF remaining bound to the ribosome (iv). A released TF is free to rebind the ribosome at the exit tunnel and assist in the folding of another (or same) emerging polypeptide (vii). Alternatively, TF might be released from the ribosome but remain bound to the growing nascent chain (v). For a long polypeptide sequence, multiple TFs on or off the ribosome may bind the chain (viii) (Agashe et al. 2004). Finally, TF may also assist in folding of a polypeptide recently released from the ribosome (ix) (Hoffmann et al. 2010).

15.2.3 Identification of Trigger Factor Substrates

Attempts have been made to identify protein substrates of TF using either co-purification with His-tagged TF or by identifying proteins that aggregate in the absence of TF but not in its presence in a Δ*dnaKJ* background strain (Martinez-Hackert and Hendrickson 2009). A total of 178 substrates were identified. Co-purification led to the identification of 42 substrates and 110 were identified by analysis of protein aggregation, while 26 substrates were identified by both techniques. Many of the identified proteins were ribosomal proteins or were part of multimeric complexes. The size distribution of proteins associating with TF was similar to the *E. coli* cytoplasmic proteome having a size range from 8 to 118 kDa with a mean of 36.5 kDa, again highlighting the promiscuity of this chaperone for its substrates.

15.3 DnaK/DnaJ/GrpE System

15.3.1 DnaK Structure and Function

DnaK is the major bacterial ortholog of the eukaryotic ATP-dependent Hsp70 chaperone. Substrates of DnaK include unfolded, misfolded and aggregated proteins (Schlecht et al. 2011). The chaperone is primarily involved in protein folding and protein disaggregation, but also has overlapping function with TF in promoting cotranslational protein folding (Deuerling et al. 1999, 2003; Teter et al. 1999; Rosenzweig et al. 2013). Structurally, like other Hsp70s, DnaK is composed of two domains (Mayer and Bukau 2005; Bertelsen et al. 2009): the N-terminal ATPase domain and the C-terminal substrate binding domain (Fig. 15.2a). DnaK function depends on a bidirectional allosteric communication between these two domains (Ung et al. 2013). The enzymatic cycle of DnaK alternates between ATP-bound open state and ADP-bound closed state (Mayer and Bukau 2005). The ATP-bound state is characterized by low affinity and fast exchange rate for substrates, while the ADP-bound state is characterized by high affinity and slow exchange rate for substrate. The hydrolysis of ATP to ADP triggers the closing of the substrate binding site resulting in locking the associated substrate to DnaK.

The functional cycle of DnaK is dependent upon DnaJ cochaperone, an Hsp40 ortholog, and the GrpE nucleotide exchange factor (Liberek et al. 1991; Szabo et al. 1994; Hartl et al. 2011). DnaJ is the major cochaperone for DnaK in *E. coli* and generally acts to stimulate the ATPase activity of DnaK. Also, DnaJ binds substrates and then transfers them to DnaK (Fig. 15.2b). The ATPase activity of DnaK is low when no substrate is bound and is stimulated two to tenfold in the presence of a substrate (Mayer and Bukau 2005). The ATPase activity is further enhanced by DnaJ. DnaJ and the DnaK-bound substrate synergistically enhance the ATPase activity of DnaK by greater than 1000-fold (Liberek et al. 1991; Karzai and McMacken 1996; Laufen et al. 1999). ATP hydrolysis then allows for a tight complex to form between the DnaK–ADP and its substrate (Kampinga and Craig 2010). The release of ADP from DnaK is slow (Brehmer et al. 2001), hence, the need for the nucleotide exchange factor GrpE which catalyzes the release of ADP from DnaK. GrpE itself dissociates from the chaperone when DnaK binds ATP, which also results in the release of the substrate protein. The substrate protein can then attempt to fold to the native state, if unsuccessful, the protein can be rebound by DnaJ or DnaK and the cycle repeated (Fig. 15.2b).

15.3.2 Interaction Network of DnaK

In order to identify how DnaK differentially recognizes its substrates in the cellular environment, the DnaK substrate binding motif was analyzed using a library of overlapping 13-mer peptides arrayed on cellulose membranes (Rudiger et al. 1997). The binding motif was found to consist of a hydrophobic core of about seven residues, enriched in leucines, flanked by basic amino acids. Based on the solved crystal structure of DnaK with a substrate peptide, the link between DnaK structure and its preferential substrates was further illustrated (Zhu et al. 1996; Mayer and Bukau 2005). Substrates interact with the substrate binding domain of the chaperone, which consists of a β-sandwich subdomain and an α-helical lid subdomain (Fig. 15.2a). The binding pocket in DnaK is composed of hydrophobic residues flanked by acidic residues contributed by both subdomains, which is consistent with the preferential DnaK binding motif identified by the peptide array analysis described above. This DnaK substrate binding motif is estimated to generally occur once every 36 residues in proteins (Rudiger et al. 1997). Indeed, it is estimated that 98 % of the *E. coli* annotated proteome would harbor potential DnaK binding sites (Srinivasan et al. 2012).

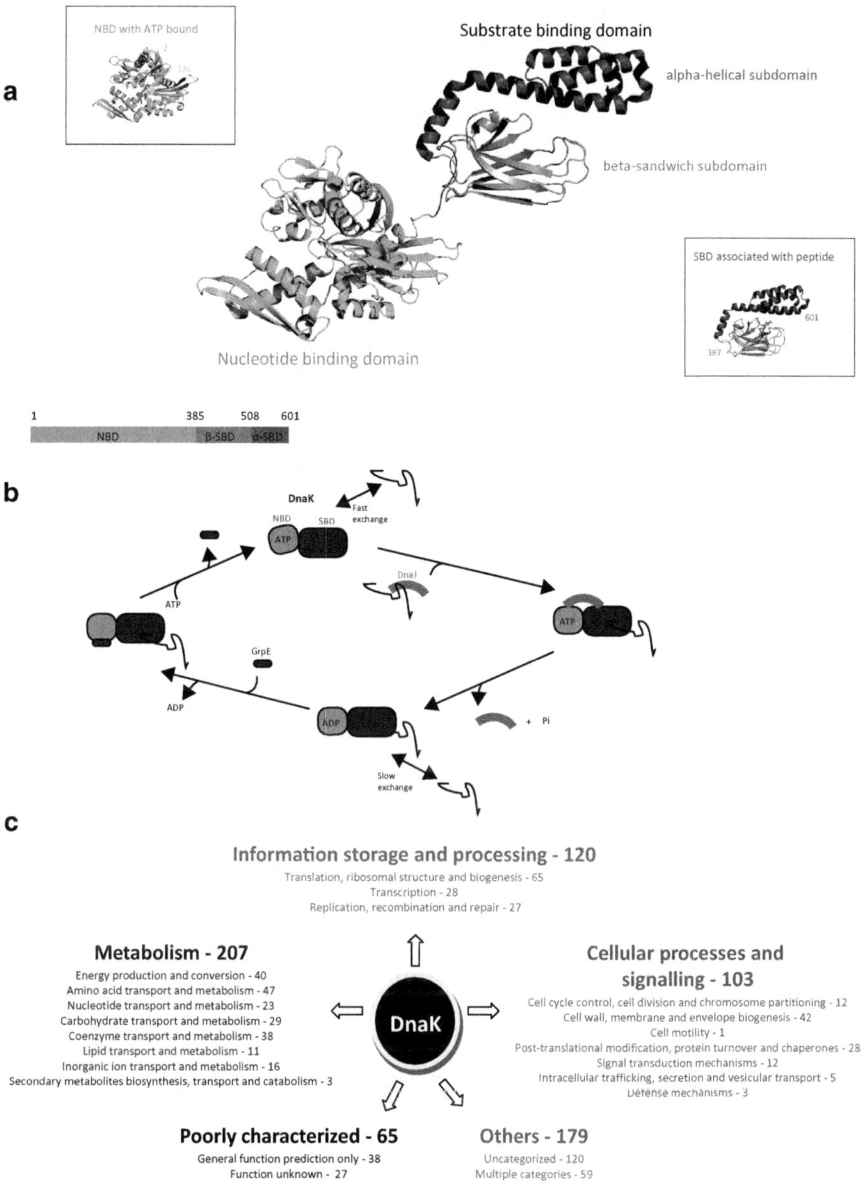

Fig. 15.2 Structure and function of DnaK. (**a**) Structure of full length DnaK [PDB ID 2KHO (Bertelsen et al. 2009)] with its nucleotide binding domain (NBD) and substrate binding domain (SBD) indicated. A bar graph of DnaK domain arrangement is shown below the structure. The inset on the *right* shows the SBD residues 387-601 with bound NRLLLTG peptide (*in blue*) [PDB ID 1DKZ (Zhu et al. 1996)]. The inset on the *left* shows residues 2-376 of NBD of DnaK with bound ATP (in *orange*) [PDB ID 4B9Q (Zhu et al. 1996; Kityk et al. 2012)]. (**b**) The functional cycle of DnaK (*red*, *green*) is shown depicting its action on its substrates with assistance from its cochaperone DnaJ (*orange*) and the nucleotide exchange factor GrpE (*violet*). As shown, the ATP-bound state of DnaK is characterized by weak binding of substrate and fast exchange rates while the ADP-bound state is characterized by strong substrate binding and slow exchange rates. (**c**) Functional categories based on Cluster of Orthologous Group (COG) for the 674 DnaK interacting proteins are shown (Tatusov et al. 2001, Calloni et al. 2012). Numbers of proteins belonging to each functional group and to each category are indicated beside the COG category name

Different approaches have been used to identify potential DnaK substrates. In one approach, protein aggregates in DnaK (and DnaJ) depleted cells with or without trigger factor, were resolved by two-dimensional gel electrophoresis (Deuerling et al. 2003). Regardless of the presence of TF, 340 major spots were identified that are representative of potential DnaK substrates. It should be noted that the levels of aggregated proteins in cells with TF were much lower, indicative of functional overlap among the two chaperones. The size range for these proteins was 16–160 kDa, but proteins larger than 60 kDa in size were found to be enriched in the aggregates compared to soluble cytoplasmic proteins. Of the distinct spots on the 2-D gel, 94 were identified by mass spectrometry (Deuerling et al. 2003). The identified proteins were all cytoplasmic and involved in different cellular processes. Though no secondary structural features or chemical features were identified to distinguish the substrates from other proteins, it was observed that a majority (~72 %) of aggregated proteins tended to be thermolabile.

More recently, in another approach to identify DnaK substrates, DnaK-substrate complexes were isolated from wild type cells or cells either lacking TF or depleted of GroEL (Calloni et al. 2012). Using endogenously expressed His-tagged DnaK, immobilized metal affinity chromatography (IMAC) was used to pulldown DnaK-interacting proteins, which were then identified and quantified by mass spectrometry. Both DnaK cochaperones, DnaJ and GrpE, were isolated in these pulldowns. In total, 674 DnaK interactors were identified belonging to diverse functional groups (Fig. 15.2c). A majority of these were predicted to be cytoplasmic (~80 %) with a significant minority of inner membrane, outer membrane and periplasmic proteins as well. Many of the interactors were involved in metabolic and cell signaling pathways (Fig. 15.2c).

Several features were observed for substrates enriched on DnaK. They were found to be more aggregation prone upon translation than less enriched DnaK substrates (Calloni et al. 2012; Niwa et al. 2012). Additionally, though enriched DnaK substrates were not more hydrophobic than the average soluble cellular protein, they were observed to be less effective in burying their hydrophobic residues from solvent (Tartaglia et al. 2010; Calloni et al. 2012). DnaK-enriched substrates were generally of low cellular abundance and of large size (Calloni et al. 2012). The negative correlation between cellular abundance and aggregation propensity was previously observed (Tartaglia et al. 2007, 2010) as folding states for abundant proteins are thought to have been evolutionarily optimized to prevent overloading chaperones. Proteins that interact extensively with DnaK were more likely to be part of hetero-oligomeric complexes. Partially structured regions of proteins that form hetero-oligomeric complexes can be shielded from the dense cellular environment through chaperone assistance (Schlecht et al. 2011). Thus, through shielding of hydrophobic charges, DnaK allows proper folding of numerous proteins.

15.3.3 DnaJ Structure and Function

DnaJ is the major cochaperone for DnaK in *E. coli* and generally acts to stimulate the ATPase activity of DnaK. DnaJ has a conserved domain of approximately 70 residues located at the N-terminus, called the J domain, that is required for DnaJ to associate with DnaK (Wall et al. 1994). There is a highly conserved His-Pro-Asp motif present in a loop between the second helix and third helix of the J domain (Fig. 15.3a), which is found to be crucial for stimulation of DnaK ATPase activity by DnaJ (Cheetham and Caplan 1998). Following the N-terminal J-domain, DnaJ is composed of a glycine/phenylalanine rich region followed by a linker region, a zinc-binding domain and a C-terminal domain (Fig. 15.3a). The C-terminus of DnaJ is known to associate with substrates in a transient fashion, binding to hydrophobic sequences that contain motifs similar to those recognized by DnaK; DnaJ then presents these substrates to DnaK (Gamer et al. 1996; Rudiger et al. 2001; Srinivasan et al. 2012) (Fig. 15.2b). Related to its importance for DnaK function, temperature sensitivity has been observed for *dnaJ* null mutant strains, as well as, a loss of bacterial motility (Sell et al. 1990; Shi et al. 1992).

DnaJ is only one of the Hsp40 paralogs present in *E. coli*; others include CbpA, DjlA, DjlB, DjlC and HscB (Ueguchi et al. 1994; Genevaux et al. 2001; Gur et al. 2005; Chenoweth et al. 2007). DjlB and DjlC are membrane associated proteins that do not associate with DnaK but interact with HscC, a specialized DnaK paralog (Kluck et al. 2002). Similarly, HscB (or Hsc20) acts as a cochaperone for another DnaK paralog termed HscA (or Hsc66) (Silberg et al. 1998; Hennessy et al. 2005; Fuzery et al. 2008). HscA/B are involved in the iron–sulfur cluster assembly pathway. Apart from DnaJ, CbpA and DjlA are the only cochaperones observed to have a significant association with DnaK (Genevaux et al. 2007). CbpA has been observed to act as a multicopy suppressor of Δ*dnaJ* mutants, though no phenotype is observed in cells lacking just CbpA (Ueguchi et al. 1994; Gur et al. 2004). DjlA, a membrane associated protein with its N-terminal anchored to the inner-membrane, has been shown to substitute for DnaJ in vitro as a cochaperone for DnaK (Genevaux et al. 2001). However, among these Hsp40 cochaperones of DnaK, DnaJ is the best characterized and functions as the premier regulator of the various DnaK activities (Kelley 1998; Gur et al. 2004).

15.3.4 GrpE Structure and Function

Along with DnaJ, DnaK works in concert with the nucleotide exchange factor GrpE. GrpE associates with DnaK and catalyzes the otherwise slow release of ADP (Packschies et al. 1997). GrpE functions as a homodimeric protein (Schonfeld et al. 1995). Structurally, the first 33 N-terminal residues of the protein are disordered, followed by a long α-helix, a short α-helix and a compact β-sheet domain (Fig. 15.3b) (Harrison et al. 1997). Much of the long α-helix forms a tail,

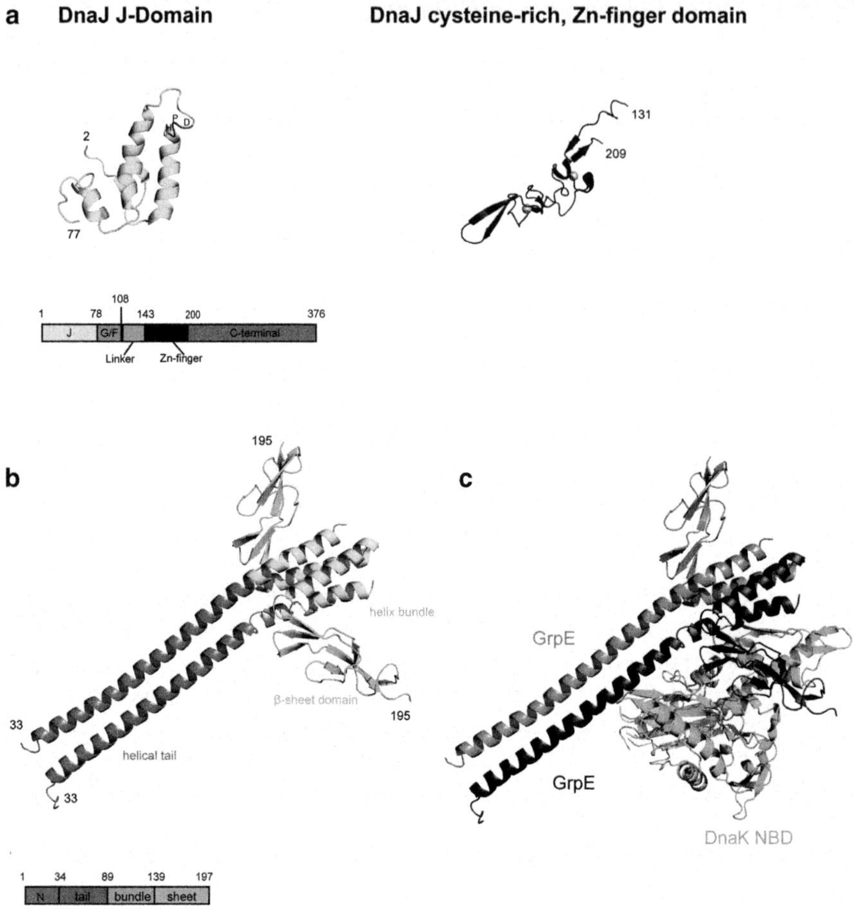

Fig. 15.3 Structure and function of DnaJ and GrpE. (**a**) Structure of DnaJ J-domain residues 2-77 (in *yellow*) with the conserved HPD motif highlighted in *red* is shown on the *left* [PDB ID 1BQ0 (Huang et al. 1999)]. DnaJ residues 131-209 (in *brown*) containing the zinc-finger domain with bound zinc depicted as *light orange spheres* is shown on the *right* [PDB ID 1EXK (Martinez-Yamout et al. 2000)]. A bar graph of DnaJ domain arrangement is shown below the structures. (**b**) Structure of GrpE [PDB ID 1DKG (Harrison et al. 1997)] highlighting its domains is shown. N-terminal amino acids 1-32 are disordered and are not shown. A bar graph of GrpE domain arrangement is shown below the structure. (**c**) Structure of the DnaK nucleotide binding domain in complex with GrpE [PDB ID 1DKG (Harrison et al. 1997)]. The GrpE subunit associating with the NBD-domain of DnaK is depicted in *dark purple* while the other GrpE in the dimer is presented in a lighter shade

with the tails in the dimer positioned parallel to each other. The remaining part of the longer helix and the shorter helix combine with their counterparts from the other GrpE in the dimer to form a helical bundle. The β-sheet domains protrude outward from the helical bundle (Fig. 15.3b). Only one of the GrpE molecules within the dimer takes part in associating with DnaK (Fig. 15.3c). The β-sheet domain makes the majority of the contacts with DnaK by binding the nucleotide binding cleft, while the helix bundle and the top of the helical tail contribute additional contacts (Harrison et al. 1997; Harrison 2003).

The interaction of GrpE with DnaK is very strong, with a K_d of 1–30 nM (Harrison et al. 1997; Packschies et al. 1997). On binding DnaK, GrpE has been shown to reduce the affinity of ADP for DnaK by 200-fold and as a result accelerating nucleotide exchange by 5000-fold (Packschies et al. 1997). This is accomplished by an associative displacement mechanism whereby the binding of the nucleotide and GrpE is not competitive and the binding sites are either distinct or only partially overlapping (Packschies et al. 1997). GrpE binding leads to a conformational change in DnaK, disrupting the contacts between the DnaK and ADP (Packschies et al. 1997; Harrison 2003).

Apart from its generalized role as a nucleotide exchange factor for DnaK, GrpE has also been found to assist in polypeptide release from the substrate binding domain of the chaperone (Mally and Witt 2001; Brehmer et al. 2004). The importance of GrpE for DnaK activity is highlighted by its temperature-dependent control of nucleotide exchange from DnaK. The long helix of GrpE was shown to undergo a reversible thermal transition above about 40 °C that leads to a decrease in the rate of ADP/ATP exchange on DnaK (Grimshaw et al. 2003). Thus, GrpE provides a thermal regulation of DnaK activity by shifting the DnaK-substrate complexes towards the ADP-associated, slow substrate exchange state at high temperatures. The slow rate of substrate release at such temperatures prevents unfolded polypeptides from accumulating in the cytoplasm where they might be susceptible to misfolding and aggregation.

15.4 The GroEL/ES Chaperone System

15.4.1 GroEL/ES Structure and Function

The GroEL protein (also called chaperonin) is the bacterial ortholog of the eukaryotic Hsp60 present in the mitochondria. GroEL (57 kDa) along with its cofactor GroES (10 kDa, Hsp10) is the only chaperone system in *E. coli* that is essential under all growth conditions (Fayet et al. 1989; Goloubinoff et al. 1989; Horwich et al. 1993). GroEL is a cylindrically-shaped oligomer composed of two rings, arranged back to back, of seven subunits each (Fig. 15.4a) (Braig et al. 1994; Horwich et al. 2006). Each GroEL subunit consists of three domains (Fig. 15.4a): the apical domain, the intermediate domain and the equatorial domain. The equatorial domain at the ring–ring interface contains the ATP-binding pocket and has been recently proposed to assist in orientation of substrate proteins as they enter the GroEL ring (Fenton et al. 1994; Weaver and Rye 2014). The apical domain at the ends of the cylinder harbors hydrophobic residues required for substrate and GroES

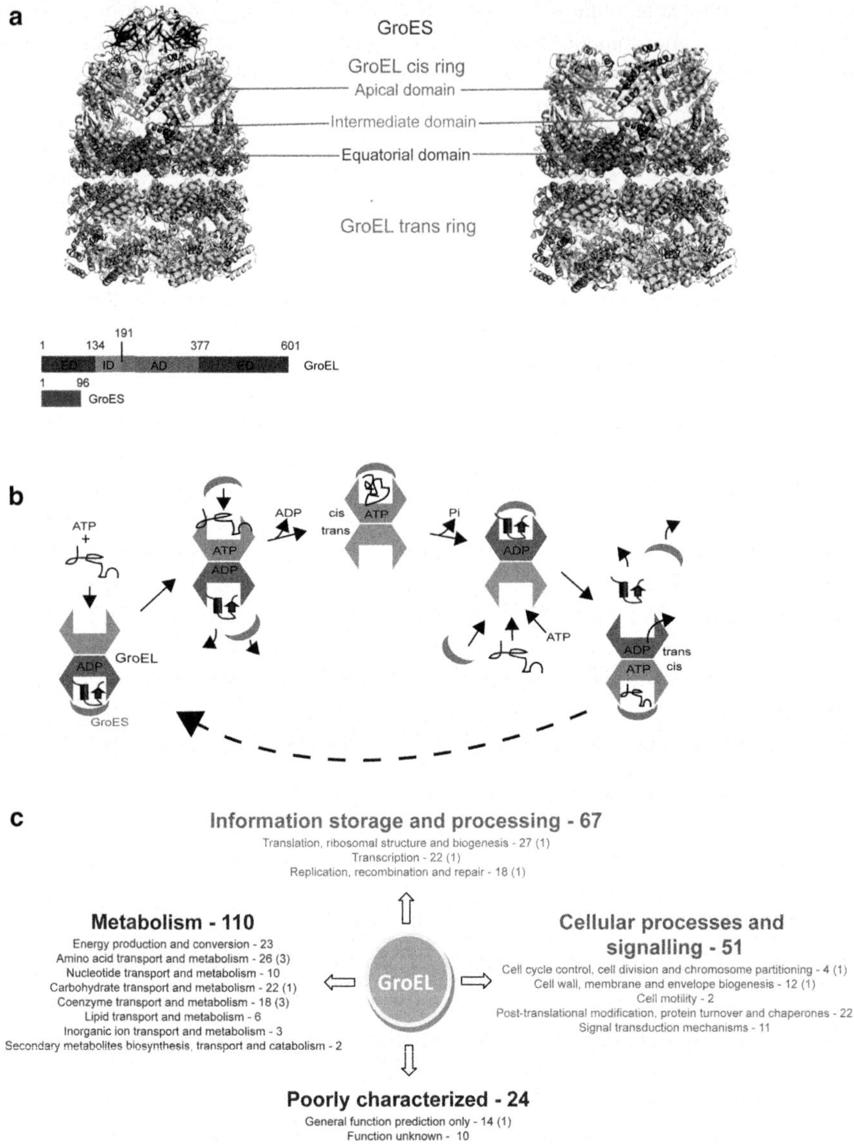

Fig. 15.4 Structure and function of the GroEL/ES chaperone system. (**a**) Shown are the structures of the GroEL–GroES complex [*left*, PDB ID 1PCQ (Chaudhry et al. 2003)] and the GroEL tetradecamer [*right*, PDB ID 1PCQ (Chaudhry et al. 2003)]. One heptameric ring is in *yellow*, while the opposite ring is in *gray* with the domains of one of the subunits colored as follows: equatorial domain in *blue*, intermediate domain in *orange* and the apical domain in *red*. GroES heptamer capping the *cis* heptameric GroEL ring is shown in *purple*. Bar graphs of GroEL and GroES domain arrangement are shown below the structures. (**b**) A cartoon representation of the nucleotide-dependent GroEL/ES functional cycle. Refer to the text for further details. (**c**) Shown are Cluster of Orthologous Group (COG) functional categories for the 252 GroEL interacting proteins (Tatusov et al. 2001; Kerner et al. 2005). The numbers between brackets indicate the GroEL substrates of the respective categories that are essential for the cell and are also obligate GroEL substrates

binding (Xu et al. 1997; Farr et al. 2000; Chaudhry et al. 2003). The intermediate domain acts as a linker between the other two domains. Each GroEL ring has a large central cavity (Fig. 15.4a).

GroES is a heptameric protein whose subunits form a ring dome-like structure (Fig. 15.4a) (Hunt et al. 1996). The sevenfold symmetry of the GroES protein complements that of GroEL and, upon association with GroEL, forms a structure analogous to a lid for the central cavity of GroEL (Langer et al. 1992; Chaudhry et al. 2003). The GroEL–GroES interaction results in the doubling of the size of the GroEL central cavity (Fig. 15.4a) due to large conformational changes in the intermediate and apical domains (Xu et al. 1997). The intermediate domain physically and functionally connects the equatorial and apical domains by transferring the energy of ATP hydrolysis in the equatorial domain with conformational changes in the apical domain (Ranson et al. 2001, 2006; Saibil et al. 2013).

The GroEL/ES functional cycle is shown schematically in Fig. 15.4b. The GroEL open ring captures non-native but compact forms of a polypeptide substrate that is exposing a hydrophobic surface, thus, GroEL prevents the substrate from misfolding or forming irreversible aggregates (Goloubinoff et al. 1989; Braig et al. 1994; Horwich et al. 2006). Mutational studies indicate that this primary association is based on hydrophobic interactions with the apical domain of the chaperone (Fenton et al. 1994; Farr et al. 2000). Binding to GroEL might result in unfolding of non-native states allowing for subsequent refolding (Lin et al. 2013). Subsequently, ATP binds cooperatively to the equatorial domains of seven subunits of one GroEL ring (Yifrach and Horovitz 1995). This allows for the association of GroES to GroEL (Chandrasekhar et al. 1986) due to large conformational shifts that release the bound substrate from its hydrophobic association with GroEL since GroES competes for the same binding sites on GroEL as the substrate. The formation of the GroEL–GroES complex results in the formation of an enclosed hydrophilic chamber that traps the substrate and promotes folding in an environment isolated from the cellular milieu (Chaudhry et al. 2003). Following ATP hydrolysis, a stable GroEL(ADP)–GroES complex is formed containing the trapped substrate (Fig. 15.4b). Subsequently, ATP binds to the opposite ring of the tetradecamer that does not contain the substrate, and, due to the negative cooperativity in nucleotide binding between the two GroEL rings (Rye et al. 1999), this leads to the release of GroES, ADP and bound substrate allowing for a new substrate interaction cycle to occur (Saibil et al. 2013).

15.4.2 The GroEL Interaction Network

Based on immunoprecipitation of GroEL and its bound substrates in pulse-chase type experiments, 10–15 % of all newly translated cytoplasmic proteins were estimated to transit through the GroEL chaperone under normal cellular conditions (Ewalt et al. 1997; Houry et al. 1999); this number increased to 30 % under heat stress of 42 °C. Combining immunoprecipitation with 2-D gel electrophoresis and mass spectrometry, 52 of the most abundant GroEL substrates were identified (Houry et al. 1999). These proteins included members of the transcription and translation machineries as well as many metabolic enzymes. To more comprehensively identify GroEL substrates, affinity chromatography was utilized to pull down proteins trapped inside the GroEL/ES chamber. Kerner et al. (2005) attempted to isolate GroEL/ES complexes formed with *E. coli* GroES-His$_6$. However, such complexes were not stable, which led the authors to replace *E. coli* GroES with GroES from *Methanosarcina mazei* (Mm). MmGroES was shown to functionally replace *E. coli* GroES but was found to bind more stably to GroEL in the presence of ADP allowing for the isolation of stable GroEL/ES complexes containing trapped substrates. Pull downs followed by mass spectrometry led to the identification of 250 GroEL/ES substrates (Fig. 15.4c). Most proteins were cytoplasmic with only eight being either periplasmic or outer membrane. Of the 250 recognized GroEL interactors, 83 were defined as obligate GroEL/ES interactors, which included 13 essential proteins (Gerdes et al. 2003; Kerner et al. 2005). Based on cellular abundance of the proteins and their GroEL dependence, about 75–80 % of cellular GroEL molecules were estimated to be occupied by 83 obligate substrates.

The identified substrates were divided into three classes based on their dependence on the GroEL/ES system for folding (Ewalt et al. 1997; Kerner et al. 2005). Class I substrates require minimal chaperone assistance to fold. Class II substrates are those that required the presence of both GroEL and GroES for folding at 37 °C but these substrates do not require GroES at lower temperatures. Furthermore, Class II substrates are not solely dependent on GroEL, as DnaK can also assist in their folding at 37 °C. Class III proteins are obligate GroEL/ES substrates (Kerner et al. 2005). Substrates belonging to Class III fail to refold in the absence of GroEL/ES even if DnaK is present; however, DnaK may be able to bind these proteins and prevent their aggregation.

Few salient characteristics were identified that differentiate a GroEL substrate from other cytosolic proteins. The GroEL-associated proteins spanned a range of sizes from 10 to 150 kDa, but they typically were of molecular weight around 20–60 kDa (Houry et al. 1999; Kerner et al. 2005), especially for class III proteins. The size range is consistent with the fact that the chamber formed upon association of GroEL with GroES can hold globular proteins with an upper size limit of 50–60 kDa (Chen et al. 1994). Considering that class I and II substrates may be assisted during their folding by chaperones other than GroEL, a size preference was not found among these proteins.

In addition to a size preference, obligate substrates had pI values around 5.5–6.5, leading to a lower net charge at physiological pH in comparison to other cytosolic proteins (Kerner et al. 2005). A lower net charge is correlated with an increased propensity to aggregate, providing an additional clue to their chaperone requirement (Chiti et al. 2002). No difference in hydrophobicity was observed for obligate GroEL substrates compared to other cytosolic proteins. Structurally, $\alpha\beta$ domains were enriched in GroEL substrates over all-α or all-β domains with a special partiality towards the $(\beta\alpha)_8$ TIM-barrel fold belonging to SCOP class c1 (Houry et al. 1999; Kerner et al. 2005; Georgescauld et al. 2014). Recently, it has been suggested the GroEL/ES can accelerate the rate of TIM-barrel domain folding (Georgescauld et al. 2014). The TIM-barrel is a common structural fold and not all proteins with such TIM-barrel need GroEL to reach their native state. Additionally, the observation that class I substrates with TIM-barrel are unable to displace class II or III substrates suggests that intermediate folded states rather than the final native state of TIM-barrel proteins may share features that favorably associate with GroEL.

15.5 Overlapping Functional Roles of the Chaperones

With trigger factor, GroEL and DnaK each responsible for the correct folding of hundreds of cellular proteins, it is interesting to note that only the GroEL/ES system is essential at all temperatures (Fayet et al. 1989). Loss of TF does not affect cell viability but DnaK is required at growth temperatures above 37 °C and below 15 °C (Bukau and Walker 1989; Deuerling et al. 1999). The indispensability of GroEL has been linked to the requirement of the chaperone to fold one or all of 13 characterized obligate GroEL substrates essential for survival of the organism (Kerner et al. 2005). The lack of essentiality for the DnaK and TF chaperones is a little more complicated, but is likely due to compensatory mechanisms and overlapping functional roles among these chaperones and GroEL.

It is well known that the viability of a Δ*tig*Δ*dnaK* mutant (*tig* is the gene for TF) can be rescued by expression of either TF or DnaK alone (Genevaux et al. 2004). Hence, the two chaperones are able to compensate for each other. In Δ*tig* cells, the DnaK interactome was found to increase by about 48 % with the chaperone associating with 998 proteins compared to 674 in wild type cells (Calloni et al. 2012). Indeed, in the absence of TF, DnaK and GroEL levels were found to increase by up to threefold compared to steady state levels in wild type cells (Deuerling et al. 2003). Also, 77 % of TF-bound peptides showed affinity for DnaK, likely based on the similarity in the binding motifs for the two chaperones, which comprise a hydrophobic core flanked by basic residues (Deuerling et al. 2003). Similarly, in the absence of DnaK and TF, an additional 150 proteins were observed to interact with GroEL compared to WT cells at 30 °C (Kerner et al. 2005). The extra burden upon the GroEL/ES system is mitigated by upregulation of its protein levels (Calloni et al. 2012).

Despite some overlap, the chaperone systems are not perfectly complementary to each other. TF has a specific role in transportation of outer membrane proteins; that role cannot be substituted by DnaK (Oh et al. 2011). As a consequence, cells lacking TF are more sensitive to the detergent deoxycholate and antibiotic vancomycin, a symptom of a weaker outer membrane (Nichols et al. 2011; Calloni et al. 2012). Similarly, when searching for their individualized importance, TF or GroEL cannot replace the function of DnaK in resolving protein aggregates in Δ*dnaK* cells (Calloni et al. 2012). Hence, while the functional overlap among chaperones ensures efficiency under stress conditions or when one of the chaperone systems is overwhelmed, each chaperone system also shows some degree of specialization in its activity.

15.6 Chaperone–Chaperone Interactions

Chaperones can be divided into two functional groups. TF and GroEL/ES belong to the group primarily involved in de novo protein folding, while a second group that is involved in refolding and protein disaggregation includes ClpB, ClpX, and some small heat shock proteins not discussed in this review (Haslbeck et al. 2005; Barends et al. 2010; Baker and Sauer 2012). ClpB acts as a disaggregase, while ClpX acts as an unfoldase and targets proteins to the ClpP protease for degradation. DnaK seems to link the two groups as it plays a major part in de novo folding and in aggregation prevention (Deuerling et al. 1999; Mogk et al. 1999; Calloni et al. 2012).

Using FRET-based analyses, utilizing fluorescence transfer from CFP to YFP, direct and indirect interactions among *E. coli* chaperones were observed (Kumar and Sourjik 2012). Consistent with the inference of DnaK as the 'chaperone' hub of the cell, it was observed that DnaK and DnaJ interact with many other chaperones including TF, the small heat shock proteins IbpA and IbpB, the Hsp100 family ATPase ClpB and Hsp90 ortholog HtpG. TF was shown to interact with DnaJ, confirming previous observation showing substrate transfer from TF to DnaJ and then DnaK (Deuerling et al. 1999; Teter et al. 1999; Kumar and Sourjik 2012). Interestingly, the DnaK nucleotide exchange factor GrpE was observed to be in close vicinity in the cell to HtpG and ClpB proteins (Genest et al. 2011; Miot et al. 2011; Kumar and Sourjik 2012). Addition of the translation inhibitor chloramphenicol, abolished these interactions suggesting that chaperone–chaperone interactions are not direct but rather are mediated by substrates. Interactions among chaperones involved in de novo folding were unaffected by similar treatment, indicative of the fact that they are substrate-independent, direct inter-chaperone interactions.

Protein–protein interactions among different chaperones have also been identified using pulldown methods. Chaperones pulled-down with GroEL included TF, DnaK, DnaJ and the redox-related chaperones Hsp33 and YegD (Kerner et al. 2005). Indeed, DnaK, DnaJ as well as TF have been noted to associate with and to deliver substrates to GroEL (Buchberger et al. 1996; Calloni et al. 2012). YegD is a member of the Hsp70/DnaK family of proteins, although its cellular function and the physiological significance of its association with GroEL is not clear. The association between GroEL and Hsp33 was previously found to occur during heat and oxidative stress and was speculated to allow GroEL to fold proteins initially interacting with Hsp33 (Echave et al. 2002; Hoffmann et al. 2004; Genevaux et al. 2007).

Among numerous proteins pulled down with DnaK were the small heat shock proteins IbpA and IbpB, whose function includes stabilization of aggregating proteins under heat stress (Laskowska et al. 1996; Calloni et al. 2012). Other chaperones identified in the pulldown with DnaK include the DnaJ paralog CbpA, the cytoplasmic chaperones TF, ClpB, HtpG, SecB, HscA, the periplasmic acid stress chaperones HdeA and HdeB, and the oxidative stress response chaperone Hsp33 (Calloni et al. 2012). ClpB and DnaK are known to act synergistically to reverse protein aggregation (Mogk et al. 1999). SecB, involved in protein export through the general secretory (Sec) system, is known to act as a chaperone in modulating folding and aggregation states of its substrates before their export (Randall and Hardy 2002; Ullers et al. 2004). DnaK and SecB might have some complementary functions but may act cooperatively as well (Wild et al. 1992; Ullers et al. 2004).

15.7 Sequence of Substrate–Chaperone Interactions

Trigger factor, DnaK and GroEL function as the main chaperone hubs for de novo protein folding. Based on the relative roles of the three chaperone systems, the most basic of models suggests a simple sequential functionality for the three chaperones (Fig. 15.5). According to such a model, TF would be the most upstream of the three systems, assisting in the folding of proteins as they emerge from the ribosome, while GroEL is thought to be utilized for proteins requiring the most chaperone assistance. Many observations support the sequential nature of the function of these chaperones. Primarily, TF is the only prokaryotic chaperone associated with the ribosome. Many GroEL substrates are directly transferred either from TF or DnaK (Kerner et al. 2005; Fujiwara et al. 2010; Calloni et al. 2012). In the absence of TF, the number of DnaK-bound GroEL substrates increases from 119 to 152 (Calloni et al. 2012). The increase suggests a shift from TF-assisted folding to DnaK-assisted folding prior to interaction with GroEL. Furthermore, although little aggregation of GroEL substrates was observed in the absence of either DnaK or TF, 70 % of GroEL substrates were found aggregated in cells lacking both DnaK and TF despite the upregulation of GroEL (Kerner et al. 2005; Calloni et al. 2012). Thus, many GroEL substrates are dependent on the 'upstream' TF and DnaK chaperones.

Such a simplistic model of Fig. 15.5 is sufficient for the description of the general de novo folding machinery based on our current knowledge of the mechanism of function of these chaperones and their interaction networks. However, as has been shown for luciferase folding, DnaK and GroEL may sometimes compete for binding to a substrate and do not always act in succession (Buchberger et al. 1996). Such cases imply that, while the simple sequential model of TF to DnaK to GroEL might be true for chaperone-assisted folding for many proteins, there are a subset of proteins that may be acted on competitively or laterally by these chaperones.

15.8 Conclusion

Numerous newly synthesized proteins rely on chaperone assistance for folding in the crowded cellular environment. Trigger factor, DnaK/DnaJ/GrpE and GroEL/ES are the three major systems that assist newly synthesized proteins. Though their mechanisms of function are well understood, their proteomic contributions are less so. Recent biochemical and structural analyses have attempted to define the

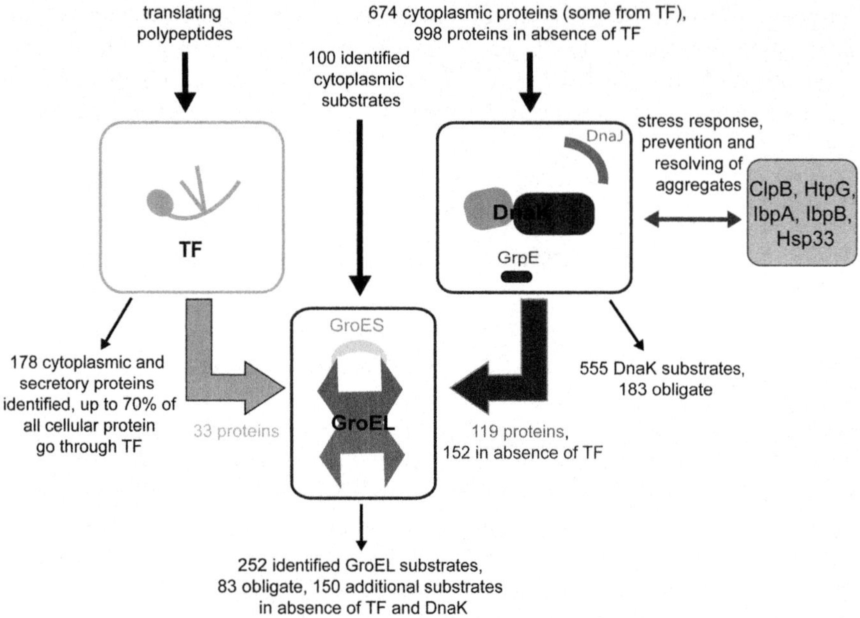

Fig. 15.5 Substrate–chaperone interaction network in *E. coli*. A schematic showing the number of substrates identified to associate with each of the three chaperone systems. Also depicted is how these substrates flux through TF, DnaK and GroEL systems. *Black arrows* leading into chaperones indicate proteins that bind to chaperones and *colored arrows* refer to the transfer of substrates between chaperone systems. TF interacts with proteins directly emerging from the ribosome. Up to 70 % of all cellular proteins are estimated to be folded by TF (Vabulas et al. 2010). Among these, 178 proteins have been identified (Martinez-Hackert and Hendrickson 2009). 674 DnaK substrates have been identified using pulldown assays (Calloni et al. 2012). Though some proteins are known to be transferred to DnaK from TF, the exact number remains unknown (Deuerling et al. 2003). TF and DnaK are thought to deliver 33 and 119 proteins to GroEL, respectively, with DnaK delivering 152 proteins in the absence of TF (Calloni et al. 2012). An additional 100 substrates are known to associate with GroEL (Ewalt et al. 1997; Kerner et al. 2005). Known increases in the number of different substrates for DnaK and GroEL are also indicated when either TF, DnaK or both are missing from the cell

interactome for each of these systems. In doing so, a better understanding is gained of the function of these chaperones in maintaining proteostasis. An understanding of the interaction networks of these chaperones can help in drug discovery efforts for pathogenic bacteria as well as in providing clues on the regulation, biogenesis and evolution of cellular proteins and protein complexes.

Acknowledgements Vaibhav Bhandari was the recipient of a Jaro Sodek Award—Ontario Student Opportunity Trust Fund (OSOTF) fellowship from the Department of Biochemistry at the University of Toronto. This work was supported by a grant from the Canadian Institutes of Health Research (MOP-130374) to WAH.

References

Agashe VR, Guha S, Chang HC, Genevaux P, Hayer-Hartl M, Stemp M et al (2004) Function of trigger factor and DnaK in multidomain protein folding: increase in yield at the expense of folding speed. Cell 117:199–209

Anfinsen CB (1973) Principles that govern the folding of protein chains. Science 181:223–230

Baker TA, Sauer RT (2012) ClpXP, an ATP-powered unfolding and protein-degradation machine. Biochim Biophys Acta 1823:15–28

Barends TR, Werbeck ND, Reinstein J (2010) Disaggregases in 4 dimensions. Curr Opin Struct Biol 20:46–53

Bertelsen EB, Chang L, Gestwicki JE, Zuiderweg ER (2009) Solution conformation of wild-type *E. coli* Hsp70 (DnaK) chaperone complexed with ADP and substrate. Proc Natl Acad Sci U S A 106:8471–8476

Braig K, Otwinowski Z, Hegde R, Boisvert DC, Joachimiak A, Horwich AL et al (1994) The crystal structure of the bacterial chaperonin GroEL at 2.8 A. Nature 371:578–586

Brehmer D, Rudiger S, Gassler CS, Klostermeier D, Packschies L, Reinstein J et al (2001) Tuning of chaperone activity of Hsp70 proteins by modulation of nucleotide exchange. Nat Struct Biol 8:427–432

Brehmer D, Gassler C, Rist W, Mayer MP, Bukau B (2004) Influence of GrpE on DnaK-substrate interactions. J Biol Chem 279:27957–27964

Buchberger A, Schroder H, Hesterkamp T, Schonfeld HJ, Bukau B (1996) Substrate shuttling between the DnaK and GroEL systems indicates a chaperone network promoting protein folding. J Mol Biol 261:328–333

Bukau B, Walker GC (1989) Cellular defects caused by deletion of the *Escherichia coli* dnaK gene indicate roles for heat shock protein in normal metabolism. J Bacteriol 171:2337–2346

Bukau B, Deuerling E, Pfund C, Craig EA (2000) Getting newly synthesized proteins into shape. Cell 101:119–122

Calloni G, Chen T, Schermann SM, Chang HC, Genevaux P, Agostini F et al (2012) DnaK functions as a central hub in the *E. coli* chaperone network. Cell Rep 1:251–264

Chandrasekhar GN, Tilly K, Woolford C, Hendrix R, Georgopoulos C (1986) Purification and properties of the groES morphogenetic protein of *Escherichia coli*. J Biol Chem 261: 12414–12419

Chaudhry C, Farr GW, Todd MJ, Rye HS, Brunger AT, Adams PD et al (2003) Role of the gamma-phosphate of ATP in triggering protein folding by GroEL-GroES: function, structure and energetics. EMBO J 22:4877–4887

Cheetham ME, Caplan AJ (1998) Structure, function and evolution of DnaJ: conservation and adaptation of chaperone function. Cell Stress Chaperones 3:28–36

Chen S, Roseman AM, Hunter AS, Wood SP, Burston SG, Ranson NA et al (1994) Location of a folding protein and shape changes in GroEL-GroES complexes imaged by cryo-electron microscopy. Nature 371:261–264

Chenoweth MR, Trun N, Wickner S (2007) *In vivo* modulation of a DnaJ homolog, CbpA, by CbpM. J Bacteriol 189:3635–3638

Chiti F, Calamai M, Taddei N, Stefani M, Ramponi G, Dobson CM (2002) Studies of the aggregation of mutant proteins in vitro provide insights into the genetics of amyloid diseases. Proc Natl Acad Sci U S A 99(Suppl 4):16419–16426

DeLano WL (2002) The PyMOL molecular graphics system, Version 1.5.0.4 Schrödinger, LLC

Deuerling E, Schulze-Specking A, Tomoyasu T, Mogk A, Bukau B (1999) Trigger factor and DnaK cooperate in folding of newly synthesized proteins. Nature 400:693–696

Deuerling E, Patzelt H, Vorderwulbecke S, Rauch T, Kramer G, Schaffitzel E et al (2003) Trigger factor and DnaK possess overlapping substrate pools and binding specificities. Mol Microbiol 47:1317–1328

Dill KA, Chan HS (1997) From Levinthal to pathways to funnels. Nat Struct Biol 4:10–19

Echave P, Esparza-Ceron MA, Cabiscol E, Tamarit J, Ros J, Membrillo-Hernandez J et al (2002) DnaK dependence of mutant ethanol oxidoreductases evolved for aerobic function and protective role of the chaperone against protein oxidative damage in *Escherichia coli*. Proc Natl Acad Sci U S A 99:4626–4631

Ellis RJ (2006) Molecular chaperones: assisting assembly in addition to folding. Trends Biochem Sci 31:395–401

Ewalt KL, Hendrick JP, Houry WA, Hartl FU (1997) *In vivo* observation of polypeptide flux through the bacterial chaperonin system. Cell 90:491–500

Farr GW, Furtak K, Rowland MB, Ranson NA, Saibil HR, Kirchhausen T et al (2000) Multivalent binding of nonnative substrate proteins by the chaperonin GroEL. Cell 100:561–573

Fayet O, Ziegelhoffer T, Georgopoulos C (1989) The GroES and GroEL heat-shock gene-products of *Escherichia-coli* are essential for bacterial-growth at all temperatures. J Bacteriol 171: 1379–1385

Fenton WA, Kashi Y, Furtak K, Horwich AL (1994) Residues in chaperonin GroEL required for polypeptide binding and release. Nature 371:614–619

Ferbitz L, Maier T, Patzelt H, Bukau B, Deuerling E, Ban N (2004) Trigger factor in complex with the ribosome forms a molecular cradle for nascent proteins. Nature 431:590–596

Fujiwara K, Ishihama Y, Nakahigashi K, Soga T, Taguchi H (2010) A systematic survey of *in vivo* obligate chaperonin-dependent substrates. EMBO J 29:1552–1564

Fuzery AK, Tonelli M, Ta DT, Cornilescu G, Vickery LE, Markley JL (2008) Solution structure of the iron-sulfur cluster cochaperone HscB and its binding surface for the iron-sulfur assembly scaffold protein IscU. Biochemistry 47:9394–9404

Gamer J, Multhaup G, Tomoyasu T, McCarty JS, Rudiger S, Schonfeld HJ et al (1996) A cycle of binding and release of the DnaK, DnaJ and GrpE chaperones regulates activity of the *Escherichia coli* heat shock transcription factor sigma32. EMBO J 15:607–617

Genest O, Hoskins JR, Camberg JL, Doyle SM, Wickner S (2011) Heat shock protein 90 from *Escherichia coli* collaborates with the DnaK chaperone system in client protein remodeling. Proc Natl Acad Sci U S A 108:8206–8211

Genevaux P, Wawrzynow A, Zylicz M, Georgopoulos C, Kelley WL (2001) DjlA is a third DnaK co-chaperone of *Escherichia coli*, and DjlA-mediated induction of colanic acid capsule requires DjlA–DnaK interaction. J Biol Chem 276:7906–7912

Genevaux P, Keppel F, Schwager F, Langendijk-Genevaux PS, Hartl FU, Georgopoulos C (2004) *In vivo* analysis of the overlapping functions of DnaK and trigger factor. EMBO Rep 5:195–200

Genevaux P, Georgopoulos C, Kelley WL (2007) The Hsp70 chaperone machines of *Escherichia coli*: a paradigm for the repartition of chaperone functions. Mol Microbiol 66:840–857

Georgescauld F, Popova K, Gupta AJ, Bracher A, Engen JR, Hayer-Hartl M et al (2014) GroEL/ES chaperonin modulates the mechanism and accelerates the rate of TIM-barrel domain folding. Cell 157:922–934

Gerdes SY, Scholle MD, Campbell JW, Balazsi G, Ravasz E, Daugherty MD et al (2003) Experimental determination and system level analysis of essential genes in *Escherichia coli* MG1655. J Bacteriol 185:5673–5684

Goloubinoff P, Christeller JT, Gatenby AA, Lorimer GH (1989) Reconstitution of active dimeric ribulose bisphosphate carboxylase from an unfolded state depends on two chaperonin proteins and Mg-ATP. Nature 342:884–889

Grimshaw JP, Jelesarov I, Siegenthaler RK, Christen P (2003) Thermosensor action of GrpE. The DnaK chaperone system at heat shock temperatures. J Biol Chem 278:19048–19053

Gur E, Biran D, Shechter N, Genevaux P, Georgopoulos C, Ron EZ (2004) The *Escherichia coli* DjlA and CbpA proteins can substitute for DnaJ in DnaK-mediated protein disaggregation. J Bacteriol 186:7236–7242

Gur E, Katz C, Ron EZ (2005) All three J-domain proteins of the *Escherichia coli* DnaK chaperone machinery are DNA binding proteins. FEBS Lett 579:1935–1939

Harrison C (2003) GrpE, a nucleotide exchange factor for DnaK. Cell Stress Chaperones 8: 218–224

Harrison CJ, Hayer-Hartl M, Di Liberto M, Hartl F, Kuriyan J (1997) Crystal structure of the nucleotide exchange factor GrpE bound to the ATPase domain of the molecular chaperone DnaK. Science 276:431–435
Hartl FU, Hayer-Hartl M (2002) Molecular chaperones in the cytosol: from nascent chain to folded protein. Science 295:1852–1858
Hartl FU, Bracher A, Hayer-Hartl M (2011) Molecular chaperones in protein folding and proteostasis. Nature 475:324–332
Haslbeck M, Franzmann T, Weinfurtner D, Buchner J (2005) Some like it hot: the structure and function of small heat-shock proteins. Nat Struct Mol Biol 12:842–846
Hennessy F, Nicoll WS, Zimmermann R, Cheetham ME, Blatch GL (2005) Not all J domains are created equal: implications for the specificity of Hsp40-Hsp70 interactions. Protein Sci 14:1697–1709
Hesterkamp T, Hauser S, Lutcke H, Bukau B (1996) *Escherichia coli* trigger factor is a prolyl isomerase that associates with nascent polypeptide chains. Proc Natl Acad Sci U S A 93: 4437–4441
Hoffmann JH, Linke K, Graf PC, Lilie H, Jakob U (2004) Identification of a redox-regulated chaperone network. EMBO J 23:160–168
Hoffmann A, Merz F, Rutkowska A, Zachmann-Brand B, Deuerling E, Bukau B (2006) Trigger factor forms a protective shield for nascent polypeptides at the ribosome. J Biol Chem 281:6539–6545
Hoffmann A, Bukau B, Kramer G (2010) Structure and function of the molecular chaperone trigger factor. Biochim Biophys Acta 1803:650–661
Horwich AL, Low KB, Fenton WA, Hirshfield IN, Furtak K (1993) Folding *in vivo* of bacterial cytoplasmic proteins: role of GroEL. Cell 74:909–917
Horwich AL, Farr GW, Fenton WA (2006) GroEL-GroES-mediated protein folding. Chem Rev 106:1917–1930
Houry WA, Frishman D, Eckerskorn C, Lottspeich F, Hartl FU (1999) Identification of *in vivo* substrates of the chaperonin GroEL. Nature 402:147–154
Huang K, Flanagan JM, Prestegard JH (1999) The influence of C-terminal extension on the structure of the "J-domain" in *E. coli* DnaJ. Protein Sci 8:203–214
Hunt JF, Weaver AJ, Landry SJ, Gierasch L, Deisenhofer J (1996) The crystal structure of the GroES co-chaperonin at 2.8 A resolution. Nature 379:37–45
Kaiser CM, Chang HC, Agashe VR, Lakshmipathy SK, Etchells SA, Hayer-Hartl M et al (2006) Real-time observation of trigger factor function on translating ribosomes. Nature 444:455–460
Kampinga HH, Craig EA (2010) The HSP70 chaperone machinery: J proteins as drivers of functional specificity. Nat Rev Mol Cell Biol 11:579–592
Karzai AW, McMacken R (1996) A bipartite signaling mechanism involved in DnaJ-mediated activation of the *Escherichia coli* DnaK protein. J Biol Chem 271:11236–11246
Kelley WL (1998) The J-domain family and the recruitment of chaperone power. Trends Biochem Sci 23:222–227
Kerner MJ, Naylor DJ, Ishihama Y, Maier T, Chang HC, Stines AP et al (2005) Proteome-wide analysis of chaperonin-dependent protein folding in *Escherichia coli*. Cell 122:209–220
Kityk R, Kopp J, Sinning I, Mayer MP (2012) Structure and dynamics of the ATP-bound open conformation of Hsp70 chaperones. Mol Cell 48:863–874
Kluck CJ, Patzelt H, Genevaux P, Brehmer D, Rist W, Schneider-Mergener J et al (2002) Structure-function analysis of HscC, the *Escherichia coli* member of a novel subfamily of specialized Hsp70 chaperones. J Biol Chem 277:41060–41069
Kramer G, Rauch T, Rist W, Vorderwulbecke S, Patzelt H, Schulze-Specking A et al (2002) L23 protein functions as a chaperone docking site on the ribosome. Nature 419:171–174
Kumar M, Sourjik V (2012) Physical map and dynamics of the chaperone network in *Escherichia coli*. Mol Microbiol 84:736–747
Lakshmipathy SK, Tomic S, Kaiser CM, Chang HC, Genevaux P, Georgopoulos C et al (2007) Identification of nascent chain interaction sites on trigger factor. J Biol Chem 282:12186–12193

Langer T, Pfeifer G, Martin J, Baumeister W, Hartl FU (1992) Chaperonin-mediated protein folding – GroES binds to one end of the GroEL cylinder, which accommodates the protein substrate within its central cavity. EMBO J 11:4757–4765

Laskowska E, Wawrzynow A, Taylor A (1996) IbpA and IbpB, the new heat-shock proteins, bind to endogenous *Escherichia coli* proteins aggregated intracellularly by heat shock. Biochimie 78:117–122

Laufen T, Mayer MP, Beisel C, Klostermeier D, Mogk A, Reinstein J et al (1999) Mechanism of regulation of hsp70 chaperones by DnaJ cochaperones. Proc Natl Acad Sci U S A 96: 5452–5457

Liberek K, Marszalek J, Ang D, Georgopoulos C, Zylicz M (1991) *Escherichia coli* DnaJ and GrpE heat shock proteins jointly stimulate ATPase activity of DnaK. Proc Natl Acad Sci U S A 88:2874–2878

Lill R, Crooke E, Guthrie B, Wickner W (1988) The "trigger factor cycle" includes ribosomes, presecretory proteins, and the plasma membrane. Cell 54:1013–1018

Lin Z, Puchalla J, Shoup D, Rye HS (2013) Repetitive protein unfolding by the *trans* ring of the GroEL-GroES chaperonin complex stimulates folding. J Biol Chem 288:30944–30955

Maier R, Eckert B, Scholz C, Lilie H, Schmid FX (2003) Interaction of trigger factor with the ribosome. J Mol Biol 326:585–592

Mally A, Witt SN (2001) GrpE accelerates peptide binding and release from the high affinity state of DnaK. Nat Struct Biol 8:254–257

Martinez-Hackert E, Hendrickson WA (2009) Promiscuous substrate recognition in folding and assembly activities of the trigger factor chaperone. Cell 138:923–934

Martinez-Yamout M, Legge GB, Zhang O, Wright PE, Dyson HJ (2000) Solution structure of the cysteine-rich domain of the *Escherichia coli* chaperone protein DnaJ. J Mol Biol 300:805–818

Mashaghi A, Kramer G, Bechtluft P, Zachmann-Brand B, Driessen AJ, Bukau B et al (2013) Reshaping of the conformational search of a protein by the chaperone trigger factor. Nature 500:98–101

Mayer MP, Bukau B (2005) Hsp70 chaperones: cellular functions and molecular mechanism. Cell Mol Life Sci 62:670–684

Mayor U, Guydosh NR, Johnson CM, Grossmann JG, Sato S, Jas GS et al (2003) The complete folding pathway of a protein from nanoseconds to microseconds. Nature 421:863–867

Merz F, Hoffmann A, Rutkowska A, Zachmann-Brand B, Bukau B, Deuerling E (2006) The C-terminal domain of *Escherichia coli* trigger factor represents the central module of its chaperone activity. J Biol Chem 281:31963–31971

Miot M, Reidy M, Doyle SM, Hoskins JR, Johnston DM, Genest O et al (2011) Species-specific collaboration of heat shock proteins (Hsp) 70 and 100 in thermotolerance and protein disaggregation. Proc Natl Acad Sci U S A 108:6915–6920

Mogk A, Tomoyasu T, Goloubinoff P, Rudiger S, Roder D, Langen H et al (1999) Identification of thermolabile *Escherichia coli* proteins: prevention and reversion of aggregation by DnaK and ClpB. EMBO J 18:6934–6949

Nichols RJ, Sen S, Choo YJ, Beltrao P, Zietek M, Chaba R et al (2011) Phenotypic landscape of a bacterial cell. Cell 144:143–156

Niwa T, Kanamori T, Ueda T, Taguchi H (2012) Global analysis of chaperone effects using a reconstituted cell-free translation system. Proc Natl Acad Sci U S A 109:8937–8942

Oh E, Becker AH, Sandikci A, Huber D, Chaba R, Gloge F et al (2011) Selective ribosome profiling reveals the cotranslational chaperone action of trigger factor *in vivo*. Cell 147:1295–1308

Packschies L, Theyssen H, Buchberger A, Bukau B, Goody RS, Reinstein J (1997) GrpE accelerates nucleotide exchange of the molecular chaperone DnaK with an associative displacement mechanism. Biochemistry 36:3417–3422

Patzelt H, Rudiger S, Brehmer D, Kramer G, Vorderwulbecke S, Schaffitzel E et al (2001) Binding specificity of *Escherichia coli* trigger factor. Proc Natl Acad Sci U S A 98:14244–14249

Raine A, Lovmar M, Wikberg J, Ehrenberg M (2006) Trigger factor binding to ribosomes with nascent peptide chains of varying lengths and sequences. J Biol Chem 281:28033–28038

Randall LL, Hardy SJ (2002) SecB, one small chaperone in the complex milieu of the cell. Cell Mol Life Sci 59:1617–1623
Ranson NA, Farr GW, Roseman AM, Gowen B, Fenton WA, Horwich AL et al (2001) ATP-bound states of GroEL captured by cryo-electron microscopy. Cell 107:869–879
Ranson NA, Clare DK, Farr GW, Houldershaw D, Horwich AL, Saibil HR (2006) Allosteric signaling of ATP hydrolysis in GroEL-GroES complexes. Nat Struct Mol Biol 13:147–152
Rosenzweig R, Moradi S, Zarrine-Afsar A, Glover JR, Kay LE (2013) Unraveling the mechanism of protein disaggregation through a ClpB-DnaK interaction. Science 339:1080–1083
Rudiger S, Germeroth L, Schneider-Mergener J, Bukau B (1997) Substrate specificity of the DnaK chaperone determined by screening cellulose-bound peptide libraries. EMBO J 16:1501–1507
Rudiger S, Schneider-Mergener J, Bukau B (2001) Its substrate specificity characterizes the DnaJ co-chaperone as a scanning factor for the DnaK chaperone. EMBO J 20:1042–1050
Rutkowska A, Mayer MP, Hoffmann A, Merz F, Zachmann-Brand B, Schaffitzel C et al (2008) Dynamics of trigger factor interaction with translating ribosomes. J Biol Chem 283:4124–4132
Rye HS, Roseman AM, Chen S, Furtak K, Fenton WA, Saibil HR et al (1999) GroEL-GroES cycling: ATP and nonnative polypeptide direct alternation of folding-active rings. Cell 97: 325–338
Saibil HR, Fenton WA, Clare DK, Horwich AL (2013) Structure and allostery of the chaperonin GroEL. J Mol Biol 425:1476–1487
Saio T, Guan X, Rossi P, Economou A, Kalodimos CG (2014) Structural basis for protein antiaggregation activity of the trigger factor chaperone. Science 344:1250494
Schlecht R, Erbse AH, Bukau B, Mayer MP (2011) Mechanics of Hsp70 chaperones enables differential interaction with client proteins. Nat Struct Mol Biol 18:345–351
Schonfeld HJ, Schmidt D, Schroder H, Bukau B (1995) The DnaK chaperone system of *Escherichia coli*: quaternary structures and interactions of the DnaK and GrpE components. J Biol Chem 270:2183–2189
Sell SM, Eisen C, Ang D, Zylicz M, Georgopoulos C (1990) Isolation and characterization of dnaJ null mutants of *Escherichia coli*. J Bacteriol 172:4827–4835
Shi W, Zhou Y, Wild J, Adler J, Gross CA (1992) DnaK, DnaJ, and GrpE are required for flagellum synthesis in *Escherichia coli*. J Bacteriol 174:6256–6263
Silberg JJ, Hoff KG, Vickery LE (1998) The Hsc66-Hsc20 chaperone system in *Escherichia coli*: chaperone activity and interactions with the DnaK-DnaJ-grpE system. J Bacteriol 180: 6617–6624
Srinivasan SR, Gillies AT, Chang L, Thompson AD, Gestwicki JE (2012) Molecular chaperones DnaK and DnaJ share predicted binding sites on most proteins in the *E. coli* proteome. Mol Biosyst 8:2323–2333
Szabo A, Langer T, Schroder H, Flanagan J, Bukau B, Hartl FU (1994) The ATP hydrolysis-dependent reaction cycle of the *Escherichia coli* Hsp70 system DnaK, DnaJ, and GrpE. Proc Natl Acad Sci U S A 91:10345–10349
Tartaglia GG, Pechmann S, Dobson CM, Vendruscolo M (2007) Life on the edge: a link between gene expression levels and aggregation rates of human proteins. Trends Biochem Sci 32: 204–206
Tartaglia GG, Dobson CM, Hartl FU, Vendruscolo M (2010) Physicochemical determinants of chaperone requirements. J Mol Biol 400:579–588
Tatusov RL, Natale DA, Garkavtsev IV, Tatusova TA, Shankavaram UT, Rao BS et al (2001) The COG database: new developments in phylogenetic classification of proteins from complete genomes. Nucleic Acids Res 29:22–28
Teter SA, Houry WA, Ang D, Tradler T, Rockabrand D, Fischer G et al (1999) Polypeptide flux through bacterial Hsp70: DnaK cooperates with trigger factor in chaperoning nascent chains. Cell 97:755–765
Ueguchi C, Kakeda M, Yamada H, Mizuno T (1994) An analogue of the DnaJ molecular chaperone in *Escherichia coli*. Proc Natl Acad Sci U S A 91:1054–1058
Ullers RS, Luirink J, Harms N, Schwager F, Georgopoulos C, Genevaux P (2004) SecB is a bona fide generalized chaperone in *Escherichia coli*. Proc Natl Acad Sci U S A 101:7583–7588

Ung PM, Thompson AD, Chang L, Gestwicki JE, Carlson HA (2013) Identification of key hinge residues important for nucleotide-dependent allostery in *E. coli* Hsp70/DnaK. PLoS Comput Biol 9:e1003279

Vabulas RM, Raychaudhuri S, Hayer-Hartl M, Hartl FU (2010) Protein folding in the cytoplasm and the heat shock response. Cold Spring Harb Perspect Biol 2:a004390

Valent QA, Kendall DA, High S, Kusters R, Oudega B, Luirink J (1995) Early events in preprotein recognition in *E. coli*: interaction of SRP and trigger factor with nascent polypeptides. EMBO J 14:5494–5505

Wall D, Zylicz M, Georgopoulos C (1994) The NH2-terminal 108 amino acids of the *Escherichia coli* DnaJ protein stimulate the ATPase activity of DnaK and are sufficient for lambda replication. J Biol Chem 269:5446–5451

Weaver J, Rye HS (2014) The C-terminal tails of the bacterial chaperonin GroEL stimulate protein folding by directly altering the conformation of a substrate protein. J Biol Chem 289: 23219–23232

Wild J, Altman E, Yura T, Gross CA (1992) DnaK and DnaJ heat shock proteins participate in protein export in *Escherichia coli*. Genes Dev 6:1165–1172

Xu Z, Horwich AL, Sigler PB (1997) The crystal structure of the asymmetric GroEL-GroES-(ADP)7 chaperonin complex. Nature 388:741–750

Yifrach O, Horovitz A (1995) Nested cooperativity in the ATPase activity of the oligomeric chaperonin GroEL. Biochemistry 34:5303–5308

Zhu X, Zhao X, Burkholder WF, Gragerov A, Ogata CM, Gottesman ME et al (1996) Structural analysis of substrate binding by the molecular chaperone DnaK. Science 272:1606–1614

Chapter 16
Genetic, Biochemical, and Structural Analyses of Bacterial Surface Polysaccharides

Colin A. Cooper, Iain L. Mainprize, and Nicholas N. Nickerson

Abstract Surface polysaccharides are an often essential component of the outer surface of bacteria. They may serve to protect organisms from harsh environmental conditions and to increase virulence. The focus of this review will be to introduce polysaccharide biosynthesis and export from the cell, and the associated techniques used to determine these glycostructures. Protein interactions and proteomics will then be discussed while introducing systems biology approaches used to determine protein–protein and protein–polysaccharide interactions. The final section will address related screening methods used to study gene regulation in bacteria relating to polysaccharide gene clusters and their associated regulators. The goal of this review will be to highlight key studies that have increased our knowledge of glycobiology and discuss novel methods that examine this field at the cellular level using systems biology.

Keywords Polysaccharide • Capsule • Lipopolysaccharide • Glycobiology • Cell envelope

C.A. Cooper (✉)
Agriculture and Food Laboratory, Laboratory Services, University of Guelph, 95 Stone Rd. W., Guelph, ON N1H 8J7, Canada
e-mail: colin.cooper@uoguelph.ca

I.L. Mainprize
Department of Molecular and Cellular Biology, University of Guelph, 95 Stone Road, Guelph, ON N1H 8J7, Canada

N.N. Nickerson
Department of Molecular and Cellular Biology, University of Guelph, 95 Stone Road, Guelph, ON N1H 8J7, Canada

Department of Infectious Diseases, Genentech Inc., South San Francisco, CA 94080, USA

N.J. Krogan, M. Babu (eds.), *Prokaryotic Systems Biology*, Advances in Experimental Medicine and Biology 883, DOI 10.1007/978-3-319-23603-2_16

16.1 Surface Expressed Polysaccharides

16.1.1 Introduction

There is a high degree of chemical and structural diversity in surface expressed glycopolymers in bacteria. A main source of this variability is in the individual sugar subunits with more than 100 distinct monosaccharides having been identified (Lindenberg 1998). Non-sugar subunits, with more than 50 observed so far, are incorporated into many bacterial glycostructures further increasing the potential complexity. The sugar subunits themselves exhibit variability with respect to constitutional (or structural) isomerism and there are multiple linkage types possible between these sugar subunits. The polymers can be linear or branched. There are also instances where the length of the glycopolymer affects its structural and biological characteristics. All of these features are considered when a glycopolymer is classified and/or serotyped.

There are a number of classical types of bacterial surface glycostructures. Peptidoglycan (PG) is a rigid layer (or sacculus) surrounding the cell and is a network of linear polymers of a disaccharide repeat unit cross-linked via short peptides. In Gram-negative bacteria, PG is a relatively thin layer found between the inner and outer membranes whilst the PG of Gram-positive bacteria is a thick layer on the exterior of the cell (reviewed in Typas et al. 2012). The majority of the diversity observed for PG is due to the amino acid composition of the peptide cross-link but there can also be chemical modifications (such as O-acetylation) of the disaccharide repeat units. Teichoic acid (TA) is a Gram-positive bacteria-specific phosphate-rich glycostructure which can be subdivided into two components: a disaccharide linkage unit and a polymer of polyol repeat units (usually containing glycerol and/or ribitol) (reviewed in Brown et al. 2013). The disaccharide linkage unit of TA is highly conserved but the repeat unit is more diverse. Wall TAs are molecules of TA that are covalently linked to PG and lipoteichoic acids are attached to the bacterial membrane via a glycolipid anchor. There is considerable diversity in PG and TA, however two glycostructures that have a much higher degree of diversity are lipopolysaccharide (LPS) and capsular polysaccharide (CPS).

Molecules of LPS are composed of three main structural components. Lipid A is a complex phosphoglycolipid, conserved across species, that anchors the LPS molecule in the outer membrane (OM) of Gram-negative bacteria. A short oligosaccharide (8–12 sugar units), referred to as core oligosaccharide (core-OS) is attached to lipid A and can be further divided into inner core-OS (closer to lipid A) and outer core-OS. O-specific polysaccharide (O-PS, also called O-antigen) is a long chain polymer of repeating units of usually 2–7 sugar residues. There are some variations in the chemical composition of lipid A and core-OS (Heinrichs et al. 1998) but the main source of variability in LPS is the O-PS. The diversity in O-PS is observed not only across species but within species as well. For instance, *Escherichia coli* has over 180 distinct O-PSs and *Vibrio cholerae* may have even more (Sozhamannan and Yildiz 2010). Some bacteria such as those from genera

Haemophilus, *Neisseria*, *Bordetella*, *Branhamella*, and *Campylobacter* express a variant of LPS, referred to as lipooligosaccharide (LOS) which lacks the long O-PS. Perhaps due to the lack of the highly variable O-PS, the core-OS of LOS has an increased chemical diversity than usually seen for the core-OS of LPS with such modifications as additional glycosyl residues (Howard et al. 2000).

Similar to O-PS, CPS molecules are composed of repeat units of mainly 2–7 monosaccharides, either as a linear polymer or with branches, but they are produced by both Gram-negative and Gram-positive bacteria. CPSs (also referred to as K-antigen) are usually much longer polymer chains than LPS, often numbering as many as several hundred repeat units per molecule, whereas LPS is usually less than 100. In most cases, the CPS extends well beyond the exterior of the LPS layer and forms a masking barrier to the environment. Another defining feature of CPSs is that they are attached to the cell surface, although the exact linkage/anchor is known for only a few bacteria. CPS-like molecules that are released into the extracellular milieu have a separate classification as exopolysaccharides (EPS). The same level of variability observed for O-PS is possible for CPS, as the monosaccharide building blocks are the same. There are several examples in which the CPS of one strain of bacteria is identical to the O-PS of another (e.g. the O-PS of *E. coli* O178 and the CPS of *E. coli* K38 have identical pentasaccharide repeat units) (Ali et al. 2005).

The distinction between CPS and LPS is not always so clear. Some bacteria produce a polysaccharide with the same repeat unit structure as the O-PS of their LPS but it is not linked to lipid A. This group of polysaccharides has been classified as O-antigen capsules (or as group 4 CPS in *E. coli*) and the biosynthetic genes are shared for the LPS and O-antigen capsule. O-antigen capsules tend to be much longer polymers than when the O-PS is part of LPS. Conversely, some bacteria produce a CPS that, under certain conditions, is ligated to lipid A and is referred to as K_{LPS}. The length of polysaccharide for K_{LPS} can be as short as a few repeat units (MacLachlan et al. 1993) or much longer (Whitfield 2006). To date, there are no known additional genes required for K_{LPS} synthesis, outside of the genes required for the biosynthesis of the 'classical' LPS and CPS produced by these bacteria.

16.1.2 Generalized Biosynthesis Pathways

Despite many significant differences, the biosynthesis of most of the known bacterial surface polysaccharides follows two major themes. In the first type of biosynthesis pathway, repeat units are assembled in the cytoplasm by glycosyltransferase enzymes (and other enzymes required for the addition of non-sugar subunits). Once a single repeat unit is complete, it is translocated out of the cytoplasm via an integral membrane protein (or complex of proteins). The repeat units are covalently linked to a growing chain of repeat units on the exocytoplasmic side of the membrane. PG and the majority of CPS and LPS are synthesized by this mode of biosynthesis (Fig. 16.1, schemes 1 and 3). For the other general mode of synthesis, the polymer is synthesized completely in the cytoplasm (Fig. 16.1, scheme 2 and 4)

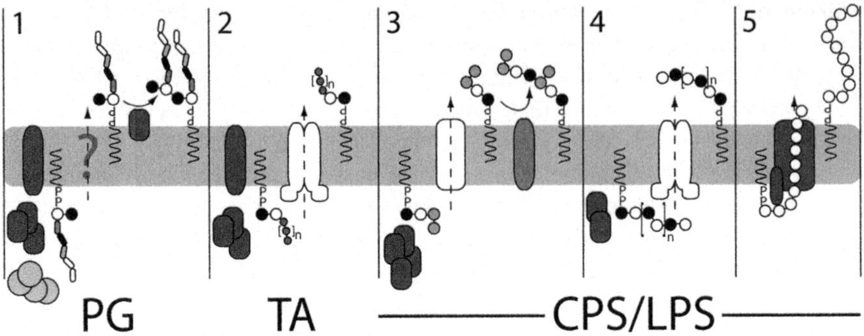

Fig. 16.1 The biosynthesis pathways for the major bacterial surface polysaccharides. One mode of synthesis involves the assembly of repeat units by glycosyltransferases (GTs) (and other types of enzymes for non-sugar additions) on the cytoplasmic face of the cytoplasmic membrane. These completed repeat units are translocated across the membrane and then polymerized. Peptidoglycan (PG, scheme 1) and most lipopolysaccharides (LPS) and capsular polysaccharides (CPS) (scheme 3) are formed by this type of assembly pathway. For the other general mode of synthesis, the polysaccharide is completely synthesized before translocation across the membrane. An ATP-binding cassette (ABC)-dependent transporter mediates the translocation of the polysaccharide. Teichoic acid (TA, scheme 2) and some LPS and CPS (scheme 4) follow this mechanism of synthesis. A slight variation of this mode of synthesis is observed for the synthase-dependent pathway in which, for a small subset of LPS and CPS, the GT that synthesizes the polysaccharide is also the conduit across the membrane. The polysaccharide is translocated across the membrane as it is being synthesized (scheme 5)

or, as in the case of synthase-dependent synthesis (Fig. 16.1, scheme 5), as it is being translocated across the membrane. Polysaccharides that use this mode of synthesis are generally simpler in structural complexity such as homopolymers (with variable linkages) or unbranched heteropolymers. TA is synthesized in this way, as are certain examples of CPS and LPS. Regardless of the synthesis mechanism used, most, if not all of the glycostructures discussed so far are synthesized initially on a common carrier lipid called undecaprenyl-phosphate (undP).

In Gram-negative bacteria, completed polysaccharides that are destined for the extracellular environment, require a second translocation complex to traverse the OM. For LPS, regardless of how it was synthesized and crossed the inner membrane (IM), the O-PS polymer is ligated to a molecule of lipid A-core-OS and transported out of the cell via the Lpt pathway (reviewed in Ruiz et al. 2009) that utilizes the β-barrel protein LptD to cross the OM (Freinkman et al. 2011). Similarly, for synthase-dependent biosynthesis of non-LPS polysaccharides, a β-barrel protein facilitates the translocation across the OM (Morgan et al. 2013). For all other exported polysaccharides, an outer membrane polysaccharide export (OPX) complex is involved. Members of the OPX family have limited sequence conservation but they have a common motif called the polysaccharide export sequence (PES) and a similar predicted secondary structure (Cuthbertson et al. 2009). X-ray crystallography revealed the structure of an octameric OPX from *E. coli* K30 to be a large periplasmic complex with a large central cavity with an α-helical barrel that crosses the OM (Dong et al. 2006) and has been proposed to act as a proteinaceous conduit for polysaccharides to cross the OM.

16.1.3 Identification and Structural Determination of Polysaccharides

Bacterial surface polysaccharides are often initially identified by serotyping using agglutination against a collection of antisera. With the increasing knowledge of the diversity of the known antigens, serotyping is becoming extremely laborious and expensive. New molecular techniques are being developed to allow faster and more reliable identification. One such technique that was designed for serotyping *Klebsiella* CPS involved PCR amplification of the CPS gene cluster followed by restriction fragment length polymorphism analysis and resulted in a fast method that was more discriminatory than classical serotyping (Brisse et al. 2004). It is expected that similar molecular techniques will be devised for other glycostructures (and other bacteria) facilitating the comprehensive identification of these important molecules.

Ultimately, solving the chemical structure of a particular polysaccharide is advantageous as physical and immunological properties can be interpreted on a molecular level. Several techniques have been used and the majority of polysaccharide structures have been determined using multiple strategies. Some of the techniques are purely chemistry-based, such as partial acid hydrolysis and Smith degradation (i.e. break-down of polysaccharides to simpler oligosaccharides) (reviewed for LPS in Banoub et al. 2010). Currently, analytical techniques such as mass spectrometry and nuclear magnetic resonance are the standard for determining the repeating structures of the polysaccharides (reviewed in Caroff and Karibian 2003). Due to the highly repetitive subunits of these glycostructures, non-stoichiometric heterogeneity and terminal modifications (if any) can require special consideration for elucidation. Also with the development of more sensitive and accurate techniques, there may be a need to revisit structures solved by more classical methods. For example, the structures of O-PS from members of the *Shigella* genus have been recently revised since their original structure determination in the 1970s (reviewed in Liu et al. 2008).

16.1.4 Glycosyltransferases: Major Determinant of Bacterial Polysaccharide Diversity

As most of the variability observed for bacterial polysaccharides resides in the actual polymer of sugar subunits, the glycosyltransferases (GTs), that determine the type of sugars added and the linkages that connect them, dictate the chemical and structural identity of the glycopolymer. There is a sequence-based family classification database called CAZy (Cantarel et al. 2009) that classifies all of the known GTs (along with other types of carbohydrate-related enzymes) based on sequence similarity. As of the beginning of 2014, there were almost 120,000 GT sequences in the database but less than 2 % had been characterized (Lombard et al. 2014). With this lack of comprehensive functional characterization, it is usually not possible to predict the substrate and product (i.e. linkage) for a particular GT *a priori*. Compounding this issue is that subtle differences in the sequence of these enzymes can have significant effects on the activity of the GT. For instance, a single amino acid substitution in *E. coli* O9 WbdA, a mannose-specific processive GT, results in seroconversion to O9a, which has one less mannose residue per repeat unit compared to the pentasaccharide of O9 O-PS (Kido and Kobayashi 2000). Also, GTs can have multiple domains with different GT activities, complicating the determination of substrate/product assignment. WbdA from *E. coli* O9a has two distinct GT domains and its homologue in *E. coli* O8 has three. Each of the five GT domains are classified to the same GT family (GT4) in the CAZy database and each add a mannose residue to the growing polymer, but there is little sequence conservation between them. It is expected that as more GTs are functionally characterized and more structures are determined, bioinformatics analyses of GTs will have more predictive accuracy.

16.1.5 Structural Determination of the CPS Anchor

The defining feature of CPSs is tight association with the bacterial cell surface, distinguishing it from loosely associated EPSs. However, in many cases the linkage or the mechanism of anchoring polysaccharide to the OM of Gram-negative or cytoplasmic membrane of Gram-positive bacteria is unknown. There is evidence to suggest that association is mediated by interactions between glycan strands and other surface molecules to generate higher-order capsule structures (Jimenez et al. 2012). Identification of the structure of the anchor is complicated by the masking effect of the long chains of polysaccharide repeat units. In *E. coli*, CPS synthesized as repeat units in the cytoplasm before translocation across the membrane (Fig. 16.1, scheme 3) has been classified as group 1 CPS. The und-P lipid acceptor of group 1 CPSs but must be removed in the periplasm prior to export to the cell surface. No information currently exists to explain how group 1 CPSs are linked to the cell surface but the association is sufficiently robust that most of it sediments with the cells during centrifugation (Whitfield 2006). Alternatively, the closely related colanic acid EPS is secreted into the growth medium (Reid and Whitfield 2005). There is no apparent explanation for the difference in cell association between these two closely related polymers.

E. coli group 2 CPSs, which are polymerized completely in the cytoplasm before export (Fig. 16.1, scheme 4), are high molecular weight polymers and 20–50 % of the chains have a diacylglycerophosphate moiety at the reducing terminus (Whitfield 2006; Jann and Jann 1990). Much of the data to support this observation is indirect and suggests that the phospholipid anchor is present in the cytoplasm before export (Gotschlich et al. 1981; Fischer et al. 1982; Tzeng et al. 2005; Schmidt and Jann 1982). Studies with *N. meningitidis* serogroup B, found that the anchor is a requirement for translocation (Frosch and Muller 1993). Further work has identified 3-deoxy-D-manno-oct-2-ulosonic acid (Kdo) as the linker between the lipid anchor and the polymer (Finke et al. 1991; Schmidt and Jann 1982). Although the evidence suggests a possible linkage of a phospholipid-Kdo anchor, these data rely on treating the CPS with extreme pH or chemical derivatization.

To gain a more complete understanding of the group 2 CPS assembly pathway it is critical to know the precise nature of the lipid terminus and its linkage to the CPS glycan. Central to addressing this question is developing an approach to isolate intact glycolipid without resorting to treatment with extreme pH or chemical derivatization. Willis et al. (2013) developed a strategy to purify CPS from the cell surface followed by depolymerization of the long-chain polymer with specific endo-acting CPS depolymerases. The CPS depolymerases are glycanases exploited from the tail-spike proteins of bacteriophages specific to the CPS of *E. coli* K1 and K5. Depolymerization of the purified CPS leaves the terminal lipid anchor and any linker domain intact, which was analyzed by mass spectrometry. Removal of most of the polysaccharide repeat units is the key to this approach as it reduces the overall contribution of CPS to the mass spectrometry signal. This work uncovered that the CPSs of group 2 *E. coli* strains K1 and K5, as well as the CPS of *N. meningitidis* group B, are attached to a lyso-phosphatidylglycerol (lyso-PG) moiety via a unique poly-Kdo linker. The lipid moiety was identical in all three bacteria, but considerable variation was observed in the number of Kdo residues, both within and between species. Identification of lyso-PG at the reducing terminus of intracellular CPS, isolated from an ABC transporter mutant, suggests that the polymer is synthesized on the lipid anchor and that the acyl chain is removed prior to export to the cell surface.

For the first time, a strategy has been designed to determine the structure of the glycolipid attached to the reducing terminus of the capsular polysaccharide. This approach omits the treatment of the polymer with harsh chemicals and exploits phage-specific glycanases to depolymerize the long-chain polymers. This strategy should provide the framework for studying other CPS systems and other polysaccharide structures to define their anchor structures.

16.2 Multi-Enzyme Complexes and Trans-Envelope Molecular Machineries

16.2.1 Introduction

New techniques and advances in microscopy have revealed that many biological processes take place in discrete subcellular locations. These complex processes take the form of larger molecular machineries, like the flagella apparatus and secretion systems, or localized areas of enzymatic activity. These loci depend on specific protein–protein interactions and identification of these interactions provides significant insights into the molecular mechanism of these processes. Numerous techniques have been developed to characterize these interactions *in vitro* and *in vivo*, however *in vivo* techniques have the advantage of characterizing the interaction in the biological context of the cell. There exists significant evidence to suggest that assembly of extracellular polysaccharides involves complex protein–protein interactions to form discrete membrane-associated multi-enzyme complexes as well as sophisticated trans-envelope export machineries. Outlined below are examples of the bacterial two-hybrid assay and *in vivo* crosslinking strategies to map specific protein–protein and protein–polysaccharide interactions to extend our understanding of polysaccharide assembly.

16.2.2 Characterization of Multi-Enzyme Complexes Using the Bacterial Two-Hybrid Assay

The bacterial two-hybrid assay offers an easy *in vivo* screening tool to identify functional interactions between two proteins. The assay depends on the functional complementation between two complementary fragments (T25 and T18) of the adenylate cyclase from *Bordetella pertussis* to restore enzymatic activity in an *E. coli* adenylate-cyclase deficient strain (*cya*) (Karimova et al. 1998; Ladant and Karimova 2000). Putative interacting proteins are genetically fused to T25 and T18, respectively (Ladant 1988; Ladant et al. 1989). Association of the proteins of interest results in functional complementation between T25 and T18 fragments leading to cAMP synthesis (Karimova et al. 1998). Cyclic AMP triggers transcriptional activation of a large number of genes, including catabolism of carbohydrates such as lactose and maltose, yielding characteristic phenotypes (Ullmann and Danchin 1983). This allows qualitative readout of interactions by colony pigmentation on indicator media or quantification of the strength of an interaction by levels of cAMP or β-galactosidase activity. The bacterial two-hybrid assay can be incorporated into large-scale screens using genetic libraries (Ladant and Karimova 2000). The major advantages of the two-hybrid system are that it allows *in vivo* selection of functional clones, interactions are spatially separated from the transcriptional activation readout, the assay exploits signal amplification as a result of signal transduction, and the assay can be redesigned with specific reporter cassettes (Karimova et al. 1998, 2000). Three examples are given below that have exploited the bacterial two-hybrid assay to demonstrate that polysaccharide biosynthesis is coordinated by coupling initiation, elongation and polysaccharide export.

As outlined in Fig. 16.1, O-PS biosynthesis can be facilitated through multiple fundamentally different assembly pathways. In one of the pathways, an ATP-binding cassette (ABC) transporter is used to translocate the completed polysaccharide across the IM and this type of pathway is called the ABC transporter-dependent pathway (Fig. 16.1, scheme 4). Prior to translocation across the IM, chain extension of the O-PS can be terminated by the addition of a non-reducing terminal residue or by interaction with the ABC transporter. Then, full length O-PS is exported across the IM by the ABC transporter where it is ligated to the lipid A-core.

The polymannose O-PS of *E. coli* O9a provides a model system for ABC transporter-dependent O-PS synthesis that requires the addition of a capping residue to terminate O-PS elongation. In this assembly pathway, three GDP-mannose dependent GTs, WbdA, WbdB and WbdC, assemble the O-PS chain (Kido et al. 1995). WbdA is the serotype-specific processive GT, or polymerase, and is solely responsible for chain extension (Kido and Kobayashi 2000; Greenfield et al. 2012). Termination is regulated by WbdD and involves methylation and phosphorylation of the non-reducing terminal residue (Clarke et al. 2004). The modification is necessary for binding of the terminated O9a O-PS to the nucleotide-binding component of the ABC transporter for export (Cuthbertson et al. 2005, 2007). Studies using the bacterial two-hybrid assay identified an interaction between WbdD (the terminator) and WbdA (the polymerase), providing evidence that synthesis and termination are coupled (Clarke et al. 2009). Furthermore, WbdD was strictly required to bring WbdA to the IM. In the absence of WbdD there was diminished mannosyltransferase activity. In this case, the bacterial two-hybrid assay established a novel mechanism to control polymerization of O-PS by coordinating the correct membrane association required for activity of the WbdA mannosyltransferase.

An additional example of O-PS biosynthesis by the ABC transporter-dependent pathway is the *K. pneumoniae* O2a polymer. The *K. pneumoniae* O2a O-PS is composed of D-galactan I, which contains the disaccharide repeat unit structure, [→3-β-D-Gal*f*-(1 → 3)-α-D-Gal*p*-(1→] (Whitfield et al. 1992). Three GTs (WbbM, WbbN and WbbO) are responsible for synthesis of the repeat unit (Guan et al. 2001). In contrast to the system described above, chain termination is determined by an interaction between the transporter and the GTs (or their product) in a manner that has not yet been determined (Kos et al. 2009). Experiments using the bacterial two-hybrid assay identified interactions between all of the GTs, suggesting that they form a multi-enzyme complex (Kos and Whitfield 2010). Although none of the GTs are integral membrane proteins, all target independently to the membrane to form a membrane located enzyme complex. These protein–protein interactions place the enzymes in proximity to the und-PP-GlcNAc acceptor and the polymer export apparatus; providing evidence that initiation, elongation, and export are coupled.

Since O-PS polymerization is completed prior to export in the ABC transporter-dependent pathway, the maintenance of GTs and modifying enzymes within a multi-protein complex is a potential mechanism to ensure efficiency and fidelity of the polymerization process.

E. coli K1 produces another well-studied example of ABC transporter-dependent (group 2) CPS that is a homopolymer of α-2,8-linked *N*-acetylneuraminic acid (NeuNAc or polysialic acid) (Barry 1958; McGuire and Binkley 1964). NeuD, NeuB, NeuC and NeuA are involved in synthesis of precursor sugar CMP-NeuNAc and the NeuS polymerase GT catalyzes the transfer of NeuNAc from CMP-NeuNAc to the non-reducing end of the polymer (Steenbergen et al. 1992; Andreishcheva and Vann 2006; Bliss and Silver 1996). KpsM and KpsT constitute the ABC transporter and form a trans-envelope complex with KpsE (localized to IM) and KpsD (localized to OM) for export of the polymer to the cell surface (Vimr and Steenbergen 2009). Exploiting the bacterial two-hybrid system revealed that KpsC (a previously uncharacterized component of the assembly pathway) interacts with itself, KpsE and the polymerase NeuS. These results suggested that KpsC is an adaptor protein to couple synthesis and export (Steenbergen and Vimr 2008; Vimr and Steenbergen 2009). Follow-up studies found that KpsC and KpsS are β-Kdo transferases and together they add a poly-Kdo linker to the lipid acceptor before polymerization of NeuNAc. KpsS adds the first Kdo residue and the reaction product is extended by KpsC (Willis and Whitfield 2013). This study extended the results of the two-hybrid assay and demonstrated that the synthesis of the linker and elongation of the polymer are coupled by forming a KpsC and NeuS complex. Furthermore, the KpsC–NeuS complex interacts with the inner membrane KpsE protein, supporting the notion that initiation, elongation and export are coupled.

The outlined studies demonstrate that the bacterial two-hybrid system is effective at identifying novel protein–protein interactions involved in the synthesis of surface polysaccharides and it enhances our understanding of the molecular mechanisms involved. However, it must be noted that when working with *in vivo* systems it is critical to maintain wild type protein levels. Overexpression of WbdD, for example, enhances chain termination frequency and reduces O9a O-PS chain length, suggesting that the stoichiometry of the biosynthesis components and their potential interactions may be important in determining function (Clarke et al. 2004). Thus, changing the stoichiometry between proteins within a complex can have significant effects on their *in vivo* function.

16.2.3 Exploring Polysaccharide Transport Using In Vivo Photocrosslinking Strategies

In Gram-negative bacteria, polysaccharides are built on either side of the IM, requiring final export to the cell surface. This can be a daunting task as these polymers can be extremely large and/or have hydrophobic anchors. While the machinery required for these processes has been identified, the molecular details are often poorly understood, in part because of the difficulty of capturing "snapshots" along the transit pathways. Major breakthroughs in this area have come as a result of *in vivo* crosslinking strategies using unnatural amino acids.

Multiple strategies exist to identify protein–protein interaction within the cell, such as the previously mentioned bacterial two-hybrid assay. Alternatively, it is more difficult to identify protein–polysaccharide interactions, especially for transient intermediates during the assembly process. Systems now exist to allow site-specific incorporation of unnatural amino acids *in vivo* with high translational fidelity (Chin et al. 2002a, b; Wang et al. 2002). An orthogonal aminoacyl-tRNA synthetase/tRNA pair has been evolved from *Methanococcus jannaschii* to reassign the amber codon (TAG) (Xie and Schultz 2006; Wang et al. 2001; Liu and Schultz 2010) for site-specific incorporation of a range of unnatural amino acids with different chemistries. Incorporation of the UV photo-crosslinkable unnatural amino acid *p*-benzoyl-L-phenylalanine (*p*Bpa) has been successfully used in many studies to characterize protein–protein interactions (Kauer et al. 1986; Chin and Schultz 2002). The benzophenone group of *p*Bpa preferentially reacts with C–H bonds upon excitation in the near-UV at 350–365 nm (Dorman and Prestwich 1994). This technique allows protein interactions to be identified in their native context, which is key to defining their cellular roles; it also allows the identification of interactions with other molecules. As described below, this strategy was successful in elucidating the transport route of LPS to the outer membrane as well as verifying the export channel for CPSs.

The Gram-negative bacterial OM is an asymmetric bilayer, with the inner leaflet composed of phospholipids and the outer leaflet composed of LPS (Raetz and Whitfield 2002). The LPS is crucial for survival in harsh environments and provides a barrier to small hydrophobic molecules, antibiotics and detergents (Nikaido 2003). LPS molecules are transported to the outer leaflet of the OM by seven essential Lpt (lipopolysaccharide transport) proteins (Silhavy et al. 2010; Polissi and Sperandeo 2014; Sperandeo et al. 2008). The ABC transporter forms a complex in the IM with a membrane bound protein, LptC (Ruiz et al. 2008; Narita and Tokuda 2009). LptC interacts with periplasmic LptA, which in turn interacts with the OM-localized LptD to form a trans-envelope bridge (Sperandeo et al. 2007, 2011). The existence of a trans-envelope bridge is still highly debated in the literature.

Using structural information, Okuda et al. (2012) incorporated *p*Bpa at 23 sites in LptC and 14 sites within LptA. Site-specific crosslinks to LPS were identified at four positions in LptC and five positions in LptA. All but one of the crosslinks were formed on the inside of the β-jellyroll structure of these proteins. Using spheroplasts, the authors demonstrated that separate rounds of ATP hydrolysis were required to first transfer LPS from the IM platform to LptC and then from LptC to LptA. These findings are significant as the *in vivo* crosslinking of LPS to LptC and LptA strengthened the model of a trans-envelope bridge as well as revealed that multiple rounds of ATP hydrolysis are required to 'push' the LPS across this periplasmic bridge.

Despite the range of known polysaccharide structures and differences in assembly pathways, CPSs share a common strategy for polymer export across the outer membrane (Whitfield 2006). The export translocon is a trans-envelope complex between the oligomeric OPX family of proteins and the oligomeric IM polysaccharide co-polymerase (PCP) protein family (Cuthbertson et al. 2009). For the group 1 CPS of *E. coli* K30, the Wza lipoprotein is the prototypical OPX protein (Drummelsmith and Whitfield 1999; Nesper et al. 2003) and it docks with the PCP family member, Wzc (Reid and Whitfield 2005; Collins et al. 2006). Together, these proteins form a contiguous molecular scaffold that spans the cell envelope (Collins et al. 2007). It has been hypothesized that the Wza-Wzc complex provides the export pathway for CPS through the OM, coupling chain growth to OM transit (Whitfield 2006).

We employed the *in vivo* crosslinking strategy to trap the nascent CPS within the lumen of the Wza octamer (Dong et al. 2006). Residues within the lumen of Wza were targeted for site-specific incorporation of the UV photo-crosslinkable unnatural amino acid *p*Bpa. Successful incorporation of *p*Bpa within the lumen of the Wza complex revealed five novel Wza-specific adducts that were dependent on synthesis of K30 CPS. Wza-specific adducts purified from the OM were crosslinked to K30 polysaccharide, providing the first definitive evidence that CPS is extruded through the Wza channel to reach the cell surface (Nickerson et al. 2014). As Wza shares many features with other OPX proteins we predict that this will be a common mechanism for translocation of high-molecular-weight polysaccharides to the cell surface.

These studies demonstrate that site-specific incorporation of the *p*Bpa unnatural amino acid and *in vivo* crosslinking provide a powerful strategy to identify protein interaction partners. The major drawbacks of this procedure are the requirement for structural information of the target protein and inability to design a feasible screen.

16.3 Genetic Regulation Associated with Bacterial Outer Envelope Structures

16.3.1 Introduction

CPS and LPS are often essential bacterial components as they may act as both a protective defence mechanism and a virulence factor for invasion and colonization. As highlighted above, many studies have elucidated detailed protein pathways and interaction maps responsible for the biosynthesis of polysaccharide precursors and their export mechanisms to the periphery of the cell; the following section will review aspects of the underlying genetic architecture for these systems. While early studies were able to characterize the coding, localization, and gene transfer between polysaccharide systems, recent advances in systems biology have been at the forefront of understanding how coding, gene regulation, acquisition and evolutionary patterns have developed to increase fitness throughout a bacterial cell.

16.3.2 Techniques Used to Understand Specific Regulatory Regions and Associated Regulatory Proteins

Since the early twentieth century, changes in the bacterial cell envelope have been a readily observable phenotype due to the qualitative observation of colony morphology (Beiser and Davis 1957; Markovitz 1964; Duguid and Wilkinson 1953). With further understanding of bacterial genetics, especially their manipulation, the ability to assay for the underlying genetic components for LPS and CPS became attainable. As modern molecular biology and microbiology fused, our knowledge of transcriptional regulation associated with these important glycostructures has excelled. Individual regulators have been identified in comprehensive analyses, DNA binding regions have been mapped, and integration of complex regulatory systems has been elucidated for genetic clusters. For the well-characterized *E. coli* K30 group 1 CPS cluster, elements including a JUMPstart sequence (Bailey et al. 1996), a putative stem-loop inhibitory region, and the requirement of RfaH anti-termination for expression of the locus have been identified. Techniques such as plasmid and chromosomal reporters were used in various mutant backgrounds to test for gene expression and ultimately for the production of the K30 CPS (Rahn and Whitfield 2003). Additional studies have utilized DNA binding techniques such as electrophoretic mobility shift assays (EMSA) to determine if a given protein has bound a purified section of DNA. This coupled with DNA footprinting can provide convincing evidence for interaction of a transcription factor at a given promoter. Such a combination of techniques was used alongside DNA reporting experiments to determine how the group 2 CPS is regulated in *E. coli* K5 (Bell et al. 2001). Furthermore, a recent study implicated a small regulatory RNA in the regulation of LPS modification. Here, a PhoPQ-dependent RNA, termed MgrR, acts to silence the gene encoding EptB, a phosphoethanolamine modifying enzyme. However, results suggested that the expression of Sigma E can out-compete the silencing effect of MgrR, allowing LPS modification under conditions such as cell envelope stress. Techniques used included: Northern blot analysis of MgrR in the presence of low Ca^{2+} (indicative of active PhoPQ conditions); 5′ RACE to determine the mRNA produced for *eptB*; and gene reporter constructs to measure accurate expression levels (Moon et al. 2013). While advanced techniques as those mentioned above are invaluable for determining genetic regulation at one (or a few) non-coding region(s) and the DNA binding proteins associated with these areas, recent advances in large scale systems biology techniques aided by bioinformatic analysis has rapidly increased our knowledge of the bacterial cellular periphery.

16.3.3 Advanced Genome Wide Assays for Genetic Regulation

In recent years, genome wide assays, genomic sequencing, and the bioinformatic tools needed to analyze these large datasets have become commonplace and are important tools for nearly all microbiology laboratories. The use of microarrays in particular can be a quick, informative, and affordable method for detecting changes in gene regulation. In a recent study of *Bacillus subtilis*, a novel RNA element termed EAR (EPS-associated RNA) was identified as a requirement for proper expression of the 16kb EPS operon. After identifying EAR through bioinformatic analyses, the region was shown to be necessary for biofilm formation using mutagenesis. Microarray analysis was performed to determine where in the 15 gene operon the anti-termination activity of EAR was occurring. Surprisingly, EAR had anti-termination effects further downstream than the gene encoded directly beside it, resulting in the last ten genes of the operon to be down-regulated in the *ear* region mutation (Irnov and Winkler 2010). An additional study sought to determine a potential cause for spontaneous capsule loss in *Pasteurella multocida* using genomic sequencing and microarray analysis. Here strains which lacked CPS production were sequenced, and while no mutations were found in the capsule biosynthesis and export loci, a point mutation in a transcription factor termed Fis was identified. Complementation with wild type Fis restored CPS production and export and microarray analysis with the *fis* point mutant strains revealed a number of genes significantly down-regulated compared to the wild type strain (Steen et al. 2010). This study illustrates an important benefit to systems biology approaches where such large quantities of data are generated that often new findings may emerge with new avenues of potential research to follow. An analysis of the regulation of Vi antigen CPS produced in *Salmonella enterica* serovar Typhi (*S.* Typhi) revealed that the regulator TviA encoded within the capsule locus (*viaB*) not only regulated the CPS biosynthesis and export genes encoded within the region (a pathogenicity island), but down-regulated both motility- and invasion-associated genes encoded elsewhere in the genome. This was observed by microarray hybridization experiments with wild type and *tviA* mutant strains (Winter et al. 2009). When the TviA regulator was expressed in a non-capsule producing serovar of *S. enterica* (Typhimurium), the same effect was observed, and subsequently as the flagellar antigen expression was decreased, the mutant *S.* Typhimurium was able to disseminate to greater levels in a chicken infection model (Winter et al. 2010).

While the use of microarray analysis has been at the forefront of genomic studies, in recent years proteomics-based analyses have been gaining favour for studying protein expression which is not only regulated at the genetic level, but at the RNA and translational levels, as well. A study making use of a technique termed iTRAQ (Isobaric tag for relative and absolute quantitation) (Gygi et al. 1999; DeSouza et al. 2005) explored the cytotoxic effect of gene mutations in LPS biosynthesis during macrophage infection by *Francisella novicida*. This strategy was used in tandem with microarray analysis but the main focus was to highlight possible changes in host protein levels and not at the genetic regulation level. A mutant strain of

F. novicida was used for macrophage infection and iTRAQ was used to identify cytoskeletal structures and the mRNA degradation regulator tristetraprolin (TTP) as potential targets during infection. These results suggested that *F. novicida* modifies host regulation of proteins such as cytokines downstream of TTP (i.e. not at the genetic level) as a method of intracellular survival within host cells (Nakayasu et al. 2013).

16.3.4 RNA Sequencing Techniques

The relatively new technique of RNA sequencing (RNAseq) has proved to be a powerful tool that is beginning to rival microarray analysis for gaining insight into biological processes (see further commentary in Shendure 2008). Not only is expressed gene content and quantity (including previously unidentified genes) achieved in an accurate manner, but the presence of non-coding RNA (ncRNA) is obtained as well. In a comprehensive study conducted with *S.* Typhi, RNAseq was compared to a traditional microarray platform for studying the differences between wild type bacteria and an *ompR* (a regulator of the *viaB* locus) mutant strain. While similar results were obtained between the microarray and RNAseq methods, a wealth of novel knowledge was obtained from the RNAseq data. The presence of hypothetical gene mRNAs was then paired to protein expression data obtained from whole proteome mass spectrometry (MS) to clarify the presence of hypothetical proteins (Perkins et al. 2009). Later, the authors expanded their OmpR regulon studies to include ChIP-seq, a technique where chromatin immunoprecipitation is conducted and the resulting DNA is subjected to sequencing (Perkins et al. 2013). A study conducted with *Yersinia enterocolitica* O:3 examined LPS substitutions and whether or not temperature had an effect on their differential biosynthesis. Here it was discovered via RNAseq that specific biosynthesis genes were expressed at lower levels at higher temperatures (Muszynski et al. 2013). RNAseq is quickly becoming a major technique in gene regulation studies, combining sequencing technology and bioinformatics in an elegant combination.

16.4 Conclusions

In summary, the study of the bacterial envelope, and the individual glycostructures, has been at the forefront of microbial research for decades. The use of novel systems biology approaches has brought this field into a new era. The ability to generate, and more importantly manage, these large datasets has been a major challenge that the scientific community has adapted to with the help of computer science. These datasets have also highlighted scientific collaboration as new methods of sharing results have been developed. As new method development continues in the future, there is no doubt that systems biology involved at the genetic, RNA, and protein levels will be a major player, and continue to bring about great discoveries and spur on new research paths.

Acknowledgements We thank Chris Whitfield, Bradley Clarke, and other members of the Whitfield lab for helpful discussions during the writing of this review.

References

Ali T, Urbina F, Weintraub A, Widmalm G (2005) Structural studies of the O-antigenic polysaccharides from the enteroaggregative *Escherichia coli* strain 522/C1 and the international type strain from *Escherichia coli* O 178. Carbohydr Res 340(12):2010–2014. doi:10.1016/j.carres.2005.06.011

Andreishcheva EN, Vann WF (2006) Gene products required for de novo synthesis of polysialic acid in *Escherichia coli* K1. J Bacteriol 188(5):1786–1797. doi:10.1128/JB.188.5.1786-1797.2006

Bailey MJ, Hughes C, Koronakis V (1996) Increased distal gene transcription by the elongation factor RfaH, a specialized homologue of NusG. Mol Microbiol 22(4):729–737

Banoub JH, El Aneed A, Cohen AM, Joly N (2010) Structural investigation of bacterial lipopolysaccharides by mass spectrometry and tandem mass spectrometry. Mass Spectrom Rev 29(4):606–650. doi:10.1002/mas.20258

Barry GT (1958) Colominic acid, a polymer of N-acetylneuraminic acid. J Exp Med 107(4): 507–521

Beiser SM, Davis BD (1957) Mucoid mutants of *Escherichia coli*. J Bacteriol 74(3):303–307

Bell CD, Kovacs K, Horvath E, Rotondo F (2001) Histologic, immunohistochemical, and ultrastructural findings in a case of minocycline-associated "black thyroid". Endocr Pathol 12(4):443–451

Bliss JM, Silver RP (1996) Coating the surface: a model for expression of capsular polysialic acid in *Escherichia coli* K1. Mol Microbiol 21(2):221–231

Brisse S, Issenhuth-Jeanjean S, Grimont PA (2004) Molecular serotyping of Klebsiella species isolates by restriction of the amplified capsular antigen gene cluster. J Clin Microbiol 42(8):3388–3398. doi:10.1128/JCM.42.8.3388-3398.2004

Brown S, Santa Maria JP Jr, Walker S (2013) Wall teichoic acids of gram-positive bacteria. Annu Rev Microbiol 67:313–336. doi:10.1146/annurev-micro-092412-155620

Cantarel BL, Coutinho PM, Rancurel C, Bernard T, Lombard V, Henrissat B (2009) The Carbohydrate-Active EnZymes database (CAZy): an expert resource for Glycogenomics. Nucleic Acids Res 37(Database issue):D233–D238. doi:10.1093/nar/gkn663

Caroff M, Karibian D (2003) Structure of bacterial lipopolysaccharides. Carbohydr Res 338(23):2431–2447

Chin JW, Schultz PG (2002) *In vivo* photocrosslinking with unnatural amino acid mutagenesis. Chembiochem 3(11):1135–1137. doi:10.1002/1439-7633(20021104)3:11<1135::AID-CBIC1135>3.0.CO;2-M

Chin JW, Martin AB, King DS, Wang L, Schultz PG (2002a) Addition of a photocrosslinking amino acid to the genetic code of *Escherichia coli*. Proc Natl Acad Sci U S A 99(17): 11020–11024. doi:10.1073/pnas.172226299

Chin JW, Santoro SW, Martin AB, King DS, Wang L, Schultz PG (2002b) Addition of p-azido-L-phenylalanine to the genetic code of *Escherichia coli*. J Am Chem Soc 124(31):9026–9027

Clarke BR, Cuthbertson L, Whitfield C (2004) Nonreducing terminal modifications determine the chain length of polymannose O antigens of *Escherichia coli* and couple chain termination to polymer export via an ATP-binding cassette transporter. J Biol Chem 279(34):35709–35718. doi:10.1074/jbc.M404738200

Clarke BR, Greenfield LK, Bouwman C, Whitfield C (2009) Coordination of polymerization, chain termination, and export in assembly of the *Escherichia coli* lipopolysaccharide O9a antigen in an ATP-binding cassette transporter-dependent pathway. J Biol Chem 284(44):30662–30672. doi:10.1074/jbc.M109.052878

Collins RF, Beis K, Clarke BR, Ford RC, Hulley M, Naismith JH, Whitfield C (2006) Periplasmic protein-protein contacts in the inner membrane protein Wzc form a tetrameric complex required for the assembly of *Escherichia coli* group 1 capsules. J Biol Chem 281(4):2144–2150. doi:10.1074/jbc.M508078200

Collins RF, Beis K, Dong C, Botting CH, McDonnell C, Ford RC, Clarke BR, Whitfield C, Naismith JH (2007) The 3D structure of a periplasm-spanning platform required for assembly of group 1 capsular polysaccharides in *Escherichia coli*. Proc Natl Acad Sci U S A 104(7):2390–2395. doi:10.1073/pnas.0607763104

Cuthbertson L, Powers J, Whitfield C (2005) The C-terminal domain of the nucleotide-binding domain protein Wzt determines substrate specificity in the ATP-binding cassette transporter for the lipopolysaccharide O-antigens in *Escherichia coli* serotypes O8 and O9a. J Biol Chem 280(34):30310–30319. doi:10.1074/jbc.M504371200

Cuthbertson L, Kimber MS, Whitfield C (2007) Substrate binding by a bacterial ABC transporter involved in polysaccharide export. Proc Natl Acad Sci U S A 104(49):19529–19534. doi:10.1073/pnas.0705709104

Cuthbertson L, Mainprize IL, Naismith JH, Whitfield C (2009) Pivotal roles of the outer membrane polysaccharide export and polysaccharide copolymerase protein families in export of extracellular polysaccharides in gram-negative bacteria. Microbiol Mol Biol Rev 73(1): 155–177. doi:10.1128/MMBR.00024-08

DeSouza L, Diehl G, Rodrigues MJ, Guo J, Romaschin AD, Colgan TJ, Siu KW (2005) Search for cancer markers from endometrial tissues using differentially labeled tags iTRAQ and cICAT with multidimensional liquid chromatography and tandem mass spectrometry. J Proteome Res 4(2):377–386. doi:10.1021/pr049821j

Dong C, Beis K, Nesper J, Brunkan-Lamontagne AL, Clarke BR, Whitfield C, Naismith JH (2006) Wza the translocon for *E. coli* capsular polysaccharides defines a new class of membrane protein. Nature 444(7116):226–229. doi:10.1038/nature05267

Dorman G, Prestwich GD (1994) Benzophenone photophores in biochemistry. Biochemistry 33(19):5661–5673

Drummelsmith J, Whitfield C (1999) Gene products required for surface expression of the capsular form of the group 1 K antigen in *Escherichia coli* (O9a:K30). Mol Microbiol 31(5):1321–1332

Duguid JP, Wilkinson JF (1953) The influence of cultural conditions on polysaccharide production by Aerobacter aerogenes. J Gen Microbiol 9(2):174–189

Finke A, Bronner D, Nikolaev AV, Jann B, Jann K (1991) Biosynthesis of the *Escherichia coli* K5 polysaccharide, a representative of group II capsular polysaccharides: polymerization in vitro and characterization of the product. J Bacteriol 173(13):4088–4094

Fischer W, Schmidt MA, Jann B, Jann K (1982) Structure of the *Escherichia coli* K2 capsular antigen. Stereochemical configuration of the glycerophosphate and distribution of galactopyranosyl and galactofuranosyl residues. Biochemistry 21(6):1279–1284

Freinkman E, Chng SS, Kahne D (2011) The complex that inserts lipopolysaccharide into the bacterial outer membrane forms a two-protein plug-and-barrel. Proc Natl Acad Sci U S A 108(6):2486–2491. doi:10.1073/pnas.1015617108

Frosch M, Muller A (1993) Phospholipid substitution of capsular polysaccharides and mechanisms of capsule formation in Neisseria meningitidis. Mol Microbiol 8(3):483–493

Gotschlich EC, Fraser BA, Nishimura O, Robbins JB, Liu TY (1981) Lipid on capsular polysaccharides of gram-negative bacteria. J Biol Chem 256(17):8915–8921

Greenfield LK, Richards MR, Li J, Wakarchuk WW, Lowary TL, Whitfield C (2012) Biosynthesis of the polymannose lipopolysaccharide O-antigens from *Escherichia coli* serotypes O8 and O9a requires a unique combination of single- and multiple-active site mannosyltransferases. J Biol Chem 287(42):35078–35091. doi:10.1074/jbc.M112.401000

Guan S, Clarke AJ, Whitfield C (2001) Functional analysis of the galactosyltransferases required for biosynthesis of D-galactan I, a component of the lipopolysaccharide O1 antigen of Klebsiella pneumoniae. J Bacteriol 183(11):3318–3327. doi:10.1128/JB.183.11.3318-3327.2001

Gygi SP, Rist B, Gerber SA, Turecek F, Gelb MH, Aebersold R (1999) Quantitative analysis of complex protein mixtures using isotope-coded affinity tags. Nat Biotechnol 17(10):994–999. doi:10.1038/13690

Heinrichs DE, Yethon JA, Whitfield C (1998) Molecular basis for structural diversity in the core regions of the lipopolysaccharides of *Escherichia coli* and Salmonella enterica. Mol Microbiol 30(2):221–232

Howard MD, Cox AD, Weiser JN, Schurig GG, Inzana TJ (2000) Antigenic diversity of Haemophilus somnus lipooligosaccharide: phase-variable accessibility of the phosphorylcholine epitope. J Clin Microbiol 38(12):4412–4419

Irnov I, Winkler WC (2010) A regulatory RNA required for antitermination of biofilm and capsular polysaccharide operons in Bacillales. Mol Microbiol 76(3):559–575. doi:10.1111/j.1365-2958.2010.07131.x

Jann K, Jann B (1990) Structure and biosynthesis of the capsular antigens of *Escherichia coli*. In: Jann K, Jann B (eds) Bacterial capsules, vol 150. Current topics in microbiology and immunology. Springer Verlag, Berlin, pp 19–42

Jimenez N, Senchenkova SN, Knirel YA, Pieretti G, Corsaro MM, Aquilini E, Regue M, Merino S, Tomas JM (2012) Effects of lipopolysaccharide biosynthesis mutations on K1 polysaccharide association with the *Escherichia coli* cell surface. J Bacteriol 194(13): 3356–3367. doi:10.1128/JB.00329-12

Karimova G, Pidoux J, Ullmann A, Ladant D (1998) A bacterial two-hybrid system based on a reconstituted signal transduction pathway. Proc Natl Acad Sci U S A 95(10):5752–5756

Karimova G, Ullmann A, Ladant D (2000) A bacterial two-hybrid system that exploits a cAMP signaling cascade in *Escherichia coli*. Methods Enzymol 328:59–73

Kauer JC, Erickson-Viitanen S, Wolfe HR Jr, DeGrado WF (1986) p-Benzoyl-L-phenylalanine, a new photoreactive amino acid. Photolabeling of calmodulin with a synthetic calmodulin-binding peptide. J Biol Chem 261(23):10695–10700

Kido N, Kobayashi H (2000) A single amino acid substitution in a mannosyltransferase, WbdA, converts the *Escherichia coli* O9 polysaccharide into O9a: generation of a new O-serotype group. J Bacteriol 182(9):2567–2573

Kido N, Torgov VI, Sugiyama T, Uchiya K, Sugihara H, Komatsu T, Kato N, Jann K (1995) Expression of the O9 polysaccharide of *Escherichia coli*: sequencing of the *E. coli* O9 rfb gene cluster, characterization of mannosyl transferases, and evidence for an ATP-binding cassette transport system. J Bacteriol 177(8):2178–2187

Kos V, Whitfield C (2010) A membrane-located glycosyltransferase complex required for biosynthesis of the D-galactan I lipopolysaccharide O antigen in Klebsiella pneumoniae. J Biol Chem 285(25):19668–19687. doi:10.1074/jbc.M110.122598

Kos V, Cuthbertson L, Whitfield C (2009) The Klebsiella pneumoniae O2a antigen defines a second mechanism for O antigen ATP-binding cassette transporters. J Biol Chem 284(5):2947–2956. doi:10.1074/jbc.M807213200

Ladant D (1988) Interaction of Bordetella pertussis adenylate cyclase with calmodulin. Identification of two separated calmodulin-binding domains. J Biol Chem 263(6):2612–2618

Ladant D, Karimova G (2000) Genetic systems for analyzing protein-protein interactions in bacteria. Res Microbiol 151(9):711–720

Ladant D, Michelson S, Sarfati R, Gilles AM, Predeleanu R, Barzu O (1989) Characterization of the calmodulin-binding and of the catalytic domains of Bordetella pertussis adenylate cyclase. J Biol Chem 264(7):4015–4020

Lindenberg B (1998) Bacterial polysaccharides: components. In: Dimitriu S (ed) Polysaccharides – structural diversity and functional versatility. Marcel Dekker, New York

Liu CC, Schultz PG (2010) Adding new chemistries to the genetic code. Annu Rev Biochem 79:413–444. doi:10.1146/annurev.biochem.052308.105824

Liu B, Knirel YA, Feng L, Perepelov AV, Senchenkova SN, Wang Q, Reeves PR, Wang L (2008) Structure and genetics of Shigella O antigens. FEMS Microbiol Rev 32(4):627–653. doi:10.1111/j.1574-6976.2008.00114.x

Lombard V, Golaconda Ramulu H, Drula E, Coutinho PM, Henrissat B (2014) The carbohydrate-active enzymes database (CAZy) in 2013. Nucleic Acids Res 42(Database issue):D490–D495. doi:10.1093/nar/gkt1178

MacLachlan PR, Keenleyside WJ, Dodgson C, Whitfield C (1993) Formation of the K30 (group I) capsule in *Escherichia coli* O9:K30 does not require attachment to lipopolysaccharide lipid A-core. J Bacteriol 175(23):7515–7522

Markovitz A (1964) Regulatory mechanisms for synthesis of capsular polysaccharide in mucoid mutants of *Escherichia coli* K12. Proc Natl Acad Sci U S A 51:239–246

McGuire EJ, Binkley SB (1964) The structure and chemistry of colominic acid. Biochemistry 3:247–251

Moon K, Six DA, Lee HJ, Raetz CR, Gottesman S (2013) Complex transcriptional and post-transcriptional regulation of an enzyme for lipopolysaccharide modification. Mol Microbiol 89(1):52–64. doi:10.1111/mmi.12257

Morgan JL, Strumillo J, Zimmer J (2013) Crystallographic snapshot of cellulose synthesis and membrane translocation. Nature 493(7431):181–186. doi:10.1038/nature11744

Muszynski A, Rabsztyn K, Knapska K, Duda KA, Duda-Grychtol K, Kasperkiewicz K, Radziejewska-Lebrecht J, Holst O, Skurnik M (2013) Enterobacterial common antigen and O-specific polysaccharide coexist in the lipopolysaccharide of Yersinia enterocolitica serotype O: 3. Microbiology 159(Pt 8):1782–1793. doi:10.1099/mic.0.066662-0

Nakayasu ES, Tempel R, Cambronne XA, Petyuk VA, Jones MB, Gritsenko MA, Monroe ME, Yang F, Smith RD, Adkins JN, Heffron F (2013) Comparative phosphoproteomics reveals components of host cell invasion and post-transcriptional regulation during Francisella infection. Mol Cell Proteomics 12(11):3297–3309. doi:10.1074/mcp.M113.029850

Narita S, Tokuda H (2009) Biochemical characterization of an ABC transporter LptBFGC complex required for the outer membrane sorting of lipopolysaccharides. FEBS Lett 583(13): 2160–2164. doi:10.1016/j.febslet.2009.05.051

Nesper J, Hill CM, Paiment A, Harauz G, Beis K, Naismith JH, Whitfield C (2003) Translocation of group 1 capsular polysaccharide in *Escherichia coli* serotype K30. Structural and functional analysis of the outer membrane lipoprotein Wza. J Biol Chem 278(50):49763–49772. doi:10.1074/jbc.M308775200

Nickerson NN, Mainprize IL, Hampton L, Jones ML, Naismith JH, Whitfield C (2014) Trapped translocation intermediates establish the route for export of capsular polysaccharides across Escherichia coli outer membranes. Proc Natl Acad Sci USA 111(22):8203–8208. doi:10.1073?pnas.1400341111

Nikaido H (2003) Molecular basis of bacterial outer membrane permeability revisited. Microbiol Mol Biol Rev 67(4):593–656

Okuda S, Freinkman E, Kahne D (2012) Cytoplasmic ATP hydrolysis powers transport of lipopolysaccharide across the periplasm in *E. coli*. Science 338(6111):1214–1217. doi:10.1126/science.1228984

Perkins TT, Kingsley RA, Fookes MC, Gardner PP, James KD, Yu L, Assefa SA, He M, Croucher NJ, Pickard DJ, Maskell DJ, Parkhill J, Choudhary J, Thomson NR, Dougan G (2009) A strand-specific RNA-Seq analysis of the transcriptome of the typhoid bacillus Salmonella typhi. PLoS Genet 5(7):e1000569. doi:10.1371/journal.pgen.1000569

Perkins TT, Davies MR, Klemm EJ, Rowley G, Wileman T, James K, Keane T, Maskell D, Hinton JC, Dougan G, Kingsley RA (2013) ChIP-seq and transcriptome analysis of the OmpR regulon of Salmonella enterica serovars Typhi and Typhimurium reveals accessory genes implicated in host colonization. Mol Microbiol 87(3):526–538. doi:10.1111/mmi.12111

Polissi A, Sperandeo P (2014) The lipopolysaccharide export pathway in *Escherichia coli*: structure, organization and regulated assembly of the Lpt machinery. Mar Drugs 12(2): 1023–1042. doi:10.3390/md12021023

Raetz CR, Whitfield C (2002) Lipopolysaccharide endotoxins. Annu Rev Biochem 71:635–700. doi:10.1146/annurev.biochem.71.110601.135414

Rahn A, Whitfield C (2003) Transcriptional organization and regulation of the *Escherichia coli* K30 group 1 capsule biosynthesis (cps) gene cluster. Mol Microbiol 47(4):1045–1060

Reid AN, Whitfield C (2005) functional analysis of conserved gene products involved in assembly of *Escherichia coli* capsules and exopolysaccharides: evidence for molecular recognition between Wza and Wzc for colanic acid biosynthesis. J Bacteriol 187(15):5470–5481. doi:10.1128/JB.187.15.5470-5481.2005

Ruiz N, Gronenberg LS, Kahne D, Silhavy TJ (2008) Identification of two inner-membrane proteins required for the transport of lipopolysaccharide to the outer membrane of *Escherichia coli*. Proc Natl Acad Sci U S A 105(14):5537–5542. doi:10.1073/pnas.0801196105

Ruiz N, Kahne D, Silhavy TJ (2009) Transport of lipopolysaccharide across the cell envelope: the long road of discovery. Nat Rev Microbiol 7(9):677–683. doi:10.1038/nrmicro2184

Schmidt MA, Jann K (1982) Phospholipid substitution of capsular (K) polysaccharide antigens from *Escherichia coli* causing extraintestinal infections. FEMS Microbiol Lett 14:69–74

Shendure J (2008) The beginning of the end for microarrays? Nat Methods 5(7):585–587. doi:10.1038/nmeth0708-585

Silhavy TJ, Kahne D, Walker S (2010) The bacterial cell envelope. Cold Spring Harb Perspect Biol 2(5):a000414. doi:10.1101/cshperspect.a000414

Sozhamannan S, Yildiz F (2010) Diversity and genetic basis of polysaccharide biosynthesis in Vibrio cholerae. In: St. Georgiev V (ed) Epidemiological and molecular aspects on cholera, infectious disease. Infectious disease series. Springer, New York, pp 129–160

Sperandeo P, Cescutti R, Villa R, Di Benedetto C, Candia D, Deho G, Polissi A (2007) Characterization of lptA and lptB, two essential genes implicated in lipopolysaccharide transport to the outer membrane of *Escherichia coli*. J Bacteriol 189(1):244–253. doi:10.1128/JB.01126-06

Sperandeo P, Lau FK, Carpentieri A, De Castro C, Molinaro A, Deho G, Silhavy TJ, Polissi A (2008) Functional analysis of the protein machinery required for transport of lipopolysaccharide to the outer membrane of *Escherichia coli*. J Bacteriol 190(13):4460–4469. doi:10.1128/JB.00270-08

Sperandeo P, Villa R, Martorana AM, Samalikova M, Grandori R, Deho G, Polissi A (2011) New insights into the Lpt machinery for lipopolysaccharide transport to the cell surface: LptA-LptC interaction and LptA stability as sensors of a properly assembled transenvelope complex. J Bacteriol 193(5):1042–1053. doi:10.1128/JB.01037-10

Steen JA, Steen JA, Harrison P, Seemann T, Wilkie I, Harper M, Adler B, Boyce JD (2010) Fis is essential for capsule production in Pasteurella multocida and regulates expression of other important virulence factors. PLoS Pathog 6(2):e1000750. doi:10.1371/journal.ppat.1000750

Steenbergen SM, Vimr ER (2008) Biosynthesis of the *Escherichia coli* K1 group 2 polysialic acid capsule occurs within a protected cytoplasmic compartment. Mol Microbiol 68(5):1252–1267. doi:10.1111/j.1365-2958.2008.06231.x

Steenbergen SM, Wrona TJ, Vimr ER (1992) Functional analysis of the sialyltransferase complexes in *Escherichia coli* K1 and K92. J Bacteriol 174(4):1099–1108

Typas A, Banzhaf M, Gross CA, Vollmer W (2012) From the regulation of peptidoglycan synthesis to bacterial growth and morphology. Nat Rev Microbiol 10(2):123–136. doi:10.1038/nrmicro2677

Tzeng YL, Datta AK, Strole CA, Lobritz MA, Carlson RW, Stephens DS (2005) Translocation and surface expression of lipidated serogroup B capsular polysaccharide in Neisseria meningitidis. Infect Immun 73(3):1491–1505. doi:10.1128/IAI.73.3.1491-1505.2005

Ullmann A, Danchin A (1983) Role of cyclic AMP in bacteria. In: Greengard P, Robinson GA (eds) Advances in cyclic nucleotide research, vol 15. Raven Press, New York, pp 1–53

Vimr ER, Steenbergen SM (2009) Early molecular-recognition events in the synthesis and export of group 2 capsular polysaccharides. Microbiology 155(Pt 1):9–15. doi:10.1099/mic.0.023564-0

Wang L, Brock A, Herberich B, Schultz PG (2001) Expanding the genetic code of *Escherichia coli*. Science 292(5516):498–500. doi:10.1126/science.1060077

Wang L, Brock A, Schultz PG (2002) Adding L-3-(2-Naphthyl)alanine to the genetic code of *E. coli*. J Am Chem Soc 124(9):1836–1837

Whitfield C (2006) Biosynthesis and assembly of capsular polysaccharides in *Escherichia coli*. Annu Rev Biochem 75:39–68. doi:10.1146/annurev.biochem.75.103004.142545

Whitfield C, Perry MB, MacLean LL, Yu SH (1992) Structural analysis of the O-antigen side chain polysaccharides in the lipopolysaccharides of Klebsiella serotypes O2(2a), O2(2a,2b), and O2(2a,2c). J Bacteriol 174(15):4913–4919

Willis LM, Whitfield C (2013) KpsC and KpsS are retaining 3-deoxy-D-manno-oct-2-ulosonic acid (Kdo) transferases involved in synthesis of bacterial capsules. Proc Natl Acad Sci U S A 110(51):20753–20758. doi:10.1073/pnas.1312637110

Willis LM, Stupak J, Richards MR, Lowary TL, Li J, Whitfield C (2013) Conserved glycolipid termini in capsular polysaccharides synthesized by ATP-binding cassette transporter-dependent pathways in Gram-negative pathogens. Proc Natl Acad Sci U S A 110(19):7868–7873. doi:10.1073/pnas.1222317110

Winter SE, Winter MG, Thiennimitr P, Gerriets VA, Nuccio SP, Russmann H, Baumler AJ (2009) The TviA auxiliary protein renders the Salmonella enterica serotype Typhi RcsB regulon responsive to changes in osmolarity. Mol Microbiol 74(1):175–193. doi:10.1111/j.1365-2958.2009.06859.x

Winter SE, Winter MG, Godinez I, Yang HJ, Russmann H, Andrews-Polymenis HL, Baumler AJ (2010) A rapid change in virulence gene expression during the transition from the intestinal lumen into tissue promotes systemic dissemination of Salmonella. PLoS Pathog 6(8):e1001060. doi:10.1371/journal.ppat.1001060

Xie J, Schultz PG (2006) A chemical toolkit for proteins—an expanded genetic code. Nat Rev Mol Cell Biol 7(10):775–782. doi:10.1038/nrm2005

Index

N.J. Krogan, M. Babu (eds.), *Prokaryotic Systems Biology*, Advances in Experimental Medicine and Biology 883, DOI 10.1007/978-3-319-23603-2

T

Y

The manufacturer's authorised representative in the EU is Springer Nature Customer Service Centre GmbH, Europaplatz 3, 69115 Heidelberg, Germany. If you have any concerns regarding our products, please contact ProductSafety@springernature.com

Printed and bound by CPI Group (UK) Ltd, Croydon, CR0 4YY
03/10/2026
02227506-0002